MODELING OF CHEMICAL PROCESS SYSTEMS

MODELING OF CHEMICAL PROCESS SYSTEMS

Edited by

SYED AHMAD IMTIAZ

Center for Risk, Integrity and Safety Engineering (C-RISE), Department of Process Engineering, Memorial University of Newfoundland, St. John's, NL, Canada

ELSEVIER

Elsevier
Radarweg 29, PO Box 211, 1000 AE Amsterdam, Netherlands
The Boulevard, Langford Lane, Kidlington, Oxford OX5 1GB, United Kingdom
50 Hampshire Street, 5th Floor, Cambridge, MA 02139, United States

ISBN: 978-0-12-823869-1

For information on all Elsevier publications
visit our website at https://www.elsevier.com/books-and-journals

Publisher: Candice G.Janco
Acquisitions Editor: Anita Koch
Editorial Project Manager: Zsereena Rose Mampusti
Production Project Manager: R. Vijay Bharath
Cover Designer: Mark Rogers

Typeset by STRAIVE, India

Contents

Part II Micro scale modeling

Contributors

S. Ahmed Center for Risk, Integrity and Safety Engineering (C-RISE), Department of Process Engineering, Memorial University of Newfoundland, St. John's, NL, Canada

Abdullah Alodhayb King Abdullah Institute for Nanotechnology; Department of Physics and Astronomy, College of Science, King Saud University, Riyadh, Saudi Arabia

Liang Cao Department of Chemical and Biological Engineering, University of British Columbia, Vancouver, BC, Canada

Yankai Cao Department of Chemical and Biological Engineering, University of British Columbia, Vancouver, BC, Canada

Paris E. Georghiou Department of Chemistry, Memorial University of Newfoundland, St. John's, Newfoundland and Labrador, Canada

R. Bhushan Gopaluni Department of Chemical and Biological Engineering, University of British Columbia, Vancouver, BC, Canada

Syed Ahmad Imtiaz Center for Risk, Integrity and Safety Engineering (C-RISE), Department of Process Engineering, Memorial University of Newfoundland, St. John's, NL, Canada

Vinod M. Janardhanan Department of Chemical Engineering, Indian Institute of Technology Hyderabad, Hyderabad, Telangana, India

Yakubu Mandafiya John Chemical Engineering, University of Bradford, Bradford, United Kingdom

Iqbal Mohammed Mujtaba Chemical Engineering, University of Bradford, Bradford, United Kingdom

E. Turco Neto Center for Risk, Integrity and Safety Engineering (C-RISE), Department of Process Engineering, Memorial University of Newfoundland, St. John's, NL, Canada

Raj Patel Chemical Engineering, University of Bradford, Bradford, United Kingdom

Prakash V. Ponugoti Department of Chemical Engineering, Indian Institute of Technology Hyderabad, Hyderabad, Telangana, India

Shofiur Rahman King Abdullah Institute for Nanotechnology, King Saud University, Riyadh, Saudi Arabia

Aminu Zakari Yusuf Nigerian Gas Marketing Limited, Abuja, Nigeria

Sohrab Zendehboudi Department of Process Engineering, Memorial University, St. John's, NL, Canada

Yan Zhang Process Engineering, Memorial University of Newfoundland, St John's, NL, Canada

Preface

Modeling and simulations are increasingly being used for design, control, optimization, fault detection and diagnosis, and various other decision-making purposes. Increasingly, models are being developed at different scales/levels, from the molecular level to the large-scale process systems scale. Though many books claim to deal with process systems modeling and simulation, most books describe modeling in the context of process control; modeling is only described using toy examples. This book covers the modeling of chemical process systems, drawing examples from traditional and contemporary areas. The book is very unique in this respect.

This book is organized to give the readers a feel for multiscale modeling. As models have emerged in various areas, a general systematic method for building a model has emerged. This book starts with the history of the model and its usefulness, then describes the modeling steps in details. A number of examples have been chosen carefully from conventional chemical process systems to contemporary systems, including fuel cells and the micro reforming process. Each chapter is accompanied with a case study explaining the step-by-step modeling methodology.

This book aims to cover the breadth of models applied in process systems. Specifically, the focus of the book is spatial resolution or the scales of the models. It covers scales from the molecular level all the way to the process systems level. The book will provide the readers a feel for multilevel modeling and their interaction. The other theme captured in the book is the progression from mechanistic to empirical modeling and the recent trend toward using machine learning techniques for process system modeling. This book can be used to teach process systems modeling to senior level undergraduate students and graduate students.

PART I

Theory and background

1

An introduction to modeling of chemical process systems

Syed Ahmad Imtiaz
Center for Risk, Integrity and Safety Engineering (C-RISE), Department of Process Engineering, Memorial University of Newfoundland, St. John's, NL, Canada

1 What is a model and modeling?

Models are an important aspect of scientific study. Models are used in a variety of scientific disciplines, ranging from physics and chemistry to ecology and Earth sciences. Application of models are not limited to only physical science but also conceptual and empirical models are common in psychology and behavioral science. However, there is no consensus on the definition of a model. Researchers from many disciplines commonly agree that a model is a "representation"; it is also emphasized that the representation is not "exact." This is nicely stated by van der Valk et al. (2007) "if a model were exactly like its target, it would not be a model but a copy." Rather, a model is a representation of a specific aspect of object, phenomena, processes, ideas, and/or their systems (Oh & Oh, 2011). The "specific aspect" is selected by the modeler depending on the purpose or modeling goal. According to Encyclopaedia Britannica, scientific modeling is "the generation of a physical, conceptual, or mathematical representation of a real phenomenon that is difficult to observe directly" (*Scientific modelling, Encyclopaedia Britannica*, 2012).

Bailer-Jones (2002) point that simplifying assumptions that are essential for models can be characterized by simplification and omissions, with the aim of capturing the essence of what is represented. Because of simplifications and omissions, models have sometimes been described as "neither true nor false" (Hutten, 1954), as has been more eloquently said "all models are wrong, some are useful" (Oh & Oh, 2011). In light of the above discussions, an encompassing definition for model can be stated as follows: "A model is a physical, mathematical, empirical, or conceptual representation of a system which helps capture the essential features of the system required to fulfil some specific purpose." Fig. 1 shows a

Modeling of Chemical Process Systems. https://doi.org/10.1016/B978-0-12-823869-1.00005-3

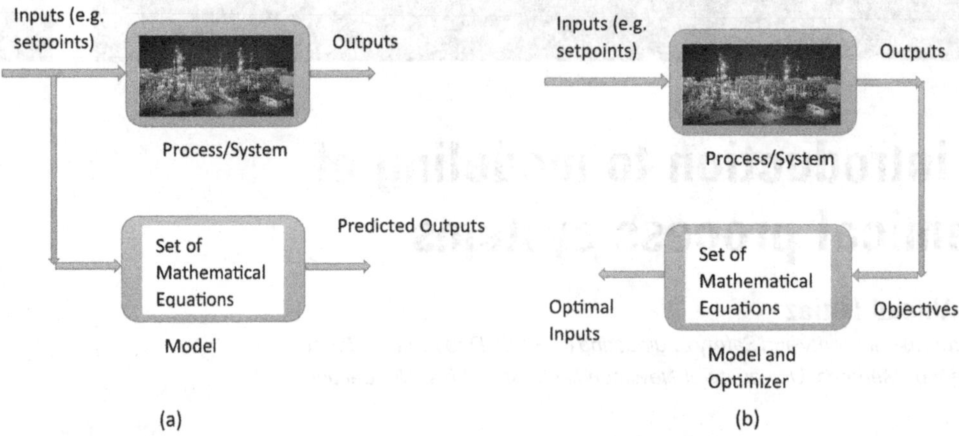

Fig. 1 A diagram showing model and its use to solve (A) direct problem and (B) inverse problem.

general depiction of a model. The common use of a model is for prediction purpose. A validated model can be used to predict the output given all the inputs and the model parameters, which is known as a forward problem or a direct problem. A model can also be used to find the desired inputs given all the output conditions and model parameters, often called "inverse problem."

2 Historical perspective of simulation, systems engineering, and process systems modeling

In the simulation world, the first breakthrough happened in the field of operation research. A number of simulation systems were developed. In 1960 at the University of Southampton, Keith Douglas Tocher developed a systematic simulation of a manufacturing line called general simulation program (GSP). The general purpose simulation system (GPSS) was developed in IBM, which is a discrete event simulator. The GPSS was designed to facilitate rapid simulation modeling of complex teleprocessing systems, for example, urban traffic control, telephone call interception and switching, airline reservation processing, and steel-mill operations. In any simulation theories, data and tools from many different disciplines are involved. To bring all these diverse elements together, a "Systems concept" was started. It is a very powerful concept in simulation that groups together interacting, interrelated, or interdependent elements, forming a complex to accomplish a defined objective(s) (INCOSE, 2018). A system interacts with its environment, which may include hardware, software, firmware, people, information, techniques, facilities, services, and other support elements. Systems concept has led to systems engineering (SE), which is a very broad, overarching idea, and generally applicable to many engineering disciplines, including biomedical systems, space vehicle systems, weapon systems, transportation systems, and so on. Systems engineering emerged as a metascience discipline to deal with system complexity at the high level when more and more segmentation was happening in different engineering and science disciplines (Klatt & Marquardt, 2009). PSE, in particular, is a narrow branch of SE associated with computers, software, and information technology (IT) (Bahill & Gissing, 1998). The application of PSE starts right at the conceptual design stage and ends at the system disposal stage. Modeling and simulation is a core component of the process systems engineering (PSE) subdiscipline of chemical engineering. PSE has its roots in systems engineering. The term PSE was first used in a special volume of the AIChE symposium series in 1961 (Grossmann & Westerberg, 2000). PSE is an interface between chemical engineering, applied mathematics, and computer science with specific model-based methods and tools as its core competencies to deal with the multiobjective design and operational challenges during the lifecycle of the manufacturing process of chemical products (Klatt & Marquardt, 2009). PSE is now a well-developed area and has several significant branches in the field of design, control, and operation.

The history of process systems modeling and simulation is intertwined with the history of simulation and the development in mathematics and computer science (Table 1). Some of the desired properties of an effective simulator are a strong graphical user interface, modular design, many functional libraries, good visualization, and minimal requirement for coding. These same goals have driven the simulation word and development of programming language. Kiviat (1967) outlined the required proficiencies for modeling and noted that it is not feasible to acquire all these proficiencies for a modeler to build a single model (Chan et al., 2017). Similarly, a process systems modeler has to rely on programmers and developers to develop user-friendly simulators so that they can focus more on the process-related aspect in the simulation. The evolution of the modeling tools has revolved around finding a delicate balance between flexibility and ease of use. External factors that have shaped the evolution of modeling and simulation are theoretical development, computer hardware, computer software, human machine interface, computer graphics etc. (Stephanopoulos & Reklaitis, 2011).

2.1 Theoretical development

On the theoretical side, the Transport Phenomena book by Bird et al. (1960) has been the major thrust in mechanistic modeling of process systems. In their seminal work, they explored the

Table 1 Evolution of process simulators with progress in computer science (Breitenecker, 2006; Nance & Sargent, 2002).

Timeline	Hardware	Software	Chemical process modeling and simulation
1950–60		FORTRAN (1956) ALGOL (1960)	General simulation program (1960) by Keith Douglas Tocher
1960–70	UNIVAC 1107 (1962)	SIMULA I (1961) SIMSCRIPT I (1963) SIMSCRIPT 67 Activity Cycle Diagram (ACD) (1964)	General purpose simulation system (GPSS) (1961) at IBM PROCESS flow sheeting (1966)—the predecessor of PRO/II DESIGN flowsheeting (1969) for oil and gas application
1970–80		Process languages based on networks of blocks (nodes) Q-GERT (Pritsker 1979) SLAM (Pegden and Pritsker 1979)	Initiation of advanced system for process engineering (ASPEN) (1976) project at MIT SPEEDUP at Imperial College in London (United Kingdom) and TISFLO at DSM in The Netherlands
1980–90	IBM PC (1981) Workstations (1985) and Operating system, UNIX	SIMAN (Pegden 1982) MAPLE (1982) MATLAB (1984)	Aspen Plus (1982)
1990–2010			HYSIM (1991) from the University of Calgary, later Aspen HYSYS DESIGN II (1995) by WinSim Inc.
2010-	Cloud-based computation	Python Virtual Reality: Simio, FlexSim, Emulate3D, Microsoft Azure Digital Twins, IBM Digital Twin Exchange	Emerson Digital Twin Honeywell Forge

interrelation between mass, energy, and momentum and studied under the term "Transport Phenomena." To this day, this remains the most used textbook in this field, and the fundamental equations have been standardized on the notations introduced in their book. The main modeling and simulation activity in the process system is "flowsheeting", and simulation programs developed for "flowsheeting" are commonly referred as "Process Simulators." The quality of a process simulator is mostly judged by its property package and equation of states (EOS). Most simulators now will have over 1000 pure components' properties in the database. A key driver specific to process system simulation was the collection, evolution, and compilation of thermophysical data. Much of these works were carried out through industrial and academic research with some government support (Rhodes, 1996).

2.2 Computer software

The machine language representations of the early 1950s gave way to the assembly language of the mid-1950s. FORTRAN has been the language of scientific and engineering computation for over five decades. In 1954, FORTRAN was developed at IBM by a group of programmers led by John Backus. The first manual of FORTRAN was published in 1956, and the compiler appeared in 1957 (Backus, 1978). Subsequent versions of FORTRAN arrived with more features for structured programming, and major versions are FORTRAN 77, Fortran90, and an object-oriented version Fortran2003. Based on FORTRAN, SIMSCRIPT was developed in 1963 for the nonprogrammers, subsequently SIMSCRIPT II emerged as a separate simulation language. SIMULA I and SIMULA 67, developed in Norwegian Computing Centre in Oslo, were the first object-oriented language. It prompted the idea of objects as elements of the simulations, and these objects can include actions described by logics that control this object and may interact synchronously or asynchronously with other objects (Chan et al., 2017). Many of the later generation object-oriented programming languages, including C++, Java, and C#, were influenced heavily by SIMULA (Goldsman et al., 2009).

2.3 Human-machine interface

The importance of visualization in modeling and simulation was not fully understood in the early days. In the 1980s, slowly the power of visualization for understanding the model output, communicating results, and fixing the errors was realized. One of the early simulation animation systems was a product called "See Why" developed in the late 1980s, which used rudimentary character-based animation. The Cinema animation system was a 2D vector-based real-time animation system for SIMAN models. In the 1990s, 3D animation capabilities began to emerge. 3D animations have been used widely for the discrete system model including manufacturing, material handling, and warehousing and to a lesser extent in the process industry. With the new trend for building digital twins for Process Plants, the 3D animation technologies are becoming more important. The goal of a digital twin is to bring all applications under one umbrella of a 3D virtual replica of the actual plant. The combination of the object-oriented framework, an immersive 3D modeling environment, and improved 3D drawing features has proven to be a very powerful combination for bringing 3D animation into the mainstream of simulation modeling. Most modern simulation tools now provide 3D as a standard capability (Chan et al., 2017).

3 Classification of models

Process models can be classified from different perspectives. In Cameron and Hangos (2001) several high-level classification criteria were stated. A modified version of the classification criteria is given in Table 2. In the context of this book, the

Table 2 Classification of process models (Cameron & Hangos, 2001).

Criteria for classification	Model type	Salient properties
Use of mechanism	Mechanistic model	Models are based on underlying physics of the system. Mass, energy, and momentum balances are the building blocks of the model
	Empirical/data-based model	Models are built from the input output data and capture the correlation between input and output data rather than the underlying mechanism
	Hybrid model	A combination of the mechanistic and empirical model
Causality	Probabilistic	Models the distribution behavior, rather than individual item. There would be some randomness in the output. The same set of inputs will not give the exact same output. Probabilistic models are widely used in reliability engineering
	Deterministic	The output of the model is fully determined by the inputs and model parameters
Spatial resolution	Lumped parameter model	Dependent variable is not a function of spatial variation. Property value is an average over the entire space, for example, continuous stirred tank reactor (CSTR)
	Distributed parameter model	Dependent variable is a function of spatial variation, for example, plug flow reactor (PFR)
Time resolution	Steady-state model	Only captures the static relationship of the system and does not cover temporal variation of the system
	Dynamic model	Captures temporal variation of the system, used primarily for controlling and monitoring process systems
Equation type	Linear model	The resulting equations are linear (i.e., superposition rule applies)
	Nonlinear model	The resulting model equations are nonlinear (i.e., superposition rule does not apply)

classification based on mechanism, classification based on spatial resolution, and scale of the model are the most relevant classifications. We will focus on these aspects of models in this book.

3.1 Use of mechanism

Mechanistic models are based on the underlying physics of the system. In the context of a process model, these types of models will be based on the mass, energy, and momentum balance of the system. All mechanistic models will have some empirical components; for example, in a reactor model, rate constants of reaction are actually fitted parameters from reaction kinetic study. One of the most widely used mechanistic model in process industries is SPYRO, which is based on fundamental reaction kinetics. In 2001, the model was reported to consist of 3288 reactions, involving 128 components and 20 radicals (Van Goethem et al., 2001). On the other hand, empirical models are not derived from the underlying mechanisms of the system. Rather, it is based on the input-output data collected from the system often through system identification experiments. Hybrid models are a combination of mechanistic and empirical models. For example, in many process control applications, running a mechanistic model on-line is not feasible due to the prohibitive computational load. In those scenarios, an empirical model is trained using the data from a mechanistic model. The empirical model runs in parallel with the mechanistic model at a higher frequency to generate outputs for the controller application. One such hybrid model is described in Sharma et al. (2018). A hybrid model was developed to predict the product sulfur content of a diesel hydrotreating (DHT) plant. The hybrid model shown in Fig. 2 is a combination of a mechanistic model based on reaction kinetics and an empirical model. The on-line empirical model is based on the support vector regression (SVR) model. It takes all the measured inputs and estimated feed sulfur concentration from the mechanistic model as its inputs. The trained SVR model predicts the product sulfur at the reactor outlet. The off-line mechanistic model on the other hand activates only when there is a significant mismatch between the predicted product sulfur and the measured sulfur from the lab or analyzer. The primary reason for such mismatch is a grade change in the feed. The mechanistic model and an optimization algorithm uses inverse modeling to calculate the feed sulfur concentration, which is then used as an input to the on-line model. The mechanistic model can also be used to generate calibration data in case the reactor is operating in a completely new operation zone.

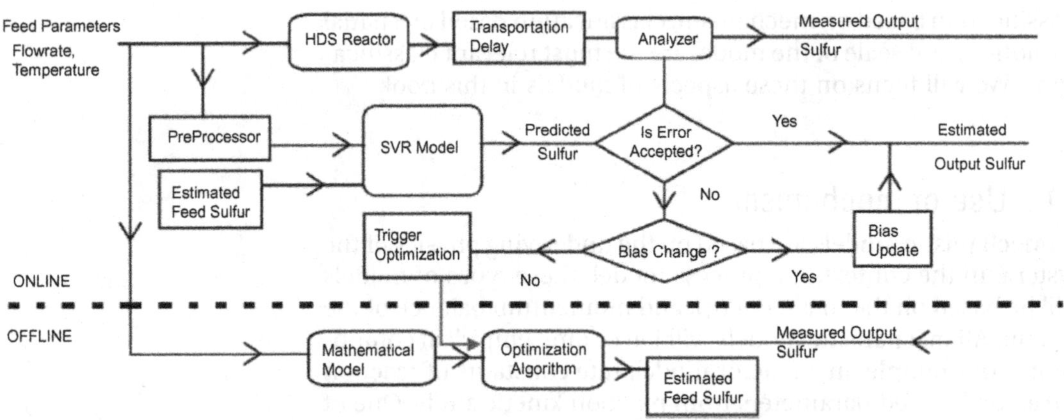

Fig. 2 A hybrid structure showing the integrated application of the mechanistic model and empirical model (Sharma et al., 2018).

3.2 Scales of models

The type of a model depends heavily on the objective or goal of the model. Developing a model requires significant time and has cost implications. A model of the same system can be very simple to very complex. It is always suggested to start with simpler models and add complexities incrementally to the extent needed. There are several dimensions to a model, Cameron and Hangos (2001) has termed this as "Hierarchies." Three dimensions/hierarchies have been identified for process models: characteristic size, level of details (complexity), and characteristic time. These three dimensions are not completely independent. Various size or scales involved in chemical engineering are pore scale (catalyst and adsorbent): 1–1000 nm; particle scale: 10 μm–1 cm; reactor/separator scale: 1–10 m. Based on level of details or complexity, Levenspiel (2002) classified models to US$10 model, US$100 model, and US$1000 model. For example, the simplest model of an adsorption column (US$10 model) assumes isothermal operation, plug fluid flow, infinitely fast mass transfer between fluid and solid phases (instantaneous equilibrium at the interface), and trace system. Model equations are the mass balance in a bed volume element and the equilibrium law at the fluid/solid interface. More details can be added to improve the accuracy of the model. Inclusion of the diffusion and intra-particle convection requires a pore-scale-level analysis, which would be considered a US$100 model. An example of a US$1000 model would be a computational fluid dynamic (CFD) model of the adsorption column describing the fluid flow pattern in the column, capturing any inefficiency in mixing from the hydrodynamics of the fluids. The model will include mechanisms that could be at the particle scale. In the last decade, increasing focus has been in the simulation of material bulk properties, equilibrium properties, and reaction rate constants from subatomic-scale simulation (Rodrigues & Minceva, 2005).

Molecular dynamic (MD) simulation and density functional theory (DFT) are two widely used techniques in chemical engineering applications. In MD simulation, the system is discretized to atomic level, and electron and nuclei are lumped together. The initial step is to assume structure. Several force fields are applied to each atom, and the total energy is calculated. The structure is optimized by minimizing the total energy of the system. Classical Newtonian mechanics is used to evaluate the position of each atom. Subsequently, the data are analyzed to relate the atomic properties to the bulk properties (Ingram et al., 2004). On the other hand, DFT is much more involved. DFT aims to find the ground state of a collection of atoms by solving Schrodinger equation. However, solving Schrodinger equation is a complex problem, and in order to solve the equation, several assumptions are made, for example, electrons are much faster than nuclei as all such calculations are carried out only on the electron, and instead of many body wave function, a one-body density is used as a fundamental variable which significantly reduces the dimension of the problem. The analysis also uses Hohenberg and Kohn theory that states the ground state energy is a functional of electron density. DFT provides important information about transition states, which can be used to confirm the reaction mechanism. The activation energy calculated from the DFT can be used to estimate rate constants. The time scale for DFT simulation is of the order of nano second (ns). Usually the DFT simulations are carried out for 100–200 atoms. Prediction of macro-scale properties from the atomic scale is computationally formidable. However, the discrete nature of systems allows multilevel and multiscale simulation. At each level, the elements and their interactions are defined, and this can be extended to the "elements" of the higher levels by substituting the atoms and molecules by higher-level elements (i.e., particles) as shown in Fig. 3. Simulations at different levels are repeated to link to the macro-scale property. For example, the discrete element method (DEM) and the smoothed particle hydrodynamics (SPH) method employ natural or fictitious material elements with Newtonian interactions for the simulation of macro-scale behavior (Ingram et al., 2004).

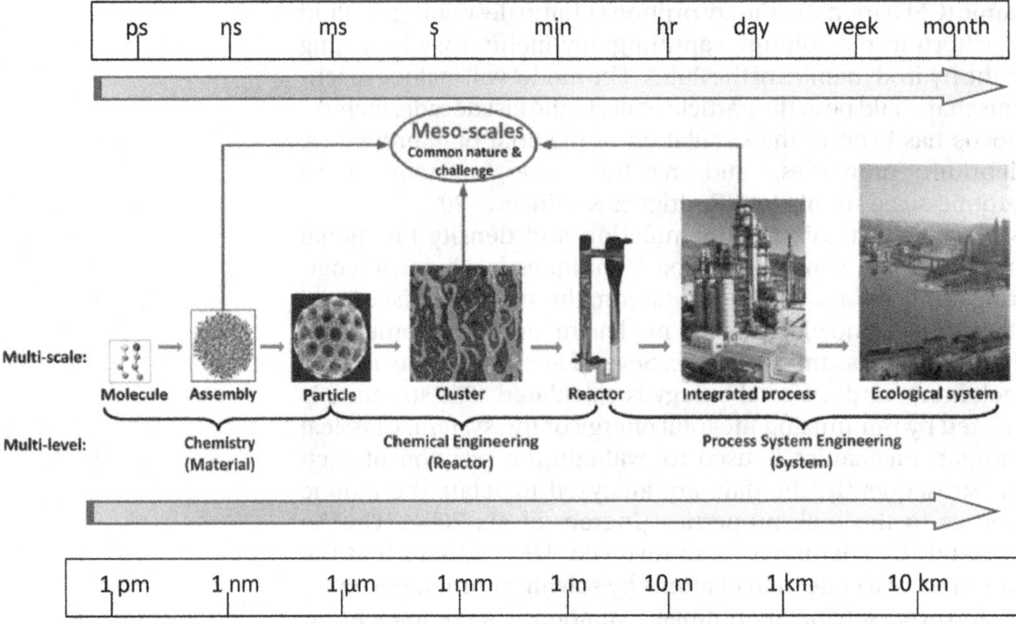

Fig. 3 Process systems in different scale. Modified from Xu, J., Li, X., Hou, C., Wang, L., Zhou, G., Ge, W., & Li, J. (2015). Engineering molecular dynamics simulation in chemical engineering. *Chemical Engineering Science, 121*, 200–216.

4 Multiscale modeling

Multiscale modeling is an important strategy to extend traditional modeling and achieve higher level of details and accuracy in the model. In van der Valk et al. (2007), a multiscale model is defined as "a composite mathematical model formed from two or more submodels that describe phenomena at different scales." Sometimes, multiscale models are viewed as software integration of models from different levels, which is not the entire picture. There are several considerations in the development and integration of multiscale models. In the development phase, the order in which the model is developed could dictate how easily the model can be developed. There are several approaches: (i) bottom-up: starting from the smaller scale models, models of increasing space and time scale are constructed. This allows building the model from fundamental mechanisms. With the increasing use of MD and DFT in chemical engineering, there has been some theoretical gain carried out in the area. However, still, more progress has to be done in order to perform a fully integrated bottom-up model from the atomic level. Also, the computational cost is formidable; (ii) top-down: starting from the simple models at the top layer and subsequently adding details, and refinements to the model. This approach is suitable at the design stage to start modeling the system with minimal information and adding details during life-time of the process; (iii) concurrent: simultaneously building models of all levels; (iv) middle-out: starting with the model for which the most information is available and then working outward from that model in smaller and larger scale (Ingram et al., 2004; van der Valk et al., 2007).

There can be many different arrangements for linking the models from different scales. Ingram and Cameron (2002) classified the approaches as follows: (i) multidomain: the domains of the model are different in this case, and the accuracy level required by these domains is different. For example, in CFD simulation, the mesh density near any boundary layer is much higher than that of the bulk phase; (ii) embedded: microscale models are embedded in the macroscale model. Some of the parameters of the macroscale model are obtained from the microscale model. One example for such models would be the reactor model where the rate constant is calculated from MD simulation; (iii) parallel: both models span the entire system domain, and the models are complementary to each other; (iv) serial: the transformation is transmitted in one way, for example, micro-scale to macro-scale; and (v) simultaneous: the entire system is built-up from the microscale models. The macro-scale properties are estimated by summation of the micro-scale properties, averaging or performing some other detailed statistical operation. No macroscale conservation equations are written.

A good example of multiscale model application can be seen in the operation of chemical processes. In chemical and petrochemical plants, operation models with different levels of details and timescale are vertically integrated to control and optimize the process as shown in Fig. 4. In a typical process plant, the first control layer above the process plant and instrumentation is a regulatory control layer, which is typically a model-free proportional-integral-derivative (PID) controller. In a process plant, PID controllers typically run in 1 s frequency. Above the regulatory layer, there will be an advanced process control (APC) layer. The APC layer typically is a model predictive controller (MPC) that performs supervisory control of unit or the entire plant. The heart of an MPC is a dynamic model that is typically built from process data using system identification techniques. The time scales for these models are of the order of minutes. On top of the APC layer, there will be a real-time optimization (RTO) layer. The RTO layer is built on a first principle/mechanistic model. These models are typically steady-state models with the time scale of the order of

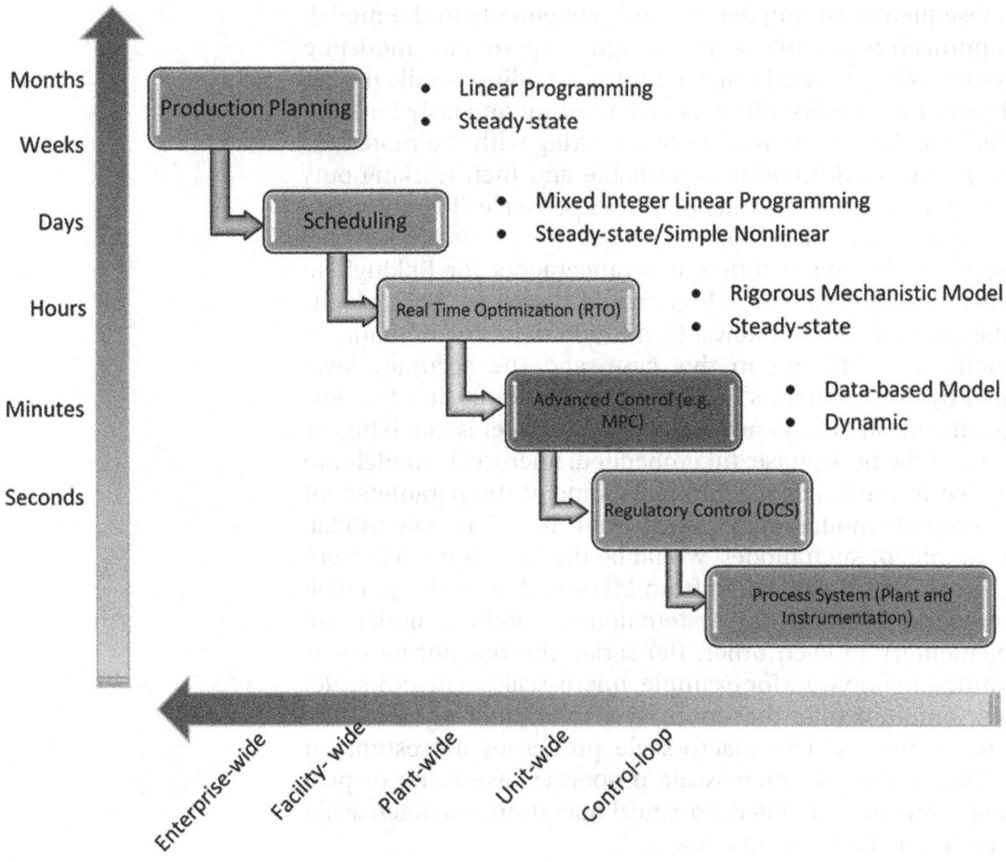

Fig. 4 Vertically integrated multiscale model for refinery operation.

hour. RTO will calculate the optimum setpoint for the MPC. Thus, the output of the RTO are inputs to the APC layer. RTO is usually used to optimize parameters which are not frequently adjusted. For example, in an ethylene cracker steam is added to the feed in order to minimize coking in the furnace tubes. Steam to hydrocarbon ratio is an important parameter to maintain optimal cracking without damaging the furnace tubes. However, it is only updated when there is change in the feed or over time as the tube gets plugged due to coke deposit. Optimizer is used to find the optimal setpoint for steam to hydrocarbon ratio, while the APC will be used to dynamically maintain the optimal steam to hydrocarbon ratio to the desired setpoint. Above the RTO layer, there are usually two other layers, a scheduling layer and a production planning layer. Production planning is the top-most layer that is used make mid-term technical decision, such as sales forecast and inventory plans for period in the weeks and months range. It typically uses the linear programming (LP) model, general plant topology, and market information to perform the forecasting. The scheduler sits in between RTO and production planning layers, which makes day to day detailed execution plan to meet the production plan. The scheduler uses the basic material balance model and operational constraints to formulate a mixed-integer linear programming (MILP) optimization problem, which relies on both continuous and discrete time representations.

5 Modeling applications in processes

Models have become an essential tool for the entire life cycle of process systems beginning from research and development, design, commissioning, operation, training, and problem solving (Dimian et al., 2014). Modeling and simulation has become an indispensable tool in all phases of a process systems.

5.1 Research and development of new product

In the 1970s, the pharmaceuticals industries started using molecular simulation to develop new products. Now, molecular simulation is in the workflow of new drug development. The practice of molecular modeling to gain new insights and aiding development of chemicals and materials started around 1990s. A combination of limited experiments and careful use of modeling and simulation tools can be very beneficial for developing polymers and chemicals (Dimian et al., 2014). For example, in Proctor and Gamble molecular dynamists study the complex interactions of billions of atoms and simulations run at the atomic level to determine how tiny nanoscale structures impact the characteristics of the ingredients in their soaps, detergents, lotions, and shampoos. Their goal is to develop more sustainable products without compromising the performance of the product. They were able to produce virtual (i.e., computer generated) "phase diagrams" that examine the behavior of materials at various concentrations. These are vital information for the formulators who can predict the concentrations at which the product will make phase change and accordingly mix and match components to create new product (Spencer & Scriba, 2008).

5.2 Design and commissioning of process systems

Applications of modeling have significantly changed the workflow and accelerated the design and development in engineering. Use of modeling and high-performance computations have cut down the time for developing new vehicle platform from an average of 5 years to 2 years. A similar impact can be seen in the process industries as well. In the design methodology, computer-aided flowsheeting is now a general norm. Immediately after a process flow diagram (PFD) has been finalized, the plant design is directly transferred to the simulator to develop the flowsheet. From that point, a simulator plays a vital role in the design and calculations. The process simulator is used to explore different process options and rank potential options, and simulators are used to find the sustainable design through heat integration and pinch analysis, to review and optimize design and operating conditions. Dynamic simulators are now widely used to assist with commissioning of a plant. Siemens have reported through use of their simulators and practicing virtual commission clients were able to speed up actual commissioning by as much as 60% and reduced unwanted standstill periods to a minimum (Chemical Engineering, 2018). Besides the flowsheeting software, CFD studies play an important role in designing several important pieces of equipment including absorber, catalytic reactor, and pipe flow. For example, the complex hydrodynamics of heterogeneous reactors are not well-understood due to complicated phenomena such as particle-particle, liquid-particle, and particle-bubble interactions. For this reason, CFD has proved to be a useful tool for understanding multiphase reactors for precise design and scale up (Panneerselvam et al., 2009).

5.3 Operator training

Human error is one of the major causes for accident. Operator training simulators (OTS) consist of dynamic models capable of running in real time and connectivity with the distributed control system (DCS). The OTS has been widely used to train inexperienced operators to practice start-up, shut-down, and handling of various operational and abnormal scenarios. Use of simulators for training allows operators to receive training on abnormal scenarios without upsetting the plant. In the process plant, the major cause for accident is human error. An OTS plays a vital role in minimizing human error-related accidents. With the advent of 3-D graphics, virtual reality (VR) is augmented with the traditional OTS, making it a more realistic environment where operators experience the anxiety, fear, and concerns similar to the real-life situation. The new generation VR-OTS are proving to be more effective in training operators (Patle et al., 2019).

5.4 Operations and debottlenecking

Different types of models are in use in the process plants to conduct different operational tasks. In the chemical and petrochemical industries, model predictive controllers (MPCs) are widely used. The core of an MPC is a dynamic model. Typically, these dynamic models are data-based models developed from carefully conducted system identification experiments in a plant. Steady-state mechanistic models are usually used for optimization purpose which calculate optimal operating conditions under various feed conditions. Models are also used to predict product quality. In many cases, a sensor is not available or too expensive to install. In those cases, operations rely on model predictions, typically known as "soft sensors." Both mechanistic and data-based models are used to develop a soft sensor. A widely used soft senor in the ethylene plants is SPYRO from TechnipFMC. In the ethylene furnaces, there is no online analyzer to analyze the cracked ethylene product. Gas chromatography in the lab usually takes about 8–12 h to get the compositional analysis, which is not suitable for model-based control. In order to predict the steam cracking yield, the plant relies on a mechanistic model. SPYRO is a highly sophisticated model that uses a very detailed kinetic scheme consisting of more than 3000 reactions involving 128 molecular and 20 radical components (Van Goethem et al., 2001). Since its introduction in 1979, SPYRO has been become an essential tool for ethylene producers worldwide. The model is also used for finding optimal operating conditions, for example, the steam to hydrocarbon ratio in the furnace feed. Careful selection of the steam to hydrocarbon ratio is important to maximize the yield and minimize coke deposition in the furnace tubes.

5.5 Abnormal situation management

Models are also the core of many activities including fault detection and diagnosis (FDD), root cause analysis, and safety and risk assessment. Models are at the center of any FDD method as shown in Fig. 5. Models are used to predict the fault-free response of the process. The predicted response is compared with the actual measurements to calculate the residuals of the system, followed by statistical test on the residuals to determine if the deviation is significant or not. Essentially, the type of the model determines the class of a FDD method. Several classifications can be found in literature. Fig. 5 shows a classification of FDD methods based upon use of different types of models (Arunthavanathan et al., 2021). In Fig. 5, all methods under "Analytical/software-based" classification uses models explicitly or implicitly. Under this classification, "Model-based method" refers to the methods that use explicit models of different type, for example, mechanistic model or well-parameterized models identified from process data. These models could be

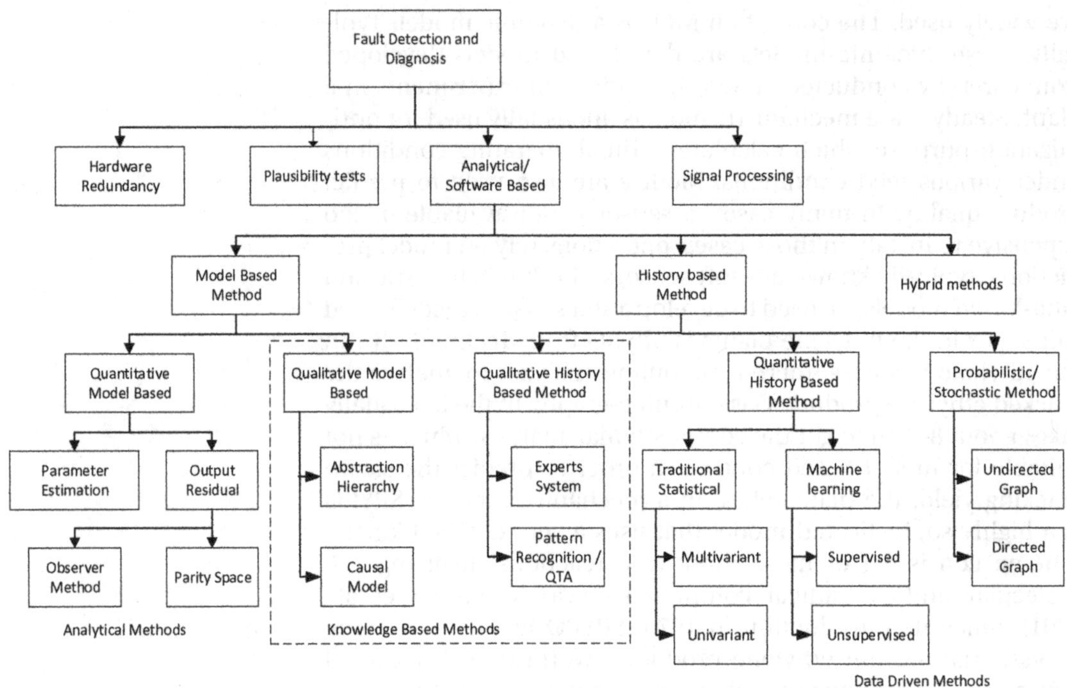

Fig. 5 Fault detection and diagnosis methodology classification (Arunthavanathan et al., 2021).

parameterized input-output model, or it could be a qualitative structural model showing the interrelation between different variables, these have been grouped under "History-based methods."

6 Scope of the book

This book aims to cover the breadth of models applied in process systems. Specifically, the focus of the book is spatial resolution or the scales of model. It covers scales from molecular level to the process systems level. The book will provide the readers a feel for multilevel modeling and their interaction. The other theme captured in the book is the progression from mechanistic to empirical modeling and the recent trend toward using machine learning techniques for process system modeling. The book can be used to teach process systems modeling to the senior-level undergraduate students and graduate students.

References

Arunthavanathan, R., Khan, F., Ahmed, S., & Imtiaz, S. (2021). An analysis of process fault diagnosis methods from safety perspectives. *Computers & Chemical Engineering, 145*, 107197.

Backus, J. (1978). The history of Fortran I, II, and III. *ACM SIGPLAN Notices, 13*(8), 165–180.

Bahill, A. T., & Gissing, B. (1998). Re-evaluating systems engineering concepts using systems thinking. *IEEE Transactions on Systems, Man and Cybernetics. Part C, Applications and Reviews, 28*(4), 516–527.

Bailer-Jones, D. M. (2002). Scientists' thoughts on scientific models. *Perspectives on Science, 10*(3), 275–301.

Bird, R. B., Stewart, W. E., & Lightfoot, E. N. (1960). *Transport phenomena* (p. 413). New York: John Wiley & Sons.

Breitenecker, F. (2006). Software for modelling and simulation-history, developments trends and challenges. In *SimVis* (pp. 7–20).

Cameron, I. T., & Hangos, K. (2001). *Process modelling and model analysis.* Elsevier.

Chan, W. K. V., D'Ambrogio, A., Zacharewicz, G., Mustafee, N., Wainer, G., Page, E., & Abeal, R. M. (2017). *Proceedings of the 2017 winter simulation conference.*

Chemical Engineering. (2018). *Simulation for faster commissioning and training.* https://www.chemengonline.com/simulation-faster-commissioning-training/.

Dimian, A. C., Bildea, C. S., & Kiss, A. A. (2014). *Integrated design and simulation of chemical processes.* Elsevier.

Goldsman, D., Nance, R. E., & Wilson, J. R. (2009). A brief history of simulation. In M. D. Rossetti, R. R. Hill, B. Johansson, A. Dunkin, & R. G. Ingalls (Eds.), *Proceedings of the 2009 winter simulation conference.*

Grossmann, I. E., & Westerberg, A. W. (2000). Research challenges in process systems engineering. *AIChE Journal, 46*(9), 1700–1703.

Hutten, E. H. (1954). The role of models in physics. *British Journal for the Philosophy of Science, 4*, 284–301.

INCOSE. (16 October 2018). *Guide to the systems engineering body of knowledge (SEBoK) version 1.9.1.* URL: https://www.sebokwiki.org/w/images/SEBoK%20v.%201.9.1.pdf.

Ingram, G., & Cameron, I. (2002). Challenges in multiscale modelling and its application to granulation systems. In *Ninth APCChE congress and CHEMECA 2002, 29 September–3 October 2002, Christchurch, NZ.*

Ingram, G. D., Cameron, I. T., & Hangos, K. M. (2004). Classification and analysis of integrating frameworks in multiscale modelling. *Chemical Engineering Science, 59*(11), 2171–2187.

Kiviat, P. J. (1967). *Digital Computer Simulation: Modeling Concepts. RM-5378-PR, August.* The Rand Corporation.

Klatt, K. U., & Marquardt, W. (2009). Perspectives for process systems engineering—Personal views from academia and industry. *Computers & Chemical Engineering, 33*(3), 536–550.

Levenspiel, O. (2002). Modeling in chemical engineering. *Chemical Engineering Science, 57*, 4691.

Nance, R. E., & Sargent, R. G. (2002). Perspectives on the evolution of simulation. *Operations Research, 50*(1), 161–172.

Oh, P. S., & Oh, S. J. (2011). What teachers of science need to know about models: An overview. *International Journal of Science Education, 33*(8), 1109–1130.

Panneerselvam, R., Savithri, S., & Surender, G. D. (2009). CFD simulation of hydrodynamics of gas–liquid–solid fluidised bed reactor. *Chemical Engineering Science, 64*(6), 1119–1135.

Patle, D. S., Manca, D., Nazir, S., & Sharma, S. (2019). Operator training simulators in virtual reality environment for process operators: A review. *Virtual Reality, 23*(3), 293–311.

Rhodes, C. L. (1996). The process simulation revolution: Thermophysical property needs and concerns. *Journal of Chemical & Engineering Data, 41*(5), 947–950.

Rodrigues, A. E., & Minceva, M. (2005). Modelling and simulation in chemical engineering: Tools for process innovation. *Computers & Chemical Engineering, 29*(6), 1167–1183.

(21 May 2012). *Scientific modelling, Encyclopaedia Britannica.* https://www.britannica.com/science/scientific-modeling. (Accessed 13 October 2020).

Sharma, P., Imtiaz, S., & Ahmed, S. (2018). A hybrid model for predicting product sulfur concentration of diesel hydrogen desulfurization process. *Chemometrics and Intelligent Laboratory Systems, 182*, 202–215.

Spencer, S., & Scriba, H. W. (2008). Best practices in process simulation for design of complex metallurgical plants. In *Metallurgical plant design and operating strategies (MetPlant 2008), 18-19.*

Stephanopoulos, G., & Reklaitis, G. V. (2011). Process systems engineering: From Solvay to modern bio-and nanotechnology: A history of development, successes and prospects for the future. *Chemical Engineering Science, 66*(19), 4272–4306.

van der Valk, T., van Driel, J. H., & de Vos, W. (2007). Common characteristics of models in present-day scientific practice. *Research in Science Education, 37*, 469–488.

Van Goethem, M. W., Kleinendorst, F. I., Van Leeuwen, C., & van Velzen, N. (2001). Equation-based SPYRO® model and solver for the simulation of the steam cracking process. *Computers & Chemical Engineering, 25*(4-6), 905–911.

2

Model equations and modeling methodology

Syed Ahmad Imtiaz

Center for Risk, Integrity and Safety Engineering (C-RISE), Department of Process Engineering, Memorial University of Newfoundland, St. John's, NL, Canada

1 Process model and model equations

Process models are widely used for a variety of applications, including design, simulation, optimization, control, and monitoring. At the heart of all these applications is the ability of the models to solve various problems. Process system modeling is a combination of many ingredients. The main building block of the model is the conservation laws of mass, energy, and momentum. In addition, to capture the various other mechanisms within the system, a host of other relations are used, for example, heat transfer equation, mass transfer equation, and reaction kinetics. All other equations besides the conservation equation are grouped under the constitutive equations. Once the equations have been formulated, usually numerical methods are used to solve the equations. Various numerical schemes can be used to numerically solve the problem. Fig. 1 schematically shows the main constituents of a mechanistic model. Typically, the size of the problems is prohibitive for hand calculation and a computer code developed in any suitable language (e.g., Matlab, Python, etc.) will be used to solve the problem. In the following sections, the main components of a model and the systematic method for building a process model will be covered.

Equations for both lumped parameter modeling and distributed parameter modeling will be described in this section. Figs. 2 and 3 show two systems representing a lumped system and a distributed parameters system, respectively. Model equations can be divided into two major categories: conservation equations and constitutive equations. Conservation equations

Modeling of Chemical Process Systems. https://doi.org/10.1016/B978-0-12-823869-1.00004-1

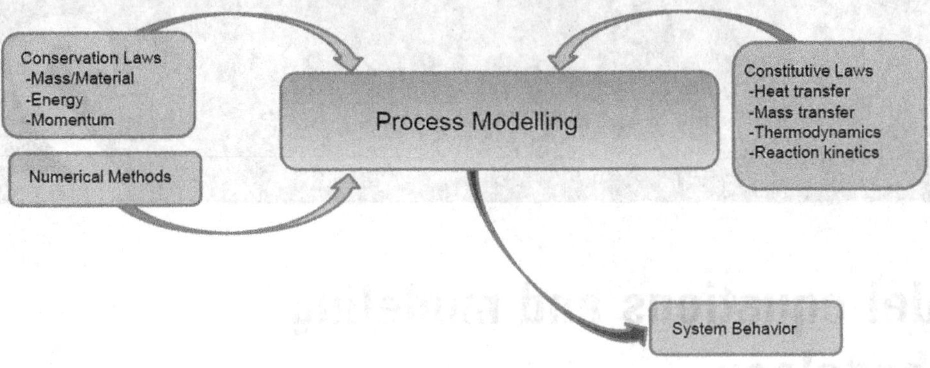

Fig. 1 Components of a process model.

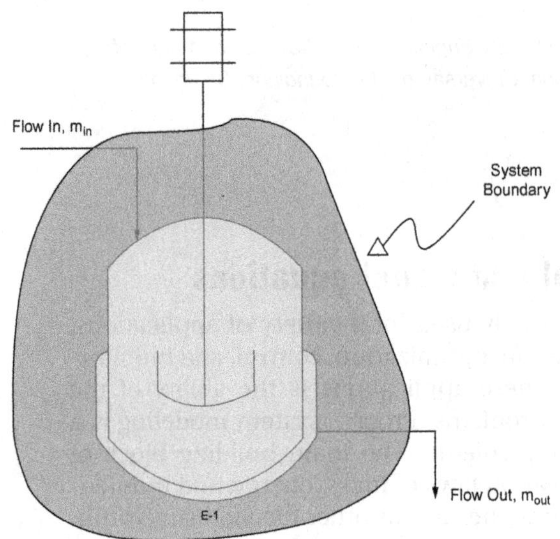

Fig. 2 A lumped parameter system.

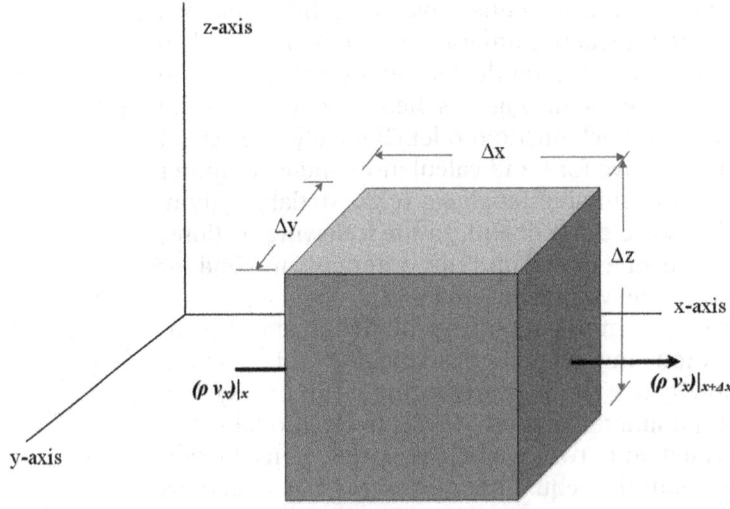

Fig. 3 A distributed parameter system.

arise from the conservation of mass, energy, and momentum; besides these equations all other equations fall under the category of constitutive equations.

2 Model equations

2.1 Conservation equations

At the heart of the conservation equations is the conservation of mass and energy described by the first law of thermodynamics. These equations can be written for the lumped system or a small discrete element. The notations of the conservation equations have been standardized based on the seminal book of Bird et al. (1960). We will use the same notations as dscribed in Bird et al. (2006) with some simplificaitons.

(i) The equation of continuity: the continuity equation is derived from mass balance. The general form of the mass balance equation is as follows:

$$\left\{\begin{matrix} \textit{rate of accumulation} \\ \textit{of mass within the} \\ \textit{system} \end{matrix}\right\} = \left\{\begin{matrix} \textit{rate of mass entering} \\ \textit{into the system} \\ \textit{through system boundary} \end{matrix}\right\} - \left\{\begin{matrix} \textit{rate of mass exiting} \\ \textit{the system} \\ \textit{through system boundary.} \end{matrix}\right\}$$

The general form of mass balance equation can be written in the system scale for lumped parameter modeling, as well as for a small control volume for distributed parameter modeling. Consider a continuous stirred tank reactor (CSTR) in Fig. 1 where a single stream enters and exits from the system.

$$\frac{dM}{dt} = m_{in} - m_{out} \tag{1}$$

Now generalizing for a system with multiple inlet and outlet streams, the general mass balance equation is given as follows:

$$\frac{dM}{dt} = \sum_{j=1}^{p} m_j - \sum_{k=1}^{q} m_k \tag{2}$$

For distributed parameter modeling, consider a small volume element, $\Delta V = \Delta x \Delta y \Delta z$ shown in Fig. 2 as the system. Fluid enters and exits the volume element through the different faces. Now writing the mass balance equation for the volume element with fluid entering and exiting the volume through the different faces results in the following total mass balance equation:

$$\Delta x \Delta y \Delta z \frac{\partial \rho}{\partial t} = \Delta y \Delta z \left[(\rho v_x)|x - (\rho v_x)|_{x+\Delta x} \right] + \Delta x \Delta z \left[(\rho v_y)|_y - (\rho v_y)|_{y+\Delta y} \right] + \Delta x \Delta y \left[(\rho v_z)|_z - (\rho v_z)|_{z+\Delta z} \right]$$

Dividing both sides by $\Delta x \Delta y \Delta z$ and taking the limits to "0," we get the following form:

$$\frac{\partial \rho}{\partial t} = -\left(\frac{\partial}{\partial x} \rho v_x + \frac{\partial}{\partial y} \rho v_y + \frac{\partial}{\partial z} \rho v_z \right) \tag{3}$$

The equation can be written more compactly using the vector notation:

$$\frac{\partial \rho}{\partial t} = -(\nabla \cdot \rho v) \tag{4}$$

where $\nabla = \frac{\partial}{\partial x} + \frac{\partial}{\partial y} + \frac{\partial}{\partial z}$ is called divergence.

For constant density, substituting $\frac{\partial \rho}{\partial t} = 0$, we obtain $\nabla \cdot v = 0$, which is a useful form of the equation and valid for incompressible fluid.

(ii) Component Mass Balance:

$$\left\{ \begin{matrix} \textit{rate of accumulation} \\ \textit{of component i within the} \\ \textit{system} \end{matrix} \right\} = \left\{ \begin{matrix} \textit{rate of component i entering} \\ \textit{into the system} \\ \textit{through system boundary} \end{matrix} \right\} - \left\{ \begin{matrix} \textit{rate of component i exiting} \\ \textit{the system} \\ \textit{through system boundary} \end{matrix} \right\} + \left\{ \begin{matrix} \textit{rate of formation or} \\ \textit{consumption of component i} \end{matrix} \right\}$$

$$\frac{dM_i}{dt} = \sum_{j=1}^{p} m_{j,i} - \sum_{k=1}^{q} m_{k,i} + g_i \tag{5}$$

(iii) Equation of Motion: Equation of motion is obtained from momentum balance over a system. The general form of the momentum balance is as follows:

$$\left\{\begin{array}{c}\textit{rate of accumulation}\\ \textit{of momentum within the}\\ \textit{system}\end{array}\right\}=\left\{\begin{array}{c}\textit{momentum entering}\\ \textit{into the system with fluid}\\ \textit{entering through system boundary}\end{array}\right\}-$$

$$\left\{\begin{array}{c}\textit{momentum exiting}\\ \textit{the system with fluid}\\ \textit{exiting through system boundary}\end{array}\right\}+\left\{\begin{array}{c}\textit{external force}\\ \textit{acting on the fluid}\end{array}\right\}$$

For the lumped system, we can write the equation of motion as follows:

$$\frac{dMv}{dt}=\sum_{j=1}^{p}(mv)_j-\sum_{k=1}^{q}(mv)_k+\sum_{l=1}^{r}F_l \tag{6}$$

For distributed parameter modeling, consider the small volume element shown in Fig. 3 as our system. The momentum entering and exiting through different faces of the system is

$$\left\{\begin{array}{c}x-\textit{component of momentum}\\ \textit{entering the system through yz face}\end{array}\right\}=\Delta y\Delta z\varphi_{xx}\Big|_{x}$$

$$\left\{\begin{array}{c}x-\textit{component of momentum}\\ \textit{exiting the system through yz face}\end{array}\right\}=\Delta y\Delta z\varphi_{xx}\Big|_{x+\Delta x}$$

$$\left\{\begin{array}{c}x-\textit{component of momentum}\\ \textit{entering the system through xz face}\end{array}\right\}=\Delta x\Delta z\varphi_{xy}\Big|_{y}$$

$$\left\{\begin{array}{c}x-\textit{component of momentum}\\ \textit{exiting the system through xz face}\end{array}\right\}=\Delta x\Delta z\varphi_{xy}\Big|_{y+\Delta y}$$

$$\left\{\begin{array}{c}x-\textit{component of momentum}\\ \textit{entering the system through xy face}\end{array}\right\}=\Delta x\Delta y\varphi_{xz}\Big|_{z}$$

$$\left\{\begin{array}{c}x-\textit{component of momentum}\\ \textit{exiting the system through xy face}\end{array}\right\}=\Delta x\Delta y\varphi_{xz}\Big|_{z+\Delta z}$$

The sum of the terms equates to the momentum accumulated within the control volume. Gathering the above components and equating to the accumulation term, we get the following form:

$$\Delta x \Delta y \Delta z \frac{\partial \rho v_x}{\partial t} = \Delta y \Delta z \varphi_{xx}|_x - \Delta y \Delta z \varphi_{xx}|_{x+\Delta x} + \Delta x \Delta z \varphi_{yx}|_y - \Delta x \Delta z \varphi_{yx}|_{y+\Delta y} + \Delta x \Delta y \varphi_{zx}|_z - \Delta x \Delta y \varphi_{zx}|_{z+\Delta z} + \Delta x \Delta y \Delta z F_x$$

where F_x is the x-component of the force acting on the fluid per unit volume. Dividing both sides by $\Delta x \Delta y \Delta z$ and taking the limits to "0," we obtain the following continuous form:

$$\frac{\partial \rho v_x}{\partial t} = -\left(\frac{\partial}{\partial x}\varphi_{xx} + \frac{\partial}{\partial y}\varphi_{yx} + \frac{\partial}{\partial z}\varphi_{zx}\right) + F_x \tag{7}$$

Similar equations can be written for y and z-components of the momentum balance.

$$\frac{\partial \rho v_y}{\partial t} = -\left(\frac{\partial}{\partial x}\varphi_{xy} + \frac{\partial}{\partial y}\varphi_{yy} + \frac{\partial}{\partial z}\varphi_{zy}\right) + F_y \tag{8}$$

$$\frac{\partial \rho v_z}{\partial t} = -\left(\frac{\partial}{\partial x}\varphi_{xz} + \frac{\partial}{\partial y}\varphi_{yz} + \frac{\partial}{\partial z}\varphi_{zz}\right) + F_z \tag{9}$$

For more details of these equations, interested readers are referred to Bird et al. (2006).

(iv) Energy balance equation: The conservation law of energy states that

$$\left\{\begin{matrix}\textit{rate of accumulation}\\ \textit{of energy within the}\\ \textit{system}\end{matrix}\right\} = \left\{\begin{matrix}\textit{rate of energy entering}\\ \textit{into the system}\\ \textit{through system boundary}\end{matrix}\right\} - \left\{\begin{matrix}\textit{rate of energy exiting}\\ \textit{the system}\\ \textit{through system boundary}\end{matrix}\right\} + \left\{\begin{matrix}\textit{Heat added to}\\ \textit{the system}\end{matrix}\right\} + \left\{\begin{matrix}\textit{Work done on}\\ \textit{the system}\end{matrix}\right\}$$

For a system with volume V, we can write the energy balance equation as follows:

$$\frac{dE}{dt} = \sum_{j=1}^{p} m_j \widehat{E}_j - \sum_{k=1}^{q} m_k \widehat{E}_k + Q + W \tag{10}$$

where total energy, $E = U + K_E + P_E$; U stands for internal energy, K_E is kinetic energy, and P_E is potential energy. Q is total heat added to the system and W is work done on the system.

The work term W can be expanded to $W = W_F + W_E + W_s$;

Flow work, $W_F = \sum_{j=1}^{p} m_j \left(P\widehat{V}\right)_j - \sum_{k=1}^{q} m_k \left(P\widehat{V}\right)_k$; W_E is expansion work and W_s is shaft work.

Expanding the energy term and the flow work, we get the following equation:

$$\frac{dE}{dt} = \sum_{j=1}^{p} m_j (\widehat{U} + \widehat{K}_E + \widehat{P}_E)_j - \sum_{k=1}^{q} m_k \left(\widehat{U} + \widehat{K}_E + \widehat{P}_E\right)_k \tag{11}$$

$$+ Q + \sum_{j=1}^{p} m_j \left(P\widehat{V}\right)_j - \sum_{k=1}^{q} m_k \left(P\widehat{V}\right)_k + W_E + W_s$$

Using $\widehat{H} = \widehat{U} + P\widehat{V}$, the above equation can be written in terms of enthalpy, which is a more useful form.

$$\frac{dE}{dt} = \sum_{j=1}^{p} m_j (\widehat{H} + \widehat{K}_E + \widehat{P}_E)_j \tag{12}$$

$$- \sum_{k=1}^{q} m_k \left(\widehat{H} + \widehat{K}_E + \widehat{P}_E\right)_k + Q + W_E + W_s$$

For distributed parameters modeling, the energy balance equations can be written for a small control volume as shown in Fig. 3.

The rate of increase of kinetic energy and internal energy within the volume element $\Delta x \Delta y \Delta z$ is

$$\Delta x \Delta y \Delta z \frac{\partial}{\partial t} \left(\frac{1}{2}\rho v_x^2 + \frac{1}{2}\rho v_y^2 + \frac{1}{2}\rho v_z^2 + \rho \widehat{U} \right)$$

Energy enters through the various faces of the control volume as convection and conductive heat transfer mechanism, which can be written in the following form;

Energy entering and exiting through the yz face of the control volume,

$$\Delta y \Delta z \left(v_x \left(\frac{1}{2} \rho v_x^2 + \widehat{U} \right) |_x - v_x \left(\frac{1}{2} \rho v_x^2 + \widehat{U} \right) |_{x+\Delta x} + \left(-K_x \frac{\partial T}{\partial x} \right) |_x - \left(-K_x \frac{\partial T}{\partial x} \right) |_{x+\Delta x} \right)$$

Energy entering and exiting through the *xz* face of the control volume,

$$\Delta x \Delta z \left(v_y \left(\frac{1}{2} \rho v_y^2 + \widehat{U} \right) |_y - v_y \left(\frac{1}{2} \rho v_y^2 + \widehat{U} \right) |_{y+\Delta y} + \left(-K_y \frac{\partial T}{\partial y} \right) |_y - \left(-K_y \frac{\partial T}{\partial y} \right) |_{y+\Delta y} \right)$$

Energy entering and exiting through the *xy* face of the control volume,

$$\Delta x \Delta y \left(v_z \left(\frac{1}{2} \rho v_z^2 + \widehat{U} \right) |_z - v_z \left(\frac{1}{2} \rho v_z^2 + \widehat{U} \right) |_{z+\Delta z} + \left(-K_z \frac{\partial T}{\partial z} \right) |_z - \left(-K_z \frac{\partial T}{\partial z} \right) |_{z+\Delta z} \right)$$

Summing the above terms and equating with the accumulation term, we get the following form:

$$\begin{aligned}
\Delta x \Delta y \Delta z \frac{\partial}{\partial t} \left(\frac{1}{2} \rho v_x^2 + \frac{1}{2} \rho v_y^2 + \frac{1}{2} \rho v_z^2 + \rho \widehat{U} \right) \\
= \Delta y \Delta z \left(v_x \left(\frac{1}{2} \rho v_x^2 + \widehat{U} \right) |_x - v_x \left(\frac{1}{2} \rho v_x^2 + \widehat{U} \right) |_{x+\Delta x} + \left(-K_x \frac{\partial T}{\partial x} \right) |_x \right. \\
\left. - \left(-K_x \frac{\partial T}{\partial x} \right) |_{x+\Delta x} \right) + \Delta x \Delta z \left(v_y \left(\frac{1}{2} \rho v_y^2 + \widehat{U} \right) |_y - v_y \left(\frac{1}{2} \rho v_y^2 + \widehat{U} \right) |_{y+\Delta y} \right. \\
\left. + \left(-K_y \frac{\partial T}{\partial y} \right) |_y - \left(-K_y \frac{\partial T}{\partial y} \right) |_{y+\Delta y} \right) \\
+ \Delta x \Delta y \left(v_z \left(\frac{1}{2} \rho v_z^2 + \widehat{U} \right) |_z - v_z \left(\frac{1}{2} \rho v_z^2 + \widehat{U} \right) |_{z+\Delta z} \right. \\
\left. + \left(-K_z \frac{\partial T}{\partial z} \right) |_z - \left(-K_z \frac{\partial T}{\partial z} \right) |_{z+\Delta z} \right)
\end{aligned}$$

After simplifying the final form,

$$\frac{\partial}{\partial t}\left(\frac{1}{2}\rho v_x^2+\frac{1}{2}\rho v_y^2+\frac{1}{2}\rho v_z^2+\rho\widehat{U}\right)$$

$$=-\left(\frac{\partial}{\partial x}v_x\left(\frac{1}{2}\rho v_x^2+\widehat{U}\right)+\frac{\partial}{\partial y}v_y\left(\frac{1}{2}\rho v_y^2+\widehat{U}\right)+\frac{\partial}{\partial z}v_z\left(\frac{1}{2}\rho v_z^2+\widehat{U}\right)\right)$$

$$+K_x\frac{\partial^2 T}{\partial x^2}+K_y\frac{\partial^2 T}{\partial y^2}+K_z\frac{\partial^2 T}{\partial z^2} \tag{13}$$

2.2 Constitutive equations

In process system modeling, a variety of other relations besides the conservation equations are used. These constitutive equations are incorporated alongside the conservation equations to describe various other mechanisms and processes within the system. The major categories of constitutive equations are as follows:

(i) Transfer relations, (ii) Reaction kinetics, (iii) Thermodynamic equilibrium relations, and (iv) Thermo-Physical property relations.

i. Transfer relations: The transfer relations include the transfer of mass, heat, and momentum. Because of the similarity, these three transfers are often studied together as transport phenomena. Below, we describe the Fick's law for diffusion, Fourier's law of conduction, and Newton's law of viscosity to explore the similarity among these three transfer relations. The general form of the transfer equations is

Rate of transfer $\propto$ gradient

For mass transfer, the transfer equation is called Fick's law, given as follows:

$j_i = -D\frac{dC_i}{dx}$, where j is the rate of mass transfer of component i, D is the diffusion coefficient, and $\frac{dC_i}{dx}$ is the concentration gradient across the boundary.

For heat transfer, the corresponding equation is the Fourier heat conduction equation

$q_x = -k\frac{dT}{dx}$, where q_x is the heat flux in x-direction, k is the conduction heat transfer coefficient, and $\frac{dT}{dx}$ is the temperature gradient across the heat transfer direction.

Finally, the momentum transfer rate is given as follows:

$\tau_{yx} = -\mu\frac{du_x}{dy}$, where τ_{yx} is the flux of momentum in y-direction due to velocity in x-direction, μ is the viscosity, and $\frac{du_x}{dy}$ is the velocity gradient of x-component of velocity in y-direction.

Besides these equations, there are relations for other mechanisms of heat, mass, and momentum transfer, and a wide range of empirical/semiempirical correlations are used to calculate the coefficients of heat, mass, and momentum transfer, for example, equations to describe heat transfer through convection and radiation mechanism, expressions to calculate the convection heat transfer coefficients (e.g., Dittus-Boelter Equation); convective mass transfer equation, and expressions to calculate the mass transfer coefficients (e.g., Wilke-Chang equation). Most of these relations are now standard and can be found in textbooks and chemical engineering hand books (Green & Southard, 2019). However, for specific systems, we may have to get the necessary empirical relations from scientific literature.

ii. Reaction Kinetics: Reactors are an essential part of process systems. Kinetic models are necessary for reactor models. Consider a reaction where reactants A and B react to form products C and D. The general form or the reaction is as follows:

$$\nu_A A + \nu_B B \rightarrow \nu_c C + \nu_D D$$

where ν_i, $i \in A, B, C,$ and D are the stoichiometric coefficients.

The reaction rate for component A defined as $r_A = -\frac{1}{V}\frac{dn_A}{dt}$ will typically be a function of the concentration of component A and B as follows,

$$r_A = k_A f\left(C_A^{\alpha}, C_B^{\beta}\right) \tag{14}$$

where C_A and C_B are concentrations of component A and B, respectively, *and* α and β determine the order of the reaction. Reaction order is given by the sum of α and β.

Reaction rate constant, k_A, is typically a function of temperature and described by Arrhenius equation:

$$k_A = k_0 e^{-E/(RT)} \tag{15}$$

where k_0 is preexponential term and E is the activation energy.

In many applications, reactions are lumped together and lumped parameters are used for modeling. However, when high degree of accuracy is desired, reaction mechanisms can be very elaborate. For example, to predict the yield of ethylene from naphtha, the commercial software SPYRO uses a very elaborate reaction kinetics consisting of 3288 reactions, involving 128 components and 20 radicals (Van Goethem et al., 2002). For the same cracking process, a lumped reaction kinetics can be found in literature with 17 reactions (Chen & Chu, 1998).

Reaction kinetics are typically obtained from lab-scale experiments. Researchers have conducted a wide range of experiments and reported in the scientific literature. Scientific literature is one of the main sources for reaction kinetics. In case kinetic data for a specific reaction are not available in literature, in-house experiments are carried out to obtain the kinetic information. More recently, with the advent of computational power, simulation methods are increasingly being used to obtain the kinetic information, especially in the pharmaceuticals industries and protein synthesis experiments. Molecular Dynamic (MD) simulations can be used to conduct the virtual experiments. A small number of molecules in the order 100–1000 are introduced in a nano-scale volume. Temperature, pressure, and other conditions are set to bring the molecules to the desired energy state. Reactive force fields are applied to the system that describes the interatomic interactions of the reactive species. At every time step, the evolution of the atoms is monitored. Some of the atoms may be short-lived and decompose quickly with or without reacting, while others may have a longer duration. Transition states of atoms are analyzed to figure out the mechanism which ultimately define the elementary reactions. Subsequently, analytical rate constants are calculated that can be used in continuum simulation in the micro-meter scale, which is further averaged to describe the system in millimeter and centimeter scale (Ilyin et al., 2019; Xie & Tinker, 2006).

Density function theory (DFT) simulations can also be used to find the transition state. In DFT simulation, transition state is determined through interpolation of the reactant and product. The orientation of geometries of the transition state is further optimized. Once the transition state has been optimized, the associated energy gives the activation energy and Gibbs energy. Subsequently, Eyring equation (Eq. 16) is used to calculate the reaction rate constant.

$$k = \frac{k_B T}{h} e^{-\frac{\Delta H}{RT}} e^{\frac{\Delta S}{R}} \tag{16}$$

where k_B is Boltzmann constant and h is Plank's constant. Gaussian9 with packages CAM-B3LYP+IDSCRF, BMK+IDSCRF, M062x +IDSCRF, and wB97xd+IDSCRF can be used to calculate energy barriers for a wide range of reactions (Joly & Miller, 2017; Li & Fang, 2014).

Though MD and DFT simulations are widely used in chemistry and pharmaceuticals for rate determination, process simulators still do not have the capability to perform MD or DFT simulation to estimate kinetic parameters. In the near future, process simulators will have to add the capability. However, researchers and practicing engineers can use these tools to get workable kinetic data that are not available otherwise.

iii. Thermodynamic equilibrium relations: In process simulations, equilibrium relation, especially vapor liquid equilibrium (VLE), is very important and appears in several unit operations (e.g., flash tank, distillation column, and liquid extraction). Cubic equation of states (EOS) is the preferred method for most simulators for equilibrium calculation. Thermodynamic state properties, such as internal energy (U), enthalpy (H), entropy (S), and Gibb's free energy (G), depend on the state variables P, V, and T, making EOS remarkably useful for thermodynamic calculations (Smith et al., 2005). Cubic equations such as Peng-Robinson (PR) and Soave-Redlich-Kwong (SRK) are very successful in predicting properties of hydrocarbon system. Cubic equations were originally developed for gas phase calculations and are not as efficient for liquid phase calculations. Various modifications were done to improve liquid phase prediction, for example, SRK equation uses an eccentric factor, which improved VLE computational accuracy considerably. Modifications have also been done for polar mixtures, for example, Peng-Robinson-Strijek-Vera for highly polar mixture, or SRK with Michelsen-Huron-Vidal mixing rules for moderately polar mixtures. A detailed discussion on the selection criteria for different EOS can be found in Wilhelmsen et al. (2017).

More recently, EOS derived from molecular modeling and statistical thermodynamics, such as, Statistical Associating Fluid Theory (SAFT) (Chapman et al., 1989) and Perturbed Chain Statistical Associating Fluid Theory (PC-SAFT) (Gross & Sadowski, 2001), have shown better prediction ability in various conditions. PC-SAFT and its variants have been successfully applied to highly asymmetric and associating mixtures to several complex polymer-solvent systems, including vapor-liquid equilibria, liquid-liquid equilibria, and gas solubility.

Simulators contain a database for the EOS parameters for pure components and also the coefficients of the mixing rules for various mixtures. Mixing rule coefficients are typically regressed using experimental data.

Fundamentally, when phases are in equilibrium, the temperature (T), pressure (P), and Gibbs free energy (G) of a component will also be in equilibrium among different phases, which leads to the following fundamental relation for phase equilibrium,

$$\mu_i^{\alpha} = \mu_i^{\beta} = \cdots = \mu_i^{\pi} \tag{17}$$

where μ_i is the chemical potential of component i defined as $\mu_i = \left[\frac{\partial(nG)}{\partial n_i}\right]_{P,T,n_j}$ $i = \alpha,\ \beta,\ \cdots\pi$ indicates different phases (Smith et al., 2005).

Since all phases are at the same temperature, an equivalent criterion in terms of fugacity (f) is widely used for phase equilibria,

$$f_i^\alpha = f_i^\beta = \cdots = f_i^\pi \tag{18}$$

The fugacity of component i in vapor phase and liquid phases is given as follows,

$$f_i^V = \varphi_{i,v} y_i P$$

$$f_i^L = \varphi_{i,l} x_i P$$

where φ_i is the fugacity coefficient of component i, x_i is composition of component i in the liquid phase, and y_i is the composition in vapor phase. Fugacity coefficients can be calculated from EOS. For evaluating mixture properties, interactions of components are captured using interaction parameters, and mixing rules are used to calculate average mixture properties.

For nonideal solution, fugacity of a component in solution deviates from pure component fugacity. The ratio of a component fugacity in solution to the pure component fugacity (f_i^o) is defined as activity,

$$a_i = \frac{f_i}{f_i^o}$$

and activity is calculated from activity coefficient,

$$a_i = \gamma_i x_i$$

Thus, activity coefficient γ_i indicates the deviation of the component activity from its measured concentration. Activity coefficients are typically calculated from semiempirical models. The most common one is the nonrandom two-liquid (NRTL) model, which can be applied to polar mixtures. Other models include Wilson, Van Laar, universal quasi-chemical functional group activity coefficients (UNIFAC), universal quasi-chemical activity coefficients (UNIQUAC), Flory Huggins, Electrolyte NRTL, and Scatchard Hildebrand models. These models indirectly use experimental data. More recently, thermodynamic tools emerged based on quantum mechanics that allow calculation of VLE from molecular theory. One such tool is COSMO (**CO**nductor-like **S**creening **MO**del) that allows predicting phase equilibrium (VLE and SLE) without any experimental data (Eckert & Klamt, 2002).

iv. Thermo-physical and transport property relations: Modeling and simulation of process systems require a wide variety of properties of pure components and mixtures as inputs. In a modern simulator, there will be a rich database of thermo-physical properties for over at least a thousand pure components. Also, a wide variety of correlations are used for properties which cannot be easily measured, and mixing rules are used to predict mixture properties. Below, we briefly describe the data sources and correlations of different thermo-physical and transport properties.

Thermo-physical properties: A rich source of thermo-physical property data and correlation is available at the DIPPR 801 database maintained by the Design Institute for Physical Properties (DIPPR) for the American Institute of Chemical Engineers. The 801 database is the largest collection of critically evaluated, pure-species, thermo-physical property data in the world (Wilding et al., 2020). REFerence fluid PROPerties (REFPROP) developed by the National Institute of Standards and Technology (NIST), calculates the thermodynamic and transport properties of industrially important fluids and their mixtures. REFPROP implements three models for the thermodynamic properties of pure fluids: equations of state explicit in Helmholtz energy, the modified Benedict-Webb-Rubin equation of state, and an extended corresponding states (ECS) model. Mixture calculations employ a model that applies mixing rules to the Helmholtz energy of the mixture components; it uses a departure function to account for the departure from ideal mixing (Lemmon et al., 2007). The other useful source is Perry's Chemical Engineers' Handbook which compiles a large data set and correlations from DIPPR, NIST, and various other scientific journals (Green & Southard, 2019).

Thermo-physical properties that are frequently used in modeling include density, vapor pressure, specific heat, solubility, enthalpy, and entropy. Most of these properties are temperature-dependent, and correlations are used to fit the properties.

The DIPPR database provides the following correlations to estimate liquid densities except those of water and *o*-terphenyl.

$$\rho = \frac{C_1}{C_2^{\left[1+\left(1-T/C_3\right)^{C_4}\right]}} \tag{19}$$

where ρ is in mol/dm^3 and T is in K. The pressure is equal to the vapor pressure for pressures greater than 1 atm and equal to 1 atm when the vapor pressure is less than 1 atm. The coefficients C_1, C_2, C_3, and C_4 are specific to components.

Vapor pressure of compounds is another important physical property. Antoine equation and several modifications of Antoine equations have been traditionally used for vapor pressure prediction. DIPPR has compiled vapor pressure data for a large set of organic and inorganic compounds. The following equation has been used for the prediction,

$$ln\ P = \beta_1 + \frac{\beta_2}{T} + \beta_3\, ln\ T + \beta_4 T^{\beta_5} \tag{20}$$

Vapor pressure P is in Pa, β_1, β_2, …, β_5 are component specific parameters.

Heat capacity is required for any enthalpy balance or energy balance. The following virial equation calculates the ideal gas heat capacity

$$C_p^0 = \gamma_1 + \gamma_2 T + \gamma_3 T^2 + \gamma_4 T^3 + \gamma_5 T^4 \tag{21}$$

where C_P^0 is in J/kmol K and T is in K.

Transport properties: The parameters of the transport equations, for example, heat conduction coefficient (k), viscosity (μ), and diffusivity (D), are calculated from various correlations. Perry's Chemical Engineer's Handbook has a collection of empirical equations and coefficients for different components to calculate the transport parameters. Below, we report few of these correlations.

$$\text{Vapor thermal conductivity, } k_v = \frac{C_1 T^{C_2}}{1 + C_3/T + C_4/T^2} \tag{22}$$

The liquid thermal conductivity correlation is expressed in a polynomial form.

$$\text{Liquid thermal conductivity, } k_l = C_1 + C_2 T + C_3 T^2 + C_4 T^3 + C_5 T^4 \tag{23}$$

where k_v and k_l are the thermal conductivity in W/(m K), T is the temperature in Kelvin, and C_1, C_2, C_3, C_4, and C_5 are coefficients specific to a component.

Similarly, correlations are also reported for viscosity,

$$\text{Vapor viscosity, } \mu^v = \frac{\alpha_1 T^{\alpha_2}}{1 + \alpha_3/T + \alpha_4/T^2} \quad \text{and} \tag{24}$$

$$\text{Liquid viscosity, } \mu^l = \exp^{\left(\alpha_1 + \alpha_2/T + \alpha_3 \ln T + \alpha_4 T^{\alpha_5}\right)} \tag{25}$$

where μ^v and μ^l are vapor and liquid viscosity is in Pa s and temperature T is in Kelvin. $\alpha_1, \alpha_2, \ldots, \alpha_5$ are coefficients for specific components (Green & Southard, 2019).

Correlations are also available to calculate diffusivity for different types of systems. Wilke and Chang (1955) proposed the following correlation for predicting diffusivity of nonelectrolytes in infinitely dilute solution.

$$\frac{D^0_{AB}\mu}{T} = \frac{7.4 \times 10^{-8}(\Phi_B M_B)^{1/2}}{V_{bA}^{0.6}} \tag{26}$$

where D^0_{AB} (cm^2/s) is the diffusivity of A in a very dilute solution of A in solvent B, μ is viscosity in (cp), M_B is molecular weight of solvent B, T is temperature in Kelvin, V_{bA} is solute molar volume in (cm^3/mol), and Φ_B is association factor for solvent B. Φ_B varies with the solvent system. Several other correlations are reported in literature to estimate the diffusion coefficient, for example, Hayduk and Minhas equation for solution and Wilke and Lee equation for gas diffusivity. These correlations are built-in in many simulators. For example, in AspenPlus under the property methods, DL01 is the commercial name for Wilke and Chang equation.

3 Systematic method for building process models

3.1 Systems thinking

Systems thinking is a holistic approach to problem solving. A system is defined as a group of components that work together in harmony to achieve a common set of objectives. Systems are often surrounded by physical or virtual boundaries and interact with the surroundings through various mechanisms. Systems approach takes into account the interaction between different elements within the system and the interaction of the system with the surroundings. This holistic approach to problem-solving brings strength to problem solving and prevents unintended negative consequences of the solution.

To describe the process of Systems Thinking, many authors, both technical and nontechnical, suggest the usage of the SIMILAR Process map as an example for tackling a problem. It comprises seven coarse-granular tasks which refer to the letters in SIMILAR (Bahill & Gissing, 1998):

I. State the problem: The objectives of the problem need to be clearly identified.
II. Investigate alternatives: Evaluation of other models based on a multicriteria decision making process including performance, cost, and risk.
III. Model the system: Capture the input-output relation of the process using mathematical equations.
IV. Integrate: Embed the designed model to the overall system, establish the link of the system with the surroundings, and use for the intended purpose.
V. Launch the system: Implementation and running of the system to output a product.
VI. Assess performance: Measurement of the system's efficiency by comparing outputs/performance with expectations.
VII. *Re*-evaluate: There should be a regular schedule to maintain the system to ensure that all information is current and up-to-date.

The above steps go sequentially from steps (i) to (vi). It all initiates with a customer demand which will determine the problem statement and goes through the steps. The output of the systematic thinking is the desired solution. Re-evaluation is done at each step to ensure the accuracy of the information. In case some discrepancy is identified, it is always advisable to fix the issue at that stage rather than progressing through the steps.

3.2 Steps for mechanistic model building

For a long time, modeling was considered more of an art, and there was no formal method for building a model. Over the past few decades, modeling has been done in many disciplines, including process systems modeling, and this has formalized the steps of modeling. Many researchers have proposed the steps for modeling, which closely follow the SIMILAR methodology. In this section, we will mainly focus on the steps for building a mechanistic model. The steps outlined by Cameron and Hangos (2001) are very general and widely followed. The steps for modeling with some modifications are shown in Fig. 4.In this section, we will briefly describe these steps.

i. Problem definition: A model is almost always built for a purpose. It is important to clearly state the problem and set the goal of the model. This will help clarify the type of the model, modeling details, and hierarchy level of the model. At this stage the modeler should clearly identify the system boundary, inputs, outputs, model type (lumped vs distributed;

steady state vs dynamic), and necessary range and accuracy of the model.

ii. Conceptualization: A model requires some degree of idealization. The next step is to conceptualize the model. This involves identifying the most important physio-chemical mechanisms of the system, making the necessary assumptions, considering the effect of these assumptions, and how it will affect the objective of the model. Some of the common controlling mechanisms in process systems are chemical reaction; diffusion of mass transfer; conductive, convective and radiative heat transfer; mixing; evaporation etc. In many cases, the mechanisms which will be included in the model are dictated by the modeling goals and level of details desired from the model. On the other hand,

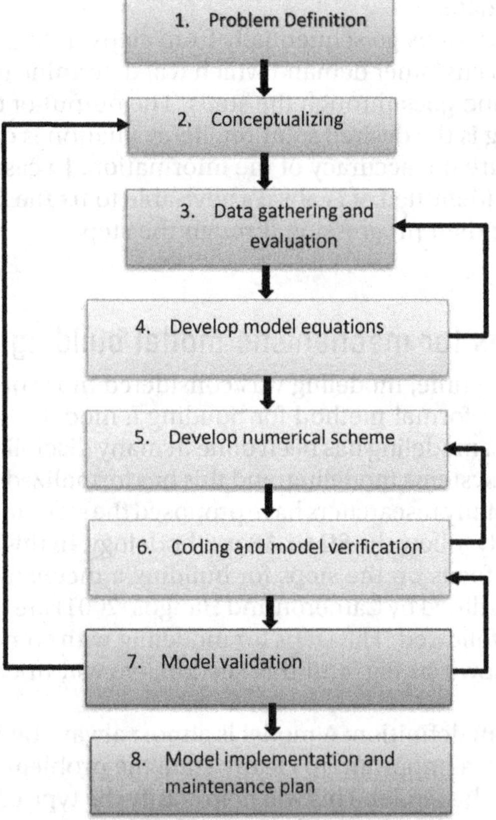

Fig. 4 Systematic method for building process model. Modified from Cameron, I. T., & Hangos, K. (2001). Process modelling and model analysis. Elsevier.

mechanisms which are seen as insignificant are not considered in the model end up as model assumptions.

iii. Data gathering and evaluation: Once the model has been conceptualized and the mechanisms have been selected, the type of data needed for building the model becomes clear. Data often have to be gathered from diverse sources, for example, scientific literature, physical property handbook, commercial simulators, and existing process system. In extreme cases, experiments have to be conducted to get the desired data, for example, conduct the kinetic study to obtain reaction rate constants. At this stage, it is also recommended to collect data to validate the model. It is also important to evaluate the accuracy of the data. There should be a clear understanding on the uncertainty involved with each data set.

iv. Develop mathematical equations: At this stage, the physical system is converted to a set of mathematical equations. The core of the mathematical equations comes from the conservation equations; in addition to this, several constitutive equations are incorporated to capture various other mechanisms in the model. Depending upon the type of model, it will lead to a set of algebraic equations, ordinary differential equations (ODEs), partial differential equations (PDEs), or set differential algebraic equations (DAEs). In writing the equations, it is important to add complexity to the model incrementally. For example, first one should write the mass balance equation, followed by energy balance, and momentum balance equation. It is important to check that the equations are well-posed and satisfies the degrees of freedom. After writing the model equations, the data requirements of the model will also become clearer. Therefore, it may be useful to revisit the data gathering step and ensure that all required data have been collected.

v. Develop numerical scheme: In some cases, it may be possible to obtain the analytical solution of the model. However, in most cases, the model size and complexity of the equations prohibit getting analytical solution, and numerical solution is the only way to solve the equations. In order to solve the problem numerically, an appropriate numerical scheme has to be selected. The first step is to determine either the problem will be solved using finite element, finite volume, or finite difference method. Accordingly, the equations will be discretized to pose the numerical problem. At this stage also appropriate solvers are selected for the equations.

vi. Coding and verification: Following the discretization of the equations, the equations are coded in a convenient computer language (e.g., C++, Matlab, and Python) for solution. In coding, it is important to follow a structured approach, for example, dividing the problem into different modules, maintaining a core code, and calling various subroutines or functions to do a specific module task. Comments should be used in the code to make it understandable to the user or developers who will be working on the code at a later stage. Codes should be checked extensively for errors. Some sanity checks, for example, making a step change in the input of the reactor and observe the output concentration should be done to ensure the correctness of the code.

vii. Validation: At this stage, the model is tested against the reality. The best way to validate the model is to compare the predicted output with measurements obtained from the actual process or experiments. It may be possible to find experimental data or plant data in the literature. The model and the inputs have to be set to the same conditions as the actual system to check against the measured output. The output should be checked to see if the predicted output is within the uncertainty of the measured data. The hypothesis test can be used to determine if the difference is statistically significant or not. In case data are not available, the model can be validated against other similar models, which may be simplified or more detailed model compared to the proposed model.

viii. Model implementation and maintenance plan: Following successful validation, the model is fully ready for the intended use. Some of the logistical issues for implementing the model results have to be sorted out at this stage, especially if the model is intended for on-line use. The model needs to be housed in a server or computer so that it provides the output at the desired time interval. For example, for a real-time optimization (RTO) application, a model output may be required at every hour; on the other hand, for an advanced process control (APC) application, model output may be required at every minute. The model may have to be optimized and further refined to fit the application need. In some cases, the mechanistic model is converted to an empirical model for on-line use and to get the output at desired frequency. Over time, the plant will change or operating conditions may change. In order to keep the model updated, a regular maintenance plan should be put in place.

3.3 Case study: PTA reactor model

Pure Terephthalic Acid (PTA) is an important raw material for many polymer products, including water bottles. A simplified process flow diagram of the PTA process is shown in Fig. 5. The quality of the product is measured by the presence of 4-CBA (4-carboxy benzyl aldehyde) which is an intermediate product. 4-CBA is not measured directly; only lab measurement is available once/twice a day. Typical concentration of 4-CBA is 2500 (±100). A historical data-based empirical model was developed to predict 4-CBA every minute, which could be used for typical feed rate of $50\,m^3/h$ Px flow. Problem aroused as the feed rate had to be changed to 60% of the max (typical) feed. The predictor was no longer valid. Also, there were not enough historical data in the 60% operating range to build a data-based model. The objective of the model is to investigate the effect of feed rate change on the product quality (4-CBA concentration) and incorporate the information in the data-based predictor for improving the prediction of the inferential predictor. Below, we use the step-by-step procedure to develop the model for the PTA reactor.

i.Problem definition:

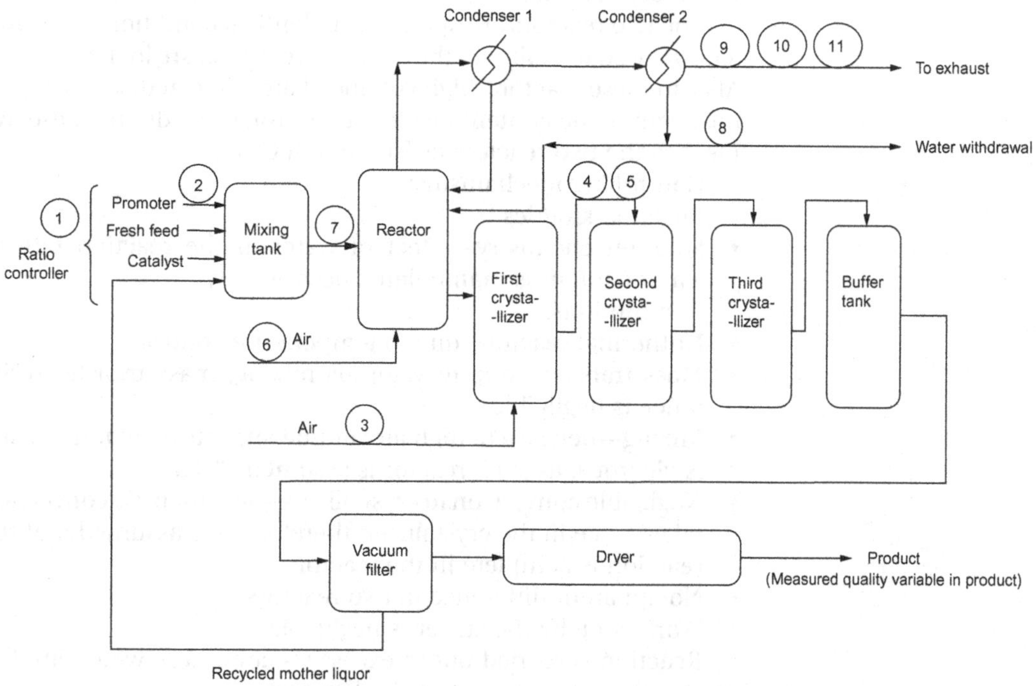

Fig. 5 A simplified process flow diagram of a PTA process (Rahman et al., 2016).

Process System: The reactor-condenser system of the PTA process needs to be modeled for the typical process operating conditions using the available kinetic information in the open literature.

Modeling Goal: A predictor needs to be developed for 4-CBA in order to control the product quality using the model predictive controller. The predictor model should be able to predict 4-CBA concentration in the product for feed rate changes which affect the residence time of the reactor. The predicted values need to be validated against the lab values.

System Boundary: Around 2 trains of Reactor Condenser

Model type: Lumped parameter modeling

Inputs: Feed, Air

Disturbances: Feed Concentration (measured)

Controller: Reactor/Condenser Pressure (fixed), Reactor Level (fixed), and Water Withdrawal (typically changed)

Output: Product Quality (4-CBA concentration)

Range of Operation: Reactor Temperature 191°C; $30 < \text{Feed rate} < 55\,m^3/h$; $2400 < \text{4-CBA concentration} < 2800\,ppm$

Desired Accuracy: 4-CBA prediction within 100 ppm of lab value

ii. Conceptualization

The two reactors are operated at identical conditions. Controllers are used to maintain the constant temperature in the reactors. Also, in these reactors, high volume of air is injected, which vigorously mixes the content inside the reactor. Considering these, we idealize the two reactors as isothermal CSTR.

Controlling mechanisms:

- Reaction kinetics
- Solution chemistry—effect of water in the reaction rate as water content is manipulated occasionally

Assumptions:

- Isothermal reactor—due to temperature control
- Mass transfer—due to vigorous mixing, mass transfer resistance is negligible
- Mixing—because of high air volume and stirrer action, mixing is vigorous, as such reactor is assumed CSTR
- Negligible conversion in crystallizers—less than 2% conversion takes place in the crystallizer; therefore, it is assumed that the reaction is complete in the reactor
- No apparent difference in two reactors
- Work done by the stirrer is negligible
- Reaction is carried out in excess O_2 condition, which implies O_2 concentration zero-order kinetics
- Constant Physio-chemical properties

iii. Data gathering and evaluation

Process Data: Temperature (accuracy ± 1°C) and Flowrates (accuracy ± 1 m^3/h) were available at 1 min sampling rate.

Lab Data: 4-CBA (accuracy of lab measurement ± 100 ppm) were available 1 or 2 times a day.

Since the frequency of the lab measurement is very sparse, we cannot do dynamic validation of the model. For steady-state validation process, data were collected for time intervals when the process was stable for a long period of time without any major disturbance or control move by the operator. Therefore, accuracy of data is not only the precision of measurement but also have to take into account the effect of the disturbances.

However, one of the objective of the model is to evaluate the effect of level on the product quality. Level affects the residence time of the reactor, which is a dynamic parameter. Therefore, initially, a dynamic mechanistic model was developed. The model will be validated only for the steady-state conditions.

The reaction kinetics (shown in Fig. 6) and the reaction kinetics data in Tables 1 and 2 were obtained from Wang et al. (2005a).

iv. Develop the model equations

Conservation equation: The component balance equation for each component is given as follows:

$$V\frac{dC_i}{dt} = f_{in}C_{i,in} - f_{out}C_i + V\sum_{j=1}^{4} \nu_{i,j}r_j \tag{27}$$

where f_{in} and f_{out} are inlet and outlet flow rate from the rector, respectively; C_i is the concentration of component i, i = 1–5; r_j is the reactor reaction rate constant; $\nu_{i,j}$ is the stoichiometric coefficient; V is the reactor volume.

Constitutive equations:

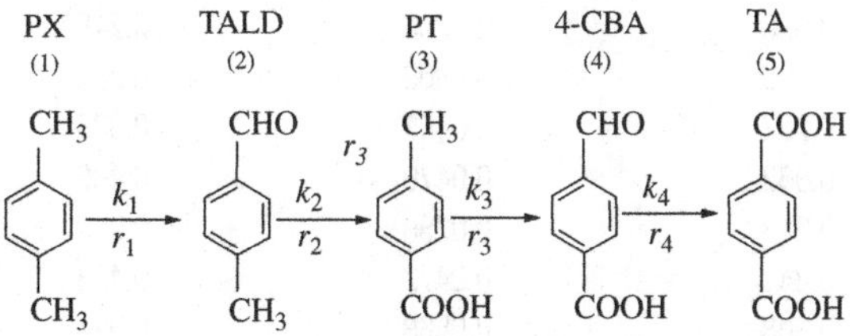

Fig. 6 Reaction pathway showing the oxidation of para xylene (PA) to terephthalic acid (TA). Reproduced with permission from Wang, Q., Li, X., Wang, L., Cheng, Y., & Xie, G. (2005). Kinetics of p-xylene liquid-phase catalytic oxidation to terephthalic acid. *Industrial & Engineering Chemistry Research, 44*(2), 261–266, Copyright 2005 American Chemical Society.

Table 1 Reaction conditions for different experimental run.

Run	T (°C)	$C_{PX,0}$ (mol/kg$_{HAc}$)	P_{O_2} (kPa)	C_{Co} (10^4 kg/kg$_{HAc}$)	C_{Mn} (10^4 kg/kg$_{HAc}$)	C_{Br} (10^4 kg/kg$_{HAc}$)
1	191	3.145	40.0	350	326	475
2	191	1.887	40.0	350	326	475
3	191	0.943	40.0	350	326	475
4	191	0.472	40.0	350	326	475
5	191	0.943	12.0	350	326	475
6	191	0.943	20.0	350	326	475
7	191	0.943	28.0	350	326	475
8	185	0.943	40.0	350	326	475
9	188	0.943	40.0	350	326	475
10	194	0.943	40.0	350	326	475
11	197	0.943	40.0	350	326	475
12	191	0.943	40.0	250	233	339
13	191	0.943	40.0	450	419	610
14	191	0.943	40.0	550	512	746
15	191	0.943	40.0	350	326	237
16	191	0.943	40.0	350	326	712
17	191	0.943	40.0	350	326	949

Table 2 Rate constants for oxidation of PX to TA.

Run	k_1 (min^{-1})	k_2 (min^{-1})	k_3 (min^{-1})	k_4 (min^{-1})
3	0.176	0.725	0.0361	0.338
5	0.138	0.547	0.0276	0.257
6	0.162	0.656	0.0329	0.307
7	0.176	0.723	0.0366	0.339
8	0.139	0.590	0.0256	0.248
9	0.153	0.650	0.0303	0.288
10	0.189	0.782	0.0409	0.386
11	0.216	0.863	0.0479	0.436
12	0.119	0.513	0.0250	0.253
13	0.211	0.841	0.0472	0.433
14	0.217	0.908	0.0568	0.521
15	0.099	0.507	0.0232	0.220
16	0.205	0.837	0.0473	0.461
17	0.219	0.855	0.0578	0.531

Reaction rate, $r_j = k_j C_i$
Effect of water concentration on reaction rate,

$$k_j = k_{j0}\left(1 + a_j\omega_{H_2O} + b_j\omega_{H_2O}^2\right) \quad (28)$$

where ω_{H_2O} is the concentration of water in the reactor (Wang et al., 2005b).

v. Develop a numerical scheme

The above set of equations lead to a set of ordinary differential equations (ODEs) and algebraic equations. In this case, no elaborate discretization scheme is required. The right hand side of the ODEs can be discretized using any central difference scheme and the left hand side of the equation using forward difference scheme.

vi. Coding and model verification

The model was coded in Matlab, and built-in ODE solver "ode23s" was used to solve the equations. The model output is given in Fig. 7. The reactant para-Xylene (PX) concentration is continuously decreasing, while the product (PTA) concentration is continually increasing. 4-CBA is an intermediate product which shows an initial increase in concentration and subsequent decrease. Clearly, the trend shows that the model is behaving as expected.

vii. Validation

The model was validated with experimental data obtained from the literature. The model showed good prediction for the experimental data. However, the match for the industrial data was poor. Therefore, the model had to be further revised. After further literature review, a correction factor for rate constant was found in the literature for industrial applications (Wang et al., 2005a). The modified rate equation is given in Eq. (29). The parameters of the rate constants are provided in Table 3.The correction factor was incorporated in the model, which significantly improved the prediction for the industrial data. Following the correction, the predicted concentration was validated with average steady-state 4-CBA concentration for two different feed. Fig. 8 shows the validation result. Since only lab data were available, the validation was done for steady-state condition only. The lab concentration is an average value. The comparison shows good agreement between the model-predicted values and the lab measurements.

$$r_j = k_j \frac{C_j}{\left(\sum_{i=1}^{4} (d_i C_i + \theta)^{\beta_j}\right)} \quad j = 1, \cdots, 4 \quad (29)$$

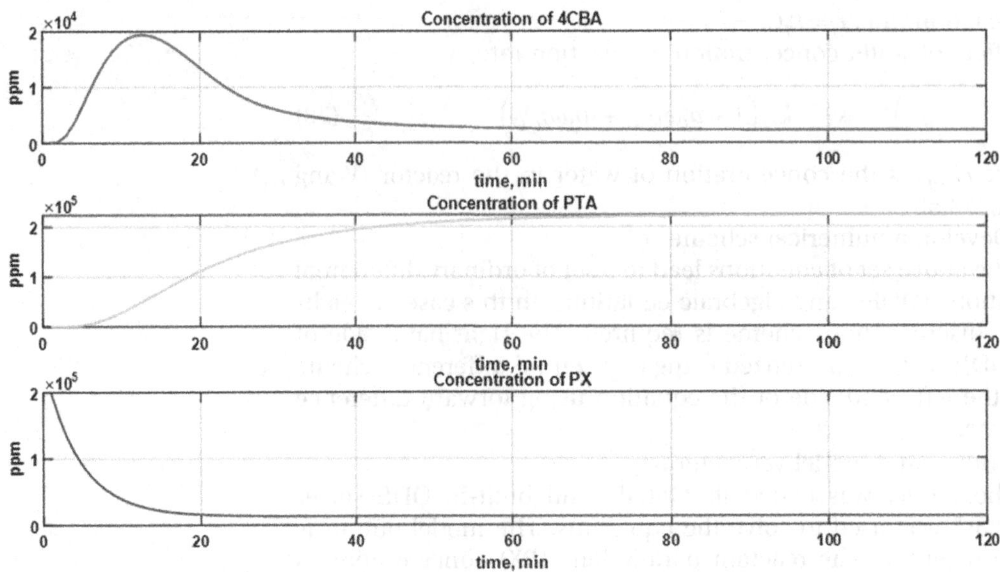

Fig. 7 Predicted concentration of 4CBA, PTA, and PX in the reactor.

viii. Model implementation and maintenance plan

In order to predict 4-CBA concentration online, a partial least squares (PLS) model was developed. There are two reasons for building a data-based model: (i) to take advantage of the existing software infrastructure. The process plant was using Aspen IQ for inferential predictor development and implementation which only supports the data-based model. Aspen IQ also has the data preprocessing module which does raw data validation, and interface with Aspen DMCplus allows seamless data transfer between the inferential predictor and advanced controller, (ii) data-based model is computationally inexpensive. It takes less than 1 s to do the calculation. Therefore, the predictions were suitable for both

Table 3 Coefficient for the rate equation in Eq. (29).

i	k_i (min^{-1})	d_i (kg acetic acid/mol)	β	θ
1	0.1716	1.4247	0	0.0146
2	0.7002	0	0.5254	
3	0.0353	0	0	
4	0.3296	4.8419	0.8111	

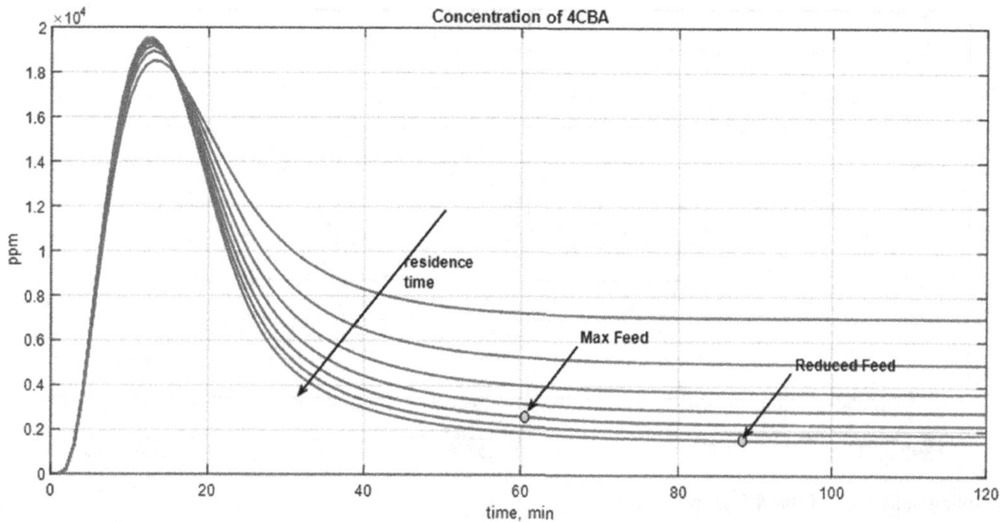

Fig. 8 Steady-state validation of model for different feed rate.

PID and model predictive controller. The PLS model was trained using historical data and the simulated data from the mechanistic model. Historical data were only available for high feed rate, and the model provided data for various flowrate. Fig. 9 compares the PLS-model prediction of 4-CBA with the lab measurements for a period of 18 days. In order to validate the model, samples were collected every 6 h which is double of the usual sampling rate. The data points marked red are validation data, while the data points marked green were training data. The prediction error is ±100 m which is comparable to the lab measurement accuracy. The prediction model was coupled with the advanced controller which provided good 4-CBA quality control in the product. A maintenance schedule was also set up to evaluate the prediction error of the model every 6 months. If the prediction error exceeds the threshold, the model will be retrained with the most recent data set.

4 Summary

A review of equations used for mechanistic modeling of process systems is provided in this chapter. It will guide readers about the different sources for collection of data. Some of the new development needs in process modeling, for example, molecular dynamic simulation and density function theory are also

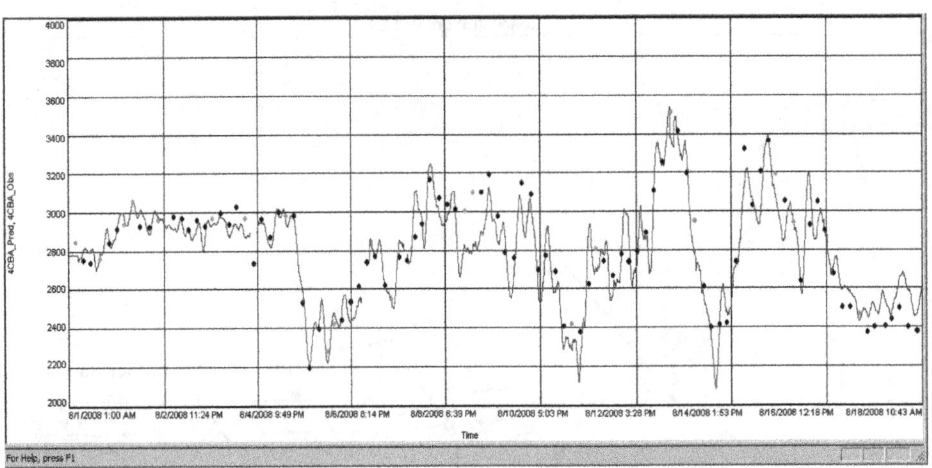

Fig. 9 Online validation of the 4-CBA predictor.

highlighted. Systems concept in process and general steps for process modeling are also covered. Finally, steps of process system modeling are described with an industrial case study.

References

Bahill, A. T., & Gissing, B. (1998). Re-evaluating systems engineering concepts using systems thinking. *IEEE Transactions on Systems, Man, and Cybernetics, Part C (Applications and Reviews)*, *28*(4), 516–527.

Bird, R. B., Stewart, W. E., & Lightfoot, E. N. (1960). *Transport phenomena. Vol. 413*. New York: John Wiley & Sons.

Bird, R. B., Stewart, W. E., & Lightfoot, E. N. (2006). *Transport phenomena. Vol. 1*. John Wiley & Sons.

Cameron, I. T., & Hangos, K. (2001). *Process modelling and model analysis*. Elsevier.

Chapman, W. G., Gubbins, K. E., Jackson, G., & Radosz, M. (1989). SAFT: Equation-of-state solution model for associating fluids. *Fluid Phase Equilibria*, *52*, 31–38.

Chen, H. J., & Chu, C. H. (1998). Modeling ethane cracker with application to flowsheet simulation of an ethylene plant. *Journal of the Chinese Institute of Chemical Engineers*, *29*(4), 275–285.

Eckert, F., & Klamt, A. (2002). Fast solvent screening via quantum chemistry: COSMO-RS approach. *AICHE Journal*, *48*(2), 369–385.

Green, D. W., & Southard, M. Z. (2019). *Perry's chemical engineers' handbook*. McGraw-Hill Education.

Gross, J., & Sadowski, G. (2001). Perturbed-chain SAFT: An equation of state based on a perturbation theory for chain molecules. *Industrial & Engineering Chemistry Research*, *40*(4), 1244–1260.

Ilyin, D. V., Goddard, W. A., Oppenheim, J. J., & Cheng, T. (2019). First-principles–based reaction kinetics from reactive molecular dynamics simulations: Application to hydrogen peroxide decomposition. *Proceedings of the National Academy of Sciences*, *116*(37), 18202–18208.

Joly, J. F., & Miller, R. E. (2017). Density functional theory rate calculation of hydrogen abstraction reactions of N-phenyl-α-naphthylamine antioxidants. *Industrial & Engineering Chemistry Research, 57*(3), 876–880.

Lemmon, E. W., Huber, M. L., & Mclinden, M. O. (2007). *NIST standard reference database 23: Reference fluid thermodynamic and transport properties—REFPROP.* Version 8.0.

Li, Y., & Fang, D. C. (2014). DFT calculations on kinetic data for some [4+ 2] reactions in solution. *Physical Chemistry Chemical Physics, 16*(29), 15224–15230.

Rahman, M. M., Imtiaz, S. A., & Hawboldt, K. (2016). A hybrid input variable selection method for building soft sensor from correlated process variables. *Chemometrics and Intelligent Laboratory Systems, 157*, 67–77.

Smith, J. M., Van Ness, H. C., & Abbott, M. M. (2005). *Introduction to chemical engineering thermodynamics.* McGraw Hill Book Company.

Van Goethem, M. W. M., Kleinendorst, F. I., van Velzen, N., Dente, M., & Ranzi, E. (2002). *Equation based SPYRO® model and optimiser for the modelling of the steam cracking process* (pp. 26–29). Escape-12 Supplementary Proceedings.

Wang, Q., Li, X., Wang, L., Cheng, Y., & Xie, G. (2005a). Kinetics of p-xylene liquid-phase catalytic oxidation to terephthalic acid. *Industrial & Engineering Chemistry Research, 44*(2), 261–266.

Wang, Q., Li, X., Wang, L., Cheng, Y., & Xie, G. (2005b). Effect of water content on the kinetics of p-xylene liquid-phase catalytic oxidation to terephthalic acid. *Industrial & Engineering Chemistry Research, 44*(13), 4518–4522.

Wilding, W. V., Knotts, T. A., Giles, N. F., & Rowley, R. L. (2020). *DIPPR® data compilation of pure chemical properties.* New York, NY: Design Institute for Physical Properties, AIChE.

Wilhelmsen, Ø., Aasen, A., Skaugen, G., Aursand, P., Austegard, A., Aursand, E., Gjennestad, M. A., & Hammer, M. (2017). Thermodynamic modeling with equations of state: Present challenges with established methods. *Industrial & Engineering Chemistry Research, 56*(13), 3503–3515.

Wilke, C. R., & Chang, P. (1955). Correlation of diffusion coefficients in dilute solutions. *AICHE Journal, 1*(2), 264–270.

Xie, Q., & Tinker, R. (2006). Molecular dynamics simulations of chemical reactions for use in education. *Journal of Chemical Education, 83*(1), 77.

PART II

Micro scale modeling

3

Density functional theory (DFT) models for the desulfurization and extraction of sulfur compounds from fuel oils using ionic liquids

Shofiur Rahman[a], Paris E. Georghiou[b], and Abdullah Alodhayb[a,c]

[a]King Abdullah Institute for Nanotechnology, King Saud University, Riyadh, Saudi Arabia. [b]Department of Chemistry, Memorial University of Newfoundland, St. John's, Newfoundland and Labrador, Canada. [c]Department of Physics and Astronomy, College of Science, King Saud University, Riyadh, Saudi Arabia

1 Introduction

The development of ionic liquids (ILs) and their applications have seen exponential growth since their potential was recognized in the late 1990s. Fewer than 40 papers involving ILs had appeared to 1999, but a recent online search of publications having the words "ionic liquids" in their titles came up with over 118,000 papers, with nearly 80,000 appearing during the last decade (2013–2023) alone.

There are excellent recent comprehensive reviews and monographs on ILs, which have been published recently, and the reader is encouraged to consult some of these for both recent developments and applications as well as for an historic overview of ILs (Freemantle, 2010; Handy, 2011; Lei et al., 2017).

In short, ILs are salts in the traditional sense, in that they are comprised of cations and anions. It is the fact that they have much lower melting points than most salts which are formed between metals and nonmetals, and that they are generally liquids at room temperatures or at temperatures below 100°C which give them their "liquid" designation. The main other characteristic of ILs

Modeling of Chemical Process Systems. https://doi.org/10.1016/B978-0-12-823869-1.00011-9

is that their cations are generally stable heteroaromatic organic ions, such as the Lewis acid *N*-alkylated pyridinium or *N*-alkylated methylimidazolium cations. As well, the IL counterions are generally from a large group of inorganic anions, most commonly tetrafluoroborate (BF_4^-) or hexafluorophosphate (PF_6^-), although the respective conjugate bases of weak and strong acids such as acetate (Ac^-) and hydrogensulfate (HSO_4^-), respectively, are also common.

ILs have received a great deal of interest across many fields of chemistry, including applications in industry (Greer et al., 2020) and very recently in the petroleum industry, where they potentially provide many advantages over conventional solvents and other technologies, particularly with respect to the removal of sulfur-containing contaminants. The most commonly employed conventional methodology for removing sulfur contaminants in petroleum products is hydrodesulfurization (HDS). HDS, however, requires both high temperatures and high pressures of hydrogen. Among several other costly limitations associated with HDS is the fact that it is not effective for the elimination of heteroaromatic sulfur-containing compounds such as thiophenes, benzothiophenes, dibenzothiophenes and their derivatives. An alternative approach to desulfurize fuel oils by avoiding the hazardous use of hydrogen or the other requirements of HDS is to instead employ oxidative desulfurization (ODS). While very effective in removing sulfur-containing compounds, the process requires the use of large amounts of solvents to selectively extract the oxidized products. The solvents needed are generally volatile, flammable, and potentially environmentally hazardous. On the other hand, the use of ILs as alternatives offers major advantages. These advantages include, among other properties, the fact that most ILs are nonvolatile having very low vapor pressures, are nonflammable, and have good thermal stability depending on reaction conditions and the nature of the anion. ILs can also be fine-tuned to increase or decrease their hydrophobicity or miscibility with hydrocarbons, or alternatively, enhance their hydrophilicity or water solubility. This can be achieved by suitable choices of their cations and/or anions, which can alter their respective coordinating abilities. ILs having chloride (Cl^-) or nitrate (NO_3^-) anions, for example, are strongly coordinating, whereas those with acidic anions, such as heptachlorodialuminate ($Al_2Cl_7^-$), are noncoordinating. On the other hand, ILs tend to be hydrophobic with weakly coordinating anions such as tetrafluororborate (BF_4^-) or hexafluorophosphate (PF_6^-). In general, as the coordinating ability of the anion decreases, the ILs become more hydrophobic (Lei et al., 2017). The cations can also be modified to enhance the hydrophobicity of ILs making them less water-miscible, for example, by increasing the alkyl chain length on the nitrogen atom of the pyridinium or the methylimidazolium cations (Bhutto et al., 2016; Lei et al., 2017).

Extensive ODS research therefore has been reported on the use of ILs as solvents, which are immiscible with fuel oils but which can both host the oxidizing agent and catalyst and also selectively extract both nonoxidized and oxidized heterosulfur contaminants. As well, ILs can subsequently be recycled and reused by removing the oxidized products, although some ILs might undergo some degradation over repeated cycles. Since the initial report by Bösmann et al. (2001), who used a series of ILs to effect deep extraction-desulfurization of fuel oils, this area of research has seen many excellent studies, and comprehensive recent reviews have recently been published (Bhutto et al., 2016; Dai et al., 2017). In summary, much of the experimental research has been conducted using model systems in which various ILs and catalysts were evaluated. In these model ODS systems, the IL served as the immiscible extractive layer of a biphasic system in which the other layer is a hydrocarbon such as *n*-octane, as a substitute for diesel oil. Dibenzothiophene (DBT) has been used most commonly as a representative of the heteroaromatic sulfur components which are usually encountered in fuel oils and which, as mentioned previously, are resistant to HDS. After the extraction into the IL, the more volatile DBT can be separated under reduced pressure, and the IL can be reused.

Many other studies have focused on a combined oxidation and extraction desulfurization [OEDS] using DBT (or other representative thiophene derivatives) with an oxidant such as hydrogen peroxide (H_2O_2) with a variety of molybdenum-, tungsten-, or vanadium-based catalysts, using an IL as both the reaction and extraction medium. The extractability of DBT and its the oxidation products, namely dibenzothiophene sulfone ($DBTO_2$), by the IL, as well as the efficiencies of the oxidant system can be experimentally determined. Bhutto et al. (2016) and more recently Dai et al. (2017) compiled much of the data obtained from such studies. There are of course many more ILs and catalysts that could be tested in model systems.

This chapter aims to show how the use of Density Functional Theory (DFT) as a quantum chemical-based tool has been used to try to determine in silico, as much as possible, the optimum ILs, which could be considered to be employed among the several thousand ILs that are potentially available or could be relatively easily synthesized to consider. There are only a few seminal papers (Gu et al., 2014; Hizaddin et al., 2014; Lin et al., 2017; Mohumed et al., 2020) that have described using DFT to computationally determine other parameters, which could be of importance in evaluating the optimum ILs to use experimentally for the desulfurization of fuel oils. These studies will be reviewed after a brief overview of the quantum chemical considerations is presented below. In this chapter, we present examples of DFT computations as an ab initio-based qualitative guide for determining the energetically most-favored noncovalent interaction energies (IEs) between a series of ILs, specifically for the fuel oil contaminants DBT and/or with $DBTO_2$. These determinations are based upon the most stable geometries generated for these molecules and their respective frontier molecular orbitals (FMOs), which are their highest occupied molecular orbitals (HOMOs) in which its electrons are located and their lowest unoccupied molecular orbitals (LUMOs). The HOMO of a chemical species is therefore nucleophilic or electron-donating, and the LUMO is electrophilic or electron-accepting. Both nonbonding or supramolecular interactions and bonding chemical reactions between a pair of reactants can be interpreted as being due to energetically favorable interactions or overlap, between a filled HOMO of the nucleophile and an empty LUMO on the electrophile. These interpretations are based upon Fukui's FMO theory (Fukui et al., 1952). For the present considerations, besides the stabilities of several combinations of cations and anions of selected ILs and the molecular interaction energies (ΔE_{INT}) between individual ILs with DBT and or $DBTO_2$, other parameters including the stabilities of the ILs themselves as ascertained by their FMO HOMO-LUMO energy (H-L) gaps can also be determined, which could also provide a guide for the individual components and also combinations thereof.

From Koopmans' theorem (Koopmans, 1933; Luo et al., 2006), the energy of the HOMO (E_{HOMO}) is related to the negative of the ionization potential (*IP*) and is characteristic for nucleophilic components. The energy of the LUMO (E_{LUMO}) is related to the negative of the electron affinity (*EA*) and is a measure of the susceptibility of the molecule or species toward reacting with electrophiles.

$$E_{HOMO} = -IP$$

$$E_{LUMO} = -EA$$

The *energy gap* ($\Delta E_{HOMO\text{-}LUMO}$) is the difference between the HOMO and LUMO energy values and is related to the *polarizability* of the species. A large HOMO-LUMO gap indicates high stability and low reactivity for the chemical species. Conversely, a large HOMO-LUMO gap is related to low chemical reactivity. A large energy gap energetically disfavors the formation of an activated complex transition state from the electron transfer between a low-energy lying HOMO of a reactant to the high-energy lying LUMO of the other reactant. Several other important and useful quantum chemical properties can be determined from the HOMO-LUMO energy values. These include global hardness (η), global softness (S), electrophilicity index (ω), electronegativity (χ), and chemical potential (μ), all of which give a measure of chemical reactivity (Gu et al., 2014; Hizaddin et al., 2014). These are defined and briefly summarized below:

1.1 Global hardness and global softness

In 1963, R.G. Pearson originally introduced the "HSAB" concept of hard and soft acids and bases to empirically account for many chemical reactions (Pearson, 1963). In this HSAB concept, small highly charged chemical species, which are also weakly polarizable, are termed "hard." Conversely, chemical species that are large, have low charge states and are strongly polarizable, are "soft." Later Parr and Pearson were able to provide a quantitative measure for both "hardness" and "softness" using quantum chemical DFT (Parr & Pearson, 1983). Thus, chemical species that have a large HOMO-LUMO energy gaps have high stability and are referred to as "hard" because they are generally resistant to changes in their electron number and distribution. The hardness value (η) is a qualitative indication of its low polarizability and can be computed using Eq. (1):

$$\eta = \left[\frac{E_{LUMO} - E_{HOMO}}{2}\right] = \left[\frac{IP - EA}{2}\right] \quad (1)$$

"Soft" molecules conversely have small HOMO-LUMO energy gaps and are highly polarizable. They only require a small amount of energy for excitation. Softness (S) is the reciprocal of the hardness and can be computed using Eq. (2):

$$S = \left[\frac{2}{E_{LUMO} - E_{HOMO}}\right] = \left[\frac{2}{IP - EA}\right] = \frac{1}{\eta} \quad (2)$$

1.2 Electronegativity

A measure of the ability of a chemical entity to attract electrons to itself and therefore can be used to make predictions about its chemical reactivity is also given by its electronegativity (χ). Electronegativity (χ) can be estimated by the average value of the HOMO and LUMO energies, as shown by Eq. (3).

$$= -\left[\frac{E_{LUMO} + E_{HOMO}}{2}\right] = \left[\frac{IP + EA}{2}\right] \tag{3}$$

1.3 Chemical potential

The tendency of an electron to be removed from a molecule is referred to as its chemical potential (μ) (Parr et al., 1978). This potential is the negative of the electronegativity and can be determined using Eq. (4):

$$\mu = \left[\frac{E_{HOMO} + E_{LUMO}}{2}\right] = -\left[\frac{IP + EA}{2}\right] \tag{4}$$

1.4 Electrophilicity index

The capability of a substance to accept electrons is quantified as its electrophilicity index (ω) and measures the energy lowering of a substance due to the electron flow between donor and acceptor. This index can be calculated by taking the square of its electronegativity divided by its chemical hardness. The mathematical expression for ω is as follows (Hizaddin et al., 2014).

$$\omega = \frac{\chi^2}{2\eta} = \left[\frac{-\left(\frac{E_{HOMO} + E_{LUMO}}{2}\right)^2}{(E_{LUMO} - E_{HOMO})}\right] = \left[\frac{\left(\frac{IP + EA}{2}\right)^2}{IP - EA}\right] \tag{5}$$

1.5 Dipole moment

The bond dipole moment $\vec{\mu}$ is a vector quantity measure of the polarity of a chemical bond within a molecule and results from the separation of opposite charges or partial charges within the bond. The dipole moment of a molecule is expressed in Debye units and is equal to the distance between the charges multiplied by the charge as shown by Eq. (6), where $\vec{\mu}$ is the dipole moment vector; q_i is the magnitude of the ith charge, and $\vec{\mu}$ is the vector representing the position of the ith charge:

$$\vec{\mu} = q_i \vec{r}_i \tag{6}$$

An early study by Anantharaj and coworkers (Hizaddin et al., 2014) employed a Hartree-Fock theory calculation with *Gaussian03* using the 6-31G* basis set to determine HOMO-LUMO interactions between the nitrogen heterocyclic compound, pyrrole with each of 18 ILs comprising 6 different cationic species (1-ethyl-3-methylimidazolium $[EMIM]^+$, *N*-ethylpyridinium $[EPY]^+$, *N*-ethyl-*N*-methylpyrrolidinium $[EMPYRO]^+$, *N*-ethyl-*N*-methylpiperidinium $[EMPIPE]^+$, *N*-ethyl-*N*-methylmorpholinium $[EMMOR]^+$, 1,2,4-trimethylpyrazolium $[TMPYRA]^+$) with three different anions (acetate, ethyl sulfonate, and methyl sulfate). Pyrrole was used as a representative nitrogen heterocycle whose elimination from fuel oils is also important for the HDS routes. To determine the optimum IL for the extraction of pyrrole, these authors included the global hardness (η) and softness (S), electronegativity (χ), chemical potential (μ), and electrophilicity index (ω) for each of the individual components and the molecular complexes formed between the ILs and pyrrole. They concluded from their analysis of all FMO and scalar quantities that $[EPY][EtSO_4]$ was the most favorable IL for the removal of pyrrole, a finding that was supported by published experimental works.

A thorough and detailed study by Gu et al. (2014) used DFT calculations for the pyridinium-based *N*-butylpyridium hydrogen sulfate IL ($[BPY][HSO_4]$) and its complexes with thiophene (TS), thiophene sulfone (TSO_2), dibenzothiophene (DBT), and dibenzothiophene sulfone ($DBTO_2$). They employed the dispersion corrected hybrid functional method ωB97X-D (Chai & Head-Gordon, 2008) with the 6-31++G(d,p) basis set with *Gaussian 09* to determine the most stable optimized structures. They also used Multiwfn software (Lu & Chen, 2012) to analyze the most stable optimized structures and several aromatic properties indices from charge decomposition analyses (Dapprich & Frenking, 1995) reduced density gradient analyses (Johnson et al., 2010), natural bond orbital NBO analyses (Reed et al., 1988), and AIM analyses (Bader, 1991). They concluded that the most stable structures of the complexes $[BPy][HSO_4]$:TS, $[BPy][HSO_4]:TSO_2$, $[BPy][HSO_4]$:DBT, and $[BPy][HSO_4]:DBTO_2$ indicated that the respective heteroaromatic rings and the pyridinium ring are *face*-to-*face* with each other, with the π-π^* stackings in $[BPy][HSO_4]$:TS and $[BPy][HSO_4]$:DBT being stronger than in the $[BPy][HSO_4]$:DBT and $[BPy][HSO_4]:DBTO_2$ complexes, respectively. TSO_2 and $DBTO_2$ are more nucleophilic than the corresponding nonoxidized precursors, TS and DBT, thus accounting for the stronger interactions between the IL and TSO_2 and $DBTO_2$ than those between IL and their corresponding nonoxidized precursors. Their results were also corroborated by their corresponding thermodynamic interaction energies and the chemical activity descriptors including chemical potential (μ), hardness (η), and electrophilicity index (ω).

Lin et al. (2017) studied the interaction between DBT and each of the following five ILs: *N*-butyl-*N*-methylimidazolium tetrafluoroborate ([BMIM][BF_4]), *N*-butyl-*N*-methylmorpholinium tetrafluoroborate, ([Bmmorpholinium][BF_4]), *N*-butyl-*N*-methylpiperdinium tetrafluoroborate ([BMPiper][BF_4]), *N*-butyl-*N*-methylpyrrolidinium tetrafluoroborate ([BMPyrro][BF_4]), and *N*-butylpyridinium tetrafluoroborate ([BPy][BF_4]). Their DFT calculations were carried out using *Gaussian 09,* and all of the ILs and the ILs-DBT structures were fully optimized using the ωB97X-D method and 6-31++G(d,p) basis set without symmetry constraints. They also included analyses of the intermolecular interactions within the IL:DBT structures. The [BMIM][BF_4]:DBT and [BPY][BF_4]:DBT complexes showed the highest interaction energies, which is consistent with the fact that they showed typical π^+-π interactions known to be key interactions between an aromatic cation and a guest molecule such as DBT. Lin et al demonstrated that the cation π^+ orbital of the IL is stacked over the middle DBT ring. From their NBO (Reed et al., 1988) reduced density gradient analyses (Johnson et al., 2010) and AIM (Bader, 1991) results, the strength of weak interactions in the IL:DBT interactions was found to be in the following order: F···H > π^+-π > N-C-H···C > anion-π.

2 Results and discussion

In 2020, we conducted a DFT structural analysis study for validation of our experimental OEDS results with a model diesel fuel and DBT using [BPy][BF_4] with hydrogen peroxide (H_2O_2) using different experimental conditions (Mohumed et al., 2020). The energetically most stable optimized structures of each of the components, [BPy][BF_4], DBT, $DBTO_2$, and their corresponding complexes ([BPy][BF_4]):DBT and ([BPy][BF_4]):$DBTO_2$ were computed and are shown below in Fig. 1. The DFT calculations were all conducted with *Gaussian 09.* In order to minimize computation time, the geometries of all the structures were initially fully optimized at the B3LYP/6-311++G(d,2p) level of theory in the gas phase, hexane, and CCl_4 solvent systems using the polarized continuum model (PCM). The structures of the [BPy] cation interacting with the BF_4^- anion at different binding sites were first optimized, and then the most stable structure of the [BPy][BF_4] salt was selected. This stable optimized structure was further optimized with DBT and also with $DBTO_2$. The interaction energies between the ILs and DBT or $DBTO_2$ were calculated using Eqs. (7)–(9):

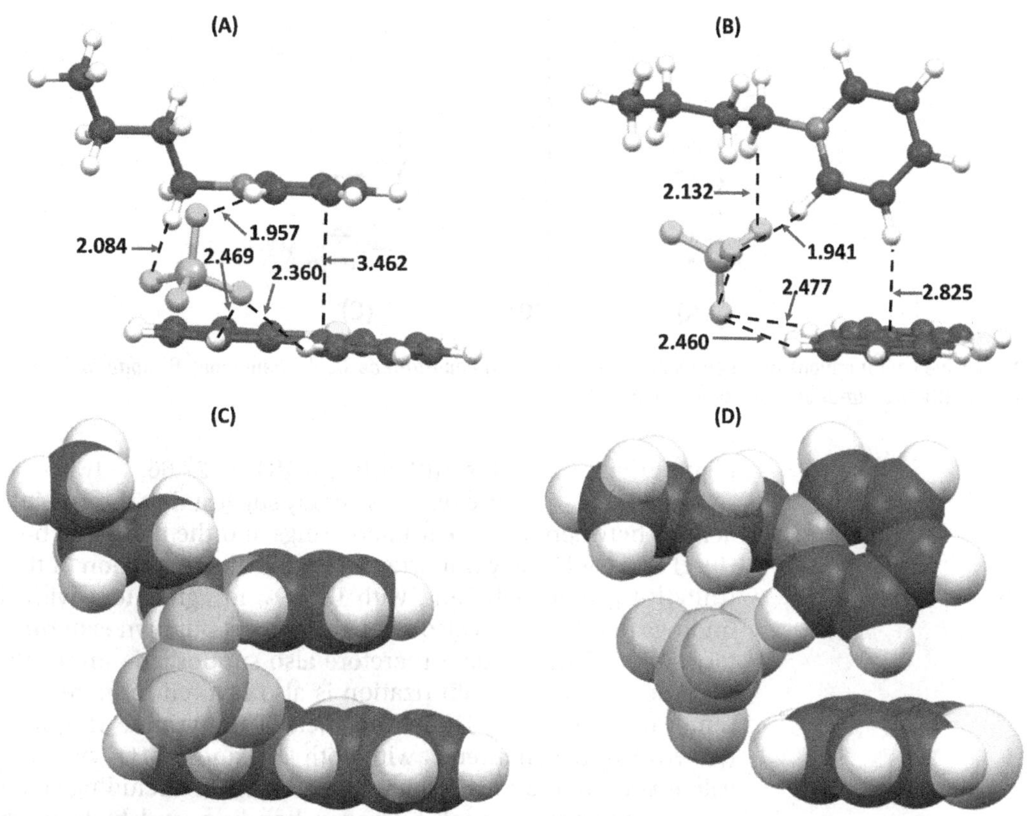

Fig. 1 The equilibrium geometry optimized structures of ([BPy][BF_4]) with DBT. (A) *Top left*: Ball-and-stick representation with selected parameter distances in Å of a *face-to-face parallel-sandwich type* structure of the IL:DBT complex; (B) *top right*: the *edge-tilted-T-shaped* structure of the IL:DBT complex; (C) *bottom left* and (D) *bottom right*: space-filling representations of the complexes depicted in (A) and (B), respectively. Color code: carbon = *purple (dark gray in the printed version)*; hydrogen = *white*; nitrogen = *blue (light gray in the printed version)*; oxygen = *red (dark gray in the printed version)*; sulfur = *yellow (light gray in the printed version)*; boron = *pink (light gray in the printed version)*; and fluorine = *green (dark gray in the printed version)*.

$$\Delta IE_{\mathrm{IL}} = E_{[\mathrm{BPy}][\mathrm{BF4}]} - \left(E_{[\mathrm{BPy}]} + E_{[\mathrm{BF4}]}\right) \tag{7}$$

$$\Delta IE_{\mathrm{IL}\supset\mathrm{DBT}} = E_{[\mathrm{BPy}][\mathrm{BF4}]):\mathrm{DBT}} - \left(E_{[\mathrm{BPy}][\mathrm{BF4}]} + E_{[\mathrm{DBT}]}\right) \tag{8}$$

$$\Delta IE_{\mathrm{IL}\supset\mathrm{DBTO2}} = E_{[\mathrm{BPy}][\mathrm{BF4}]):\mathrm{DBTO2}} - \left(E_{[\mathrm{BPy}][\mathrm{BF4}]} + E_{[\mathrm{DBTO2}]}\right) \tag{9}$$

$$\Delta G_{\mathrm{IL}\supset\mathrm{DBTO2}} = G_{[\mathrm{BPy}][\mathrm{BF4}]):\mathrm{DBTO2}} - \left(G_{[\mathrm{BPy}][\mathrm{BF4}]} + G_{[\mathrm{DBTO2}]}\right) \tag{10}$$

The calculated interaction energies in the gas phase, hexane, and CCl_4 solvent systems, respectively, for ([BPy][BF_4]):$DBTO_2$ (−51.63, −32.67, and −28.91 kJ mol^{-1}) are almost two times

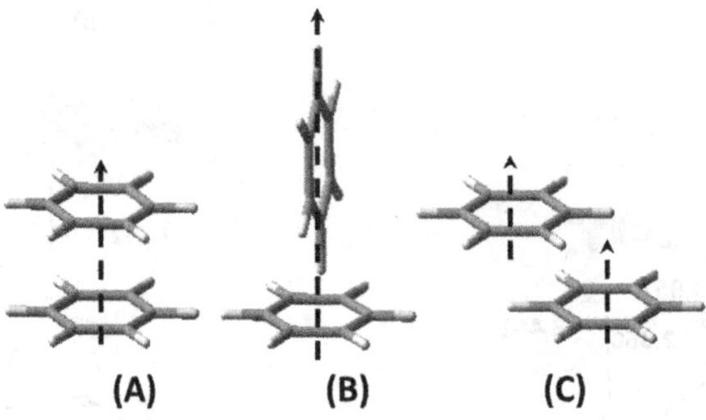

Fig. 2 Possible conformations between two aromatic rings: (A) *Face-to-face* stack orientation; (B) *edge-to-face* stack orientation; (C) *edge-to-edge* stack orientation.

higher than those for ([BPy][BF_4]):DBT (−27.66, −16.15, and −14.69 kJ mol^{-1}). These results strongly suggest that the π-π^* interactions between the two aromatic rings and the hydrogen bonds (F···H and O···H) play important roles for the interaction of the IL ([BPy][BF_4]) with DBT and with $DBTO_2$, in agreement with the findings of others, and also strongly support our own experimental results. The DFT data therefore also support the implication that the oxidative desulfurization is also favored to increase the removal efficiency of DBT from fuels. The pyridine ring in the ([BPy][BF_4]) IL can interact with both DBT and $DBTO_2$ via noncovalent π-π^* interactions. Such weak interactions including van der Waals interactions, weak hydrogen bonding, and hydrophobic lipophilic interactions all play important roles in noncovalent or supramolecular interactions and could account for the extraction by pyridinium-based ILs of the aromatic sulfur-containing compounds from fuel oils such as diesel and fuel oil models. Aromatic–aromatic π-π^* stacking interactions are the attractive noncovalent interactions between the π-electronic clouds of aromatic systems in *parallel, face-to-face*, or *edge-to-face* orientations (Fig. 2) (Lehn, 2007). The presence of a nitrogen atom on an aromatic ring is also known to have large effects on similar *stacked-*, *parallel-displaced-*, and *T-shaped* dimeric structures.

Fig. 1 shows the most stable optimized structures of the DFT-generated ([BPy][BF_4]):DBT and ([BPy][BF_4]):$DBTO_2$ complexes. Fig. 1A and C *(top & bottom left)* andFig. 1B and D *(top & bottom right)*, respectively, show the *face-to-face parallel-sandwich* and *edge-tilted-T-shaped* structures for the computed [BPy][BF_4]: DBT structures. In the *face-to-face parallel-sandwich* complex (Fig. 1A), two hydrogen bonds from the pyridinium cation to the closest BF_4^- fluorine atoms can be seen, with bond distances of 1.957 and 2.084 Å. The DBT also has closest hydrogen bond contacts with the BF_4^- anion at bond distances of 2.360 and 2.469 Å. Furthermore, the plane of the pyridinium ring is slightly displaced and is nearly parallel to that of the DBT plane.

In the *edge-tilted-T-shaped* structure (Fig. 1B and D), two hydrogen bonds from the cation to the closest BF_4^- fluorine atoms can be seen, with bond distances of 1.941 and 2.132 Å. A pyridinium C—H···π interaction with DBT, which is almost perpendicular to the DBT plane, forms the "*T-shaped*" arrangement with a C—H bond distance of 2.825 Å. The DBT also shows a weak H—F bond from the BF_4^- anion with bond distances of 2.460 and 2.477 Å. The calculated interaction energies (Δ*IE*s) for these two geometries, respectively, are −28.99 and −27.66 kJ mol^{-1} in the gas-phase, −16.48 and −16.15 kJ mol^{-1} in the hexane solvent system, and −14.37 and −14.69 kJ mol^{-1} in the CCl_4 solvent system. The Δ*IE*s in the CCl_4 solvent system are slightly anomalous with the trend seen for the other two solvent systems.

On the other hand, the [BPy][BF_4]:$DBTO_2$ structures, which are respectively shown in Fig. 3A and C *(top & bottom left)* and Fig. 3B and D *(top & bottom right)*, reveal that the π-π* interactions lead to *face-to-face parallel-displaced* and *edge-tilted-T-shaped* structures. In Fig. 3A, the three shortest hydrogen bonds with bond distances of 1.940 and 2.362 Å from the pyridinium cation to the fluorine atoms of the anion, and a close contact of 2.292 Å from a pyridinium hydrogen to an oxygen atom of the $DBTO_2$ can be seen. The $DBTO_2$ also has closest hydrogen bond contacts with the BF_4^- anion with a bond distances of 2.565 and 2.741 Å, which are larger than the corresponding closest contacts seen in the DBT:IL complex. The planes of the pyridinum ring and that of the $DBTO_2$ are slightly displaced and form the *face-to-face parallel-displaced*-type arrangement.

In the *edge-tilted-T-shaped* structure (Fig. 3B), two pyridinium hydrogen bonds to the anion with bond distances of 1.992 and 2.221 Å and a close contact of 3.434 Å from a pyridinium hydrogen to an oxygen atom of the $DBTO_2$ can be seen. This is closer than the corresponding distance in the corresponding parallel arrangement. The pyridinium C—H···π interaction with $DBTO_2$, which is almost perpendicular to the DBT plane in the "*T-shaped*" arrangement, has a hydrogen atom to a $DBTO_2$-carbon atom bond distance of 3.796 Å, which is also larger than the corresponding distance in the IL:DBT structure. The shortest fluorine to $DBTO_2$ hydrogen bonding distance is 2.349 Å. The interaction energies of ([BPy][BF_4^-]):$DBTO_2$ for *sandwich* and *T-shaped* geometries are −50.26 and −51.63 kJ mol^{-1} in the gas-phase calculations, −31.16 and −32.77 kJ mol^{-1} in the hexane solvent system,

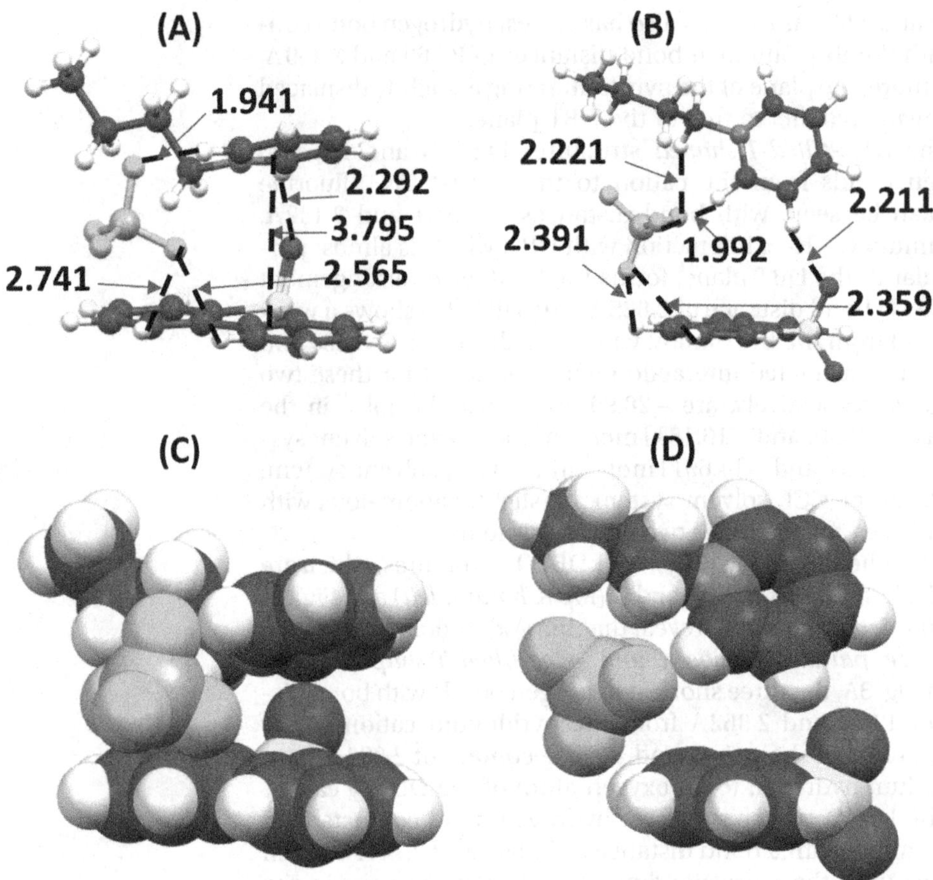

Fig. 3 The equilibrium geometry optimized structures of ([BPy][BF_4]) with $DBTO_2$. (A) *Top left*: Ball-and-stick representation with selected parameter distances in Å of a *face-to-face parallel-sandwich type* structure of the IL: $DBTO_2$ complex; (B) *top right*: the *edge-tilted-T-shaped* structure of the IL:DBTO complex; (C) *bottom left* and (D) *bottom right*: space-filling representations of the complexes depicted in (A) and (B) respectively. Color code: carbon = *purple (dark gray in the printed version)*; hydrogen = *white*; nitrogen = *blue (light gray in the printed version)*; oxygen = *red (dark gray in the printed version)*; sulfur = *yellow (light gray in the printed version)*; boron = *pink (light gray in the printed version)*; and fluorine = *green (dark gray in the printed version)*.

−29.07 and −28.91 kJ mol^{-1} in the CCl_4 solvent system, respectively. As with the DBT, slightly anomalous values are seen in the interaction energies with the CCl_4. The hydrogen bonding between the oxygen atom (S=O) of the $DBTO_2$ with the pyridinium hydrogen and the higher polarity of the $DBTO_2$ could be the possible reasons for the higher interaction energies of ([BPy][BF_4]):$DBTO_2$ versus those of ([BPy][BF_4]):DBT. The dipole moments of $DBTO_2$ and DBT are 6.699 and 0.869 D, respectively, in the hexane solvent system. Due to the higher polarity of the

$DBTO_2$, it is more likely to be solubilized in the ionic-liquid phase rather than the model-fuel phase based upon these calculations. These postulates were further evaluated using the quantum chemical properties that can be extracted by examination of the FMOs of the ([BPy][BF_4]) with DBT and $DBTO_2$, and these will be presented below, along with findings with other ILs. The HOMOs and LUMOs of the complexes depicted in Figs. 1 and 3 were then analyzed.

Their corresponding orbital representations are shown in Figs. 4 and 5. It can be visually ascertained how the favored orbital overlaps between the LUMOs of the IL conformations shown in Fig. 4A and B, respectively, with the HOMO of the DBT being formed. Fig. 5A and B shows the corresponding HOMOs and LUMOs of the same IL with $DBTO_2$. To account for the observed IEs, it is useful to examine the HOMOs and LUMOs of the DBT, $DBTO_2$, and IL, which yield the chemical activity parameters described previously. Table 1 lists these values in eV. Although we used a different basis set to compute these values, our findings are in basic agreement with those of Gu et al. (2014). Similarly, we found that the chemical potential (m) of $DBTO_2$ was more negative than that of DBT, and also that the electrophilicity index (w) of $DBTO_2$ was more positive than that of DBT. Thus, $DBTO_2$ is more nucleophilic than DBT.

The chemical hardness (h) values of DBT and $DBTO_2$ are not much different. The larger negative chemical potential (m) and positive electrophilicity index (w) of the [BPy][BF_4] IL confirm that it clearly acts as the electron acceptor in the interaction process with the DBT and $DBTO_2$. Furthermore, we similarly found that the interaction energy (IE) between the IL and $DBTO_2$ is more favorable than that between the IL and DBT. However, as shown in Table 2, the *edge tilted-T-shaped* [BPy][BF_4]:$DBTO_2$ complex structures shown in Figs. 3 and 5 are more energetically favored than the corresponding *face-to-face parallel sandwich* structures, in both the gas and hexane solvent phases.

To extend these studies and further illustrate the application of DFT for evaluating potential ILs, we present additional preliminary results, which we have conducted with *Gaussian 16* (Frisch et al., 2019) on 35 ILs (shown in Fig. 6) using 6-311++G(d,2p) with ωB97X-D to more-closely compare with the results that Gu et al. (2014) employed. The ILs formed from four *N*-alkylated pyridinium cations *N*-ethyl ([EPy^+]), *N*-butyl ([BPy^+]), *N*-hexyl ([HPy^+]), *N*-octyl ([OPy^+]), and three carboxy pyridinium cations *N*-carbomethoxy ([$CMePy^+$]), *N*-carboethoxy ([$CEtPy^+$]), and *N*-carbopropoxy ([$CEtPy^+$]) were examined in silico by combinations of each of these cations with each of the following anions: BF_4^-, and four conjugate

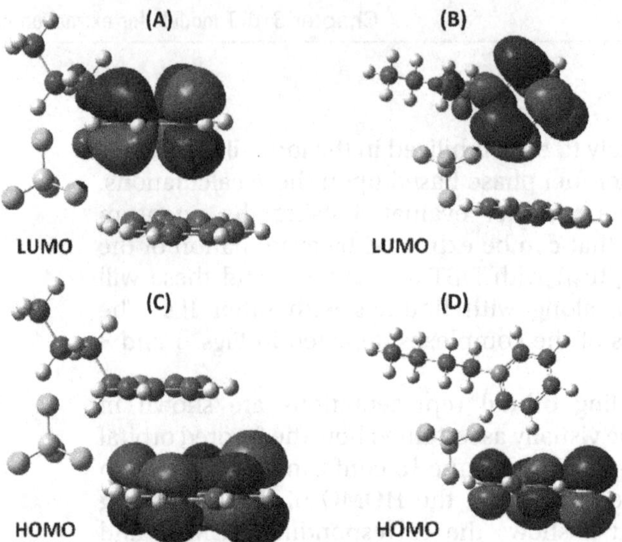

Fig. 4 The HOMO and LUMO distributions of the ([BPy][BF_4]):DBT complex structures. (A) *Top left: face-to-face parallel-sandwich type* LUMO of the complex; (B) *top right:* LUMO of the *edge-tilted-T-shaped* complex; (C) *bottom left:* HOMO of the *face-to-face parallel-sandwich type* structure of complex; and (D) *bottom right:* LUMO of the *edge-tilted-T-shaped* structure of the complex. Color code: carbon = *purple (dark gray in the printed version)*; hydrogen = *white*; nitrogen = *blue (light gray in the printed version)*; oxygen = *red (dark gray in the printed version)*; sulfur = *yellow (light gray in the printed version)*; boron = *pink (light gray in the printed version)*; and fluorine = *green (dark gray in the printed version)*.

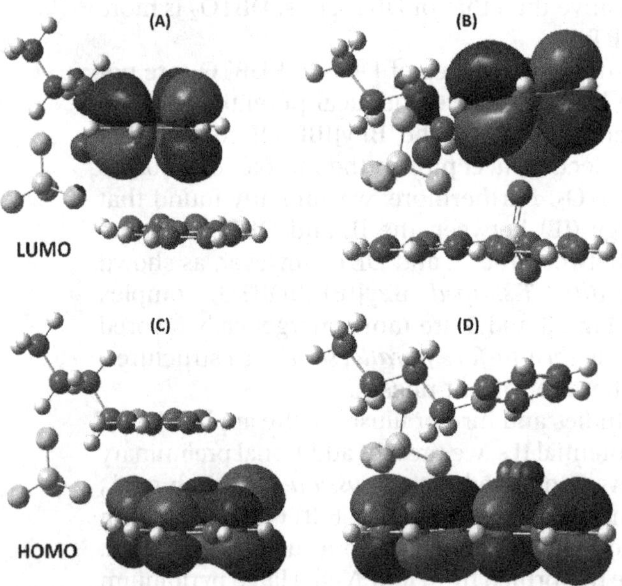

Fig. 5 The HOMO and LUMO distributions of the ([BPy][BF_4]):$DBTO_2$ complex structures. (A) *Top left: face-to-face parallel-sandwich type* LUMO of the complex; (B) *top right:* LUMO of the *edge-tilted-T-shaped* complex; (C) *bottom left:* HOMO of the *face-to-face parallel-sandwich type* structure of complex; and (D) *bottom right:* LUMO of the *edge-tilted-T-shaped* structure of the complex. Color code: carbon = *purple (dark gray in the printed version)*; hydrogen = *white*; nitrogen = *blue (light gray in the printed version)*; oxygen = *red (dark gray in the printed version)*; sulfur = *yellow (light gray in the printed version)*; boron = *pink (light gray in the printed version)*; and fluorine = *green (dark gray in the printed version)*.

Table 1 HOMO-LUMO gap (ΔE_g), ionization potential (IP), electron affinity (EA), electronegativity (χ), chemical potential (μ), hardness (η), softness (S), electrophilicity index (ω), and dipole moments of DBT, $DBTO_2$, and [BPy][BF_4] using B3LYP/6-311++G(d,p) basis set in the gas phase.

	HOMO energy (eV)	LUMO energy(eV)	HOMO-LUMO gap (ΔE_g) (eV)	Ionization potential (IP) (eV)	Electron affinity (EA) (eV)	Electro-negativity (c) (eV)	Chemical potential (m) (eV)	Hardness (h) (eV)	Softness (s) (eV)	Electro-philicity Index (ω) (eV)	Dipole moment (Debye)
DBT	−6.1163	−1.4074	4.7089	6.1163	1.4074	3.7618	−1.8809	2.3545	0.4247	0.7513	0.7676
$DBTO_2$	−7.0222	−2.2275	4.7946	7.0222	2.2275	4.6248	−2.3124	2.3973	0.4171	1.1153	5.7590
[BPy] [BF_4]	−8.5136	−3.3467	5.1669	8.5136	3.3467	5.9302	−5.9302	2.5835	0.3871	6.8062	14.7340

Table 2 Calculated interaction energies (ΔIE kJ mol^{-1}) for the ionic liquid [BPy][BF$_4$] with DBT and DBTO$_2$ sulfur-containing compounds in model fuel (hexane) and CCl$_4$ solvent system.

	ΔIE (kJ mol^{-1})		
	Gas phase	*Hexane*	*CCl$_4$*
[BPy]:[BF$_4$]	−336.97	−175.78	−148.43
[BPy][BF$_4$]:DBT *Face to face parallel sandwich*	−28.98	−16.48	−14.37
[BPy][BF$_4$]:DBT *Edge tilted-T-shaped*	−27.66	−16.15	−14.69
[BPy][BF$_4$]:DBTO$_2$ *Face to face parallel-displaced*	−50.26	−31.16	−29.07
[BPy][BF$_4$]:DBTO$_2$ *Edge tilted-T-shaped*	−51.63	−32.67	−28.91

base anions, acetate [Ac$^-$], dihydrogen phosphate, [$H_2PO_4^-$], hydrogen sulfate [HSO_4^-], and trifluoroacetate [TFA$^-$]. The four conjugate base anions were chosen in order to ascertain whether their respective pK_b values were potentially correlated with their DFT-determined stabilities and/or reactivities, and their corresponding ILs toward interacting with DBT and DBTO$_2$, in light also of the experimental findings of Fang et al. (2014). These authors studied the H_2O_2-mediated OEDS of thiophene using a model oil with five 3-*N*-methylimmidazolium ILs with the different anions shown and reported the order of reactivity to be as follows: [HMIM$^+$][TFA$^-$] > [HMIM$^+$][HSO_4^-] > [HMIM$^+$][HCO_2^-] > [HMIM$^+$][$AlCl_4^-$] > [HMIM$^+$][Ac$^-$]. The three *N*-pyridinium carboxyalkyl cations were chosen for examination in light of the experimental OEDS findings of Zhang et al. (2012) and earlier, by Gui et al. (2010), who showed better activities for the IL-mediated H_2O_2 oxidation of model oils containing benzothiophene (BT), DBT or 4,6-dimethyl-DBT (DMDBT) with ILs [CMePy][HSO_4] and [CEtPy][HSO_4] in methanol solutions. It is presumed that under the acidic conditions, the carboxylic acids were converted with the hydrogen peroxide to their corresponding better oxidizing peroxycarboxylic acids in situ. The shorter-chain [CMePy][HSO_4] IL was both the better oxidant and extractant in effecting the sulfur removal, and it was concluded that this was partly due to the relatively stronger acidity of the IL.

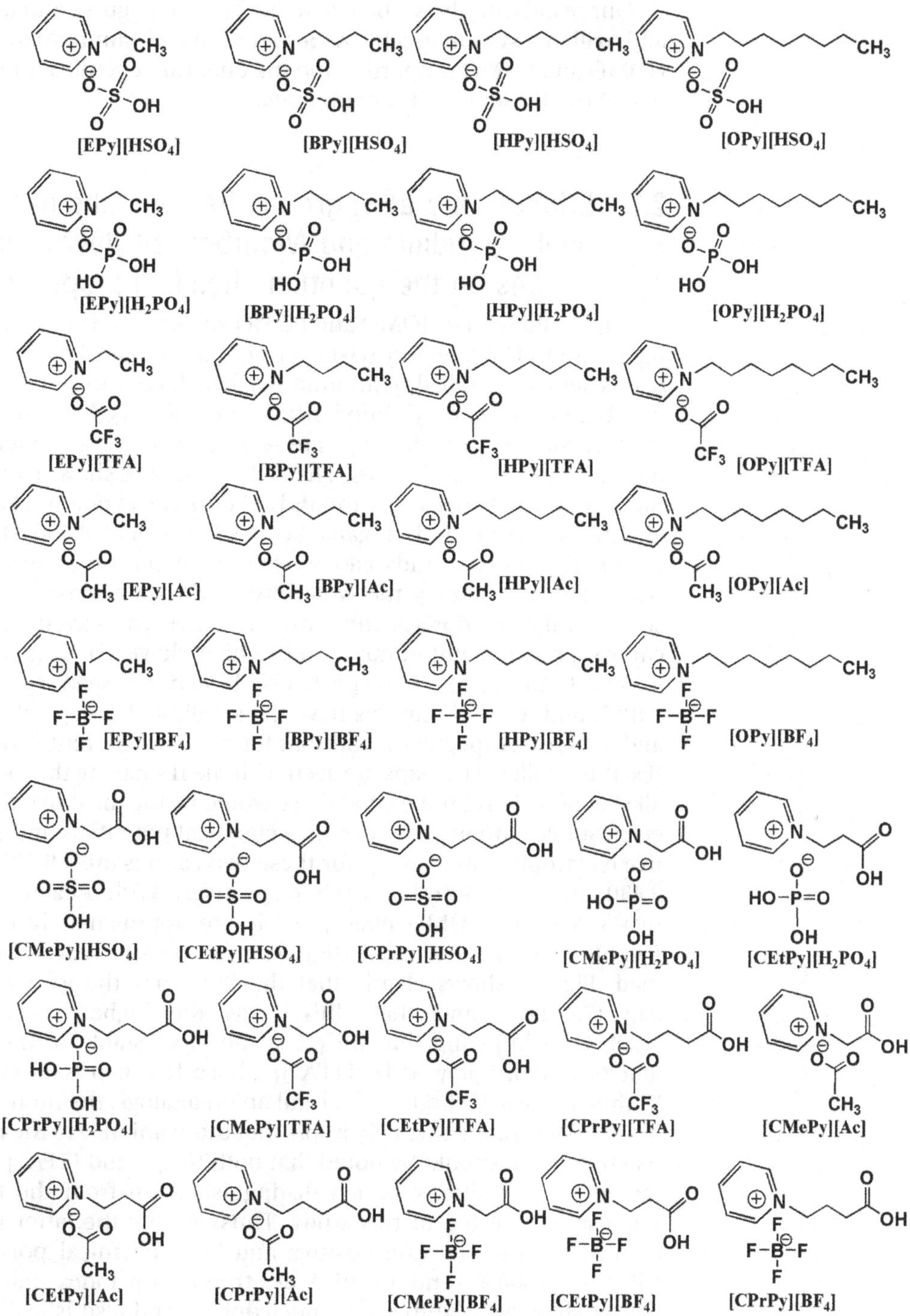

[EPy][HSO$_4$] [BPy][HSO$_4$] [HPy][HSO$_4$] [OPy][HSO$_4$]

[EPy][H$_2$PO$_4$] [BPy][H$_2$PO$_4$] [HPy][H$_2$PO$_4$] [OPy][H$_2$PO$_4$]

[EPy][TFA] [BPy][TFA] [HPy][TFA] [OPy][TFA]

[EPy][Ac] [BPy][Ac] [HPy][Ac] [OPy][Ac]

[EPy][BF$_4$] [BPy][BF$_4$] [HPy][BF$_4$] [OPy][BF$_4$]

[CMePy][HSO$_4$] [CEtPy][HSO$_4$] [CPrPy][HSO$_4$] [CMePy][H$_2$PO$_4$] [CEtPy][H$_2$PO$_4$]

[CPrPy][H$_2$PO$_4$] [CMePy][TFA] [CEtPy][TFA] [CPrPy][TFA] [CMePy][Ac]

[CEtPy][Ac] [CPrPy][Ac] [CMePy][BF$_4$] [CEtPy][BF$_4$] [CPrPy][BF$_4$]

Fig. 6 Pyridinium-based ionic liquids (ILs) used this present work for DFT calculation.

Our results are shown below in the following figures and tables and will be systematically reviewed. Table 3 summarizes the HOMO and LUMO properties and the quantum chemical properties of the ILs and their components.

2.1 Effect of the alkyl group chain lengths of the *N*-alkylpyridium and *N*-carboxyalkylpyridinium cations on the quantum chemical properties

Fig. 7A shows the HOMO and LUMO energies and corresponding HOMO-LUMO energy (H-L) gaps of the cations of the homologous series of *N*-alkylpyridinium ILs, which comprise the even-numbered carbon ethyl, butyl, hexyl, and octyl alkyl groups. The HOMO energies and H-L gaps are seen to decrease with increasing alkyl group chain lengths. The LUMO energies show relatively much smaller changes. The trends in the chemical potentials parallel the trend in the H-L gaps: i.e., 9.938 > 9.765 > 8.756 > 8.039 (Table 1). Similar trends can be seen with the other series of ILs in Fig. 7B, namely the *N*-carboxyalkylpyridinium-based ILs series, which comprises the carboxymethyl, carboxyethyl, and carboxypropyl compounds, and which have H-L gaps of 9.489 > 8.515 > 8.199 eV, respectively. It can be seen that the [OPy^+] and [$CPrPy^+$] cations have the smallest H-L gaps of 8.039 and 8.199 eV, respectively, and that for the two different series of ILs, the smallest H-L gaps are seen with the ILs having the longest alkyl chains. This implies that these would be the most reactive or electrophilic cations. The negative chemical potentials and positive electrophilicity indexes for these two cations are −8.707 and 9.430, and −8.905 and 9.672 eV, respectively. Within each of the two series, the LUMO energy levels are approximately equal. Fig. 6B shows the corresponding data for the four anions examined. Fig. 6B shows clearly that the [Ac^-] has the lowest H-L gap (7.491 eV) and that [BF_4^-] has the highest H-L gap (12.854 eV) implying that [BF_4^-] is the most stable of the five anions tested, followed by [TFA^-], whose H-L gap is 11.466 eV. Within a series of similar structural anion analogs, the more positive LUMO energies usually imply that those anions are the more reactive, but it should be noted that both [BF_4^-] and [TFA^-] contained fluorine atoms, which distinguish them from the three other anions tested in this study. Thus, among the latter three anions, the [Ac^-] whose positive and large chemical potential value of 0.608 eV and positive electrophilicity index value of 0.0494 show it to be the most nucleophilic and also is expected

Table 3 Calculated global scalar properties of DBT, $DBTO_2$, cations, anions, and their combined ionic liquids computed at the DFT ωb97xd/6-311++g(d,2p) level of theory in the gas phase (Frisch et al., 2019).

	HOMO energy (eV)	LUMO energy (eV)	H-L gap (eV)	Ionization potential (IP) (eV)	Electron affinity (EA) (eV)	Electro-negativity (χ) (eV)	Chemical potential (μ) (eV)	Hardness (η) (eV)	Softness (S) (eV)	Electro-philicity index (ω) (eV)	Dipole moment (Debye)
DBT	−7.9642	−0.4063	7.5580	7.9642	0.4063	4.1852	−4.1852	3.7790	0.2646	2.3176	0.8635
DBTO2	−8.9079	−0.3880	8.5199	8.9079	0.3880	4.6480	−4.6480	4.2599	0.2347	2.5357	5.7373
[EPy][HSO_4]	−8.4902	−1.3195	7.1707	8.4902	1.3195	4.9049	−4.9049	3.5854	0.2789	3.3550	12.5195
[BPy][HSO_4]	−8.4823	−1.3442	7.1381	8.4823	1.3442	4.9133	−4.9133	3.5690	0.2802	3.3819	13.8968
[HPy][HSO_4]	−8.4796	−1.3641	7.1155	8.4796	1.3641	4.9219	−4.9219	3.5578	0.2811	3.4045	13.9498
[OPy][HSO_4]	−8.4793	−1.3663	7.1131	8.4793	1.3663	4.9228	−4.9228	3.5565	0.2812	3.4070	13.9351
[EPy][H_2PO_4]	−8.3855	−1.1388	7.2467	8.3855	1.1388	4.7621	−4.7621	3.6233	0.2760	3.1294	12.1520
[BPy][H_2PO_4]	−8.3830	−1.1396	7.2434	8.3830	1.1396	4.7613	−4.7613	3.6217	0.2761	3.1298	11.8345
[HPy][H_2PO_4]	−8.3800	−1.1407	7.2393	8.3800	1.1407	4.7604	−4.7604	3.6197	0.2763	3.1303	11.8148
[OPy][H_2PO_4]	−8.3776	−1.1415	7.2361	8.3776	1.1415	4.7595	−4.7595	3.6180	0.2764	3.1306	11.8290
[EPy][TFA]	−8.2472	−1.1331	7.1141	8.2472	1.1331	4.6902	−4.6902	3.5571	0.2811	3.0921	12.8833
[BPy][TFA]	−8.2206	−1.1633	7.0573	8.2206	1.1633	4.6919	−4.6919	3.5286	0.2834	3.1194	14.4045
[HPy][TFA]	−8.2000	−1.1867	7.0133	8.2000	1.1867	4.6933	−4.6933	3.5066	0.2852	3.1408	14.2935
[OPy][TFA]	−8.1520	−1.2365	6.9155	8.1520	1.2365	4.6942	−4.6942	3.4578	0.2892	3.1864	14.2977

Continued

Table 3 Calculated global scalar properties of DBT, $DBTO_2$, cations, anions, and their combined ionic liquids computed at the DFT ωb97xd/6-311++g(d,2p) level of theory in the gas phase—cont'd

	HOMO energy (eV)	LUMO energy (eV)	H-L gap (eV)	Ionization potential (IP) (eV)	Electron affinity (EA) (eV)	Electro-negativity (χ) (eV)	Chemical potential (μ) (eV)	Hardness (η) (eV)	Softness (S) (eV)	Electro-philicity index (ω) (eV)	Dipole moment (Debye)
[EPy][Ac]	−7.5098	−0.9108	6.5990	7.5098	0.9108	4.2103	−4.2103	3.2995	0.3031	2.6862	9.9931
[BPy][Ac]	−7.5123	−0.9388	6.5735	7.5123	0.9388	4.2255	−4.2255	3.2867	0.3043	2.7162	10.4885
[HPy][Ac]	−7.5169	−0.9543	6.5626	7.5169	0.9543	4.2356	−4.2356	3.2813	0.3048	2.7337	10.5736
[OPy][Ac]	−7.5220	−0.9782	6.5438	7.5220	0.9782	4.2501	−4.2501	3.2719	0.3056	2.7604	10.7684
[EPy][BF_4]	10.7572	−1.3481	9.4092	10.7572	1.3481	6.0526	−6.0526	4.7046	0.2126	3.8935	14.3221
[BPy][BF_4]	10.7953	−1.3508	9.4445	10.7953	1.3508	6.0730	−6.0730	4.7223	0.2118	3.9051	14.3336
[HPy][BF_4]	10.8271	−1.3551	9.4720	10.8271	1.3551	6.0911	−6.0911	4.7360	0.2111	3.9170	14.2357
[OPy][BF_4]	10.8620	−1.3630	9.4990	10.8620	1.3630	6.1125	−6.1125	4.7495	0.2105	3.9333	14.2230
[CMePy][HSO_4]	−8.6734	−1.5755	7.0978	8.6734	1.5755	5.1245	−5.1245	3.5489	0.2818	3.6997	8.1161
[CEtPy][HSO_4]	−8.6633	−1.4849	7.1784	8.6633	1.4849	5.0741	−5.0741	3.5892	0.2786	3.5867	14.6351
[CPrPy][HSO_4]	−8.6554	−1.4177	7.2377	8.6554	1.4177	5.0366	−5.0366	3.6188	0.2763	3.5048	12.3471
[CMePy][H_2PO_4]	−8.6437	−1.2564	7.3874	8.6437	1.2564	4.9500	−4.9500	3.6937	0.2707	3.3169	10.7597
[CEtPy][H_2PO_4]	−8.6227	−1.2275	7.3952	8.6227	1.2275	4.9251	−4.9251	3.6976	0.2704	3.2801	12.2583
[CPrPy][H_2PO_4]	−8.6179	−1.1970	7.4208	8.6179	1.1970	4.9074	−4.9074	3.7104	0.2695	3.2453	11.6303
[CMePy][TFA]	−8.4244	−1.2308	7.1936	8.4244	1.2308	4.8276	−4.8276	3.5968	0.2780	3.2397	13.1132

[CEtPy] [TFA]	−8.3841	−1.1274	7.2567	8.3841	1.1274	4.7557	−4.7557	3.6284	0.2756	3.1167	13.0962
[CPrPy] [TFA]	−8.1079	−1.0855	7.0224	8.1079	1.0855	4.5967	−4.5967	3.5112	0.2848	3.0089	14.2420
[CMePy] [Ac]	−7.7283	−1.0574	6.6709	7.7283	1.0574	4.3929	−4.3929	3.3354	0.2998	2.8928	9.3708
[CEtPy] [Ac]	−7.7468	−1.0093	6.7375	7.7468	1.0093	4.3780	−4.3780	3.3688	0.2968	2.8448	9.5695
[CPrPy] [Ac]	−7.6845	−0.9404	6.7441	7.6845	0.9404	4.3125	−4.3125	3.3720	0.2966	2.7576	8.8505
[CMePy] [BF_4]	10.3213	−1.5832	8.7381	10.3213	1.5832	5.9522	−5.9522	4.3691	0.2289	4.0545	12.9776
[CEtPy] [BF_4]	10.2767	−1.5293	8.7474	10.2767	1.5293	5.9030	−5.9030	4.3737	0.2286	3.9835	15.4958
[CPrPy] [BF_4]	10.2179	−1.4553	8.7626	10.2179	1.4553	5.8366	−5.8366	4.3813	0.2282	3.8876	13.8422
[EPy]	14.7717	−4.8333	9.9384	14.7717	4.8333	9.8025	−9.8025	4.9692	0.2012	9.6684	0.5510
[BPy]	14.4974	−4.7323	9.7651	14.4974	4.7323	9.6149	−9.6149	4.8825	0.2048	9.4670	4.4670
[HPy]	13.4563	−4.7000	8.7564	13.4563	4.7000	9.0781	−9.0781	4.3782	0.2284	9.4117	9.3016
[OPy]	12.7265	−4.6874	8.0391	12.7265	4.6874	8.7070	−8.7070	4.0195	0.2488	9.4304	14.5631
[CMePy]	14.4316	−4.9424	9.4892	14.4316	4.9424	9.6870	−9.6870	4.7446	0.2108	9.8889	3.3901
[CEtPy]	13.4920	−4.9767	8.5153	13.4920	4.9767	9.2343	−9.2343	4.2576	0.2349	10.0141	7.2509
[CPrPy]	13.0049	−4.8055	8.1993	13.0049	4.8055	8.9052	−8.9052	4.0997	0.2439	9.6718	8.1029
[HSO_4]	−4.5487	4.6308	9.1795	4.5487	−4.6308	−0.0411	0.0411	4.5897	0.2179	0.0002	2.5965
[H_2PO_4]	−4.5892	4.6542	9.2434	4.5892	−4.6542	−0.0325	0.0325	4.6217	0.2164	0.0001	3.1533
[TFA]	−9.2832	2.1832	11.463	9.2832	−2.1832	3.5500	−3.5500	5.7332	0.1744	1.0991	4.8212
[Ac]	−3.1369	4.3538	7.4908	3.1369	−4.3538	−0.6084	0.6084	3.7454	0.2670	0.0494	3.9180
[BF_4]	−6.843	6.0110	12.854	6.843	−6.011	0.4161	−0.4161	6.4271	0.1556	0.0135	0.0048

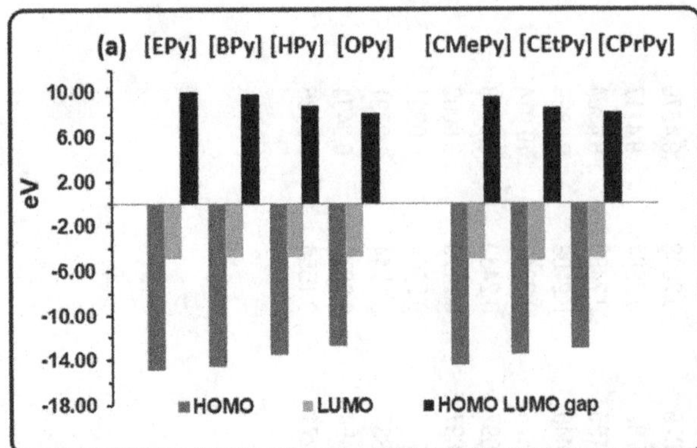

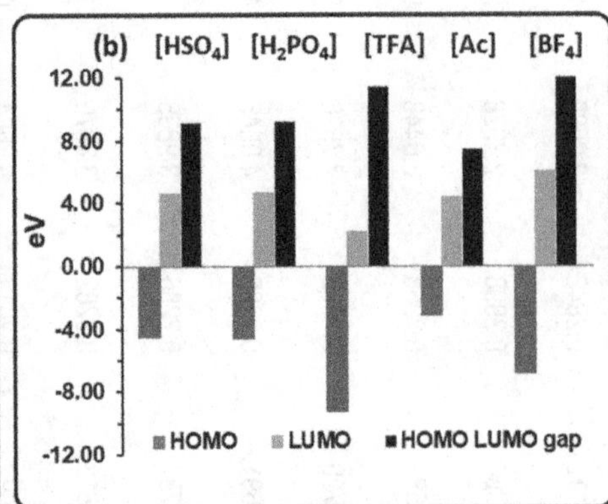

Fig. 7 Histograms showing the HOMO, LUMO energies, and HOMO-LUMO energy gaps of (A) *left:* the IL cations examined; and (B) *right:* the IL anions examined.

to be the more reactive one to form the more stable IL with the cations shown in Fig. 7A.

2.2 Effect of alkyl group chain lengths on the quantum chemical properties of the *N*-alkylpyridium ILs

As can be seen in Fig. 8, there are only very slight variations in the selected quantum chemical properties for the ILs based on the homologous series of the *N*-alkypyridinium ILs as shown here, for example, with *N*-alkypyridinium [HSO_4^-] ILs (Gu et al., 2014), and Zhao et al. (2010) had used [BPy][HSO_4] IL in their studies.

2.3 Effect of the five anions on the quantum chemical properties of the *N*-alkylpyridium ILs

Fig. 9 shows the effect of the five different anions with the *N*-butylpyridinium ILs as representative ILs and reveals that the [BPy][BF_4] has the largest H-L energy gap, but also has the larger negative chemical potential and highest electrophilicity in this series. These would have predicted it to be the more electrophilic IL acceptor in the interactions with DBT and $DBTO_2$. However, the interaction energies and thermodynamic data suggest a different predictive outcome.

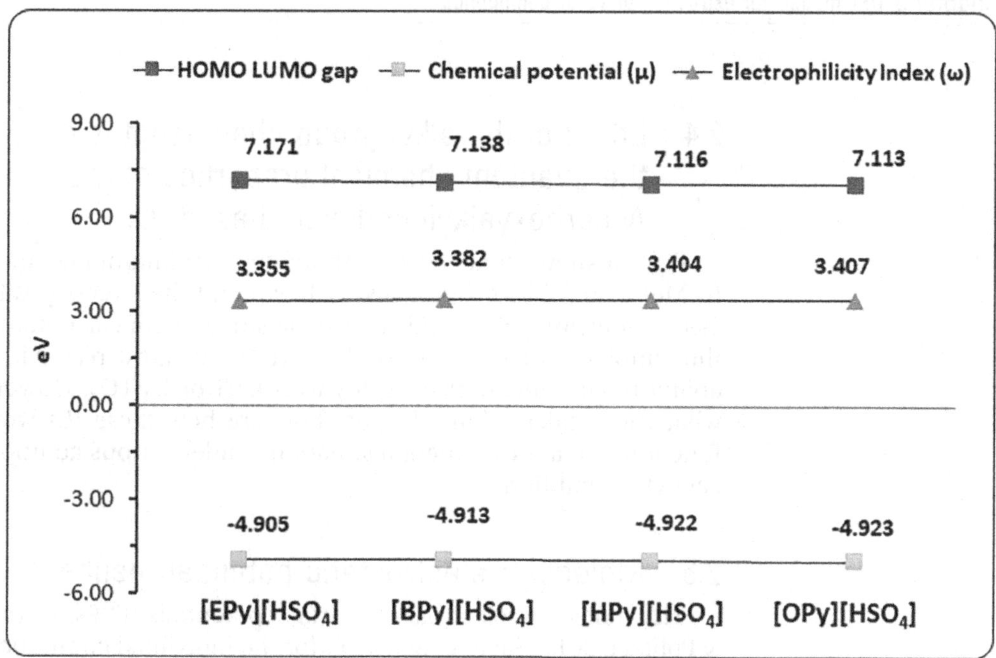

Fig. 8 Variations in the H-L gaps, chemical potential (μ) and electrophilicity indexes (ω) with increasing alkyl group chain lengths for the HSO_4^-, ILs.

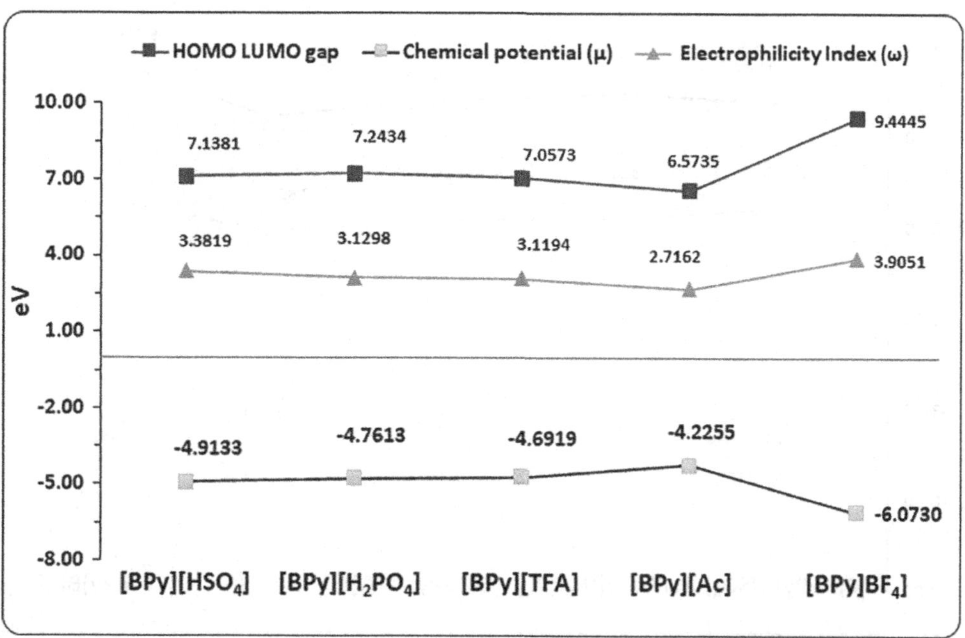

Fig. 9 Variations in the H-L, chemical potential (μ) and electrophilicity indexes (ω) for the five different counter ions with *N*-butylpyridinium ILs.

2.4 Effect of the alkyl group chain lengths on the quantum chemical properties of the *N*-carboxyalkylpyridinium-based ILs

Fig. 10 shows that the *N*-carbomethylpyridinium acetate IL, ([CMePy][Ac]) has the smallest H-L gap and the largest positive electrophilicity index, and largest negative chemical potential, thus implying that this IL would have the greatest reactivity or ability to coordinate or complex with DBT or $DBTO_2$. However, what these calculations do not show are how these ILs would function in real experimental situations under various additional catalytic conditions.

2.5 Molecular electrostatic potential results

The molecular electrostatic surface potentials (ESPs) (Murray & Politzer, 2011; Savin et al., 1998) for the individual components and the complexes formed from the ILs [BuPy][HSO_4] and [CMePy][HSO_4], with each of DBT and $DBTO_2$, are shown in

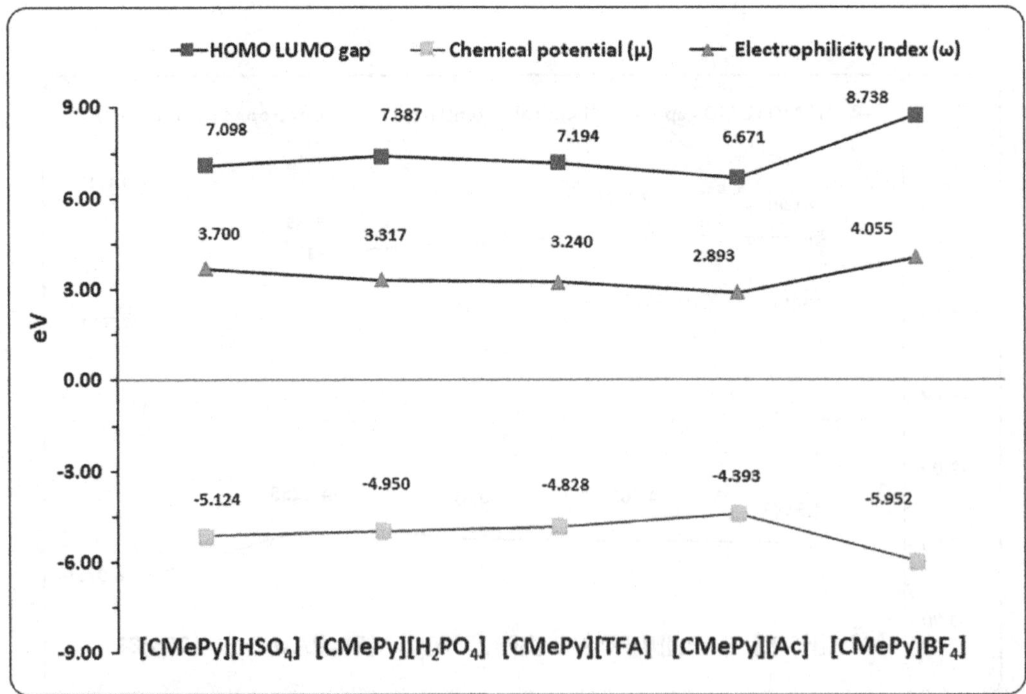

Fig. 10 Variations in the H-L, chemical potential (μ) and electrophilicity indexes (ω) for the five different counter ions with the corresponding *N*-carboxymethylpyridinium based ILs.

Figs. 11–16. These were generated using the *GaussView 6.0.16* software, and the surfaces were formed by mapping the electrostatic potentials (EPs) onto their HOMO electron density surfaces. The EPs show the relative polarities and thus the reactive sites of the species: the negative EPs are shown in red (dark gray in the printed version), and the order of increasing electrostatic potentials (i.e., highest -ve value) is: red (dark gray in the printed version) > orange (light gray in the printed version) < yellow (light gray in the printed version) < green (dark gray in the printed version) < blue (light gray in the printed version). From Figs. 11–16, it is clearly evident that the hydrogen atoms in pyridine, DBT and $DBTO_2$ rings are the most nucleophilic sites, which tend to interact with the electrophilic HSO_4^- anion of the IL molecule. Figs. 12 and 13 show the ESP plots of the ILs [BPy][HSO_4] and its complexes with DBT and $DBTO_2$ ([BPy][HSO_4]:[DBT] and [BPy]

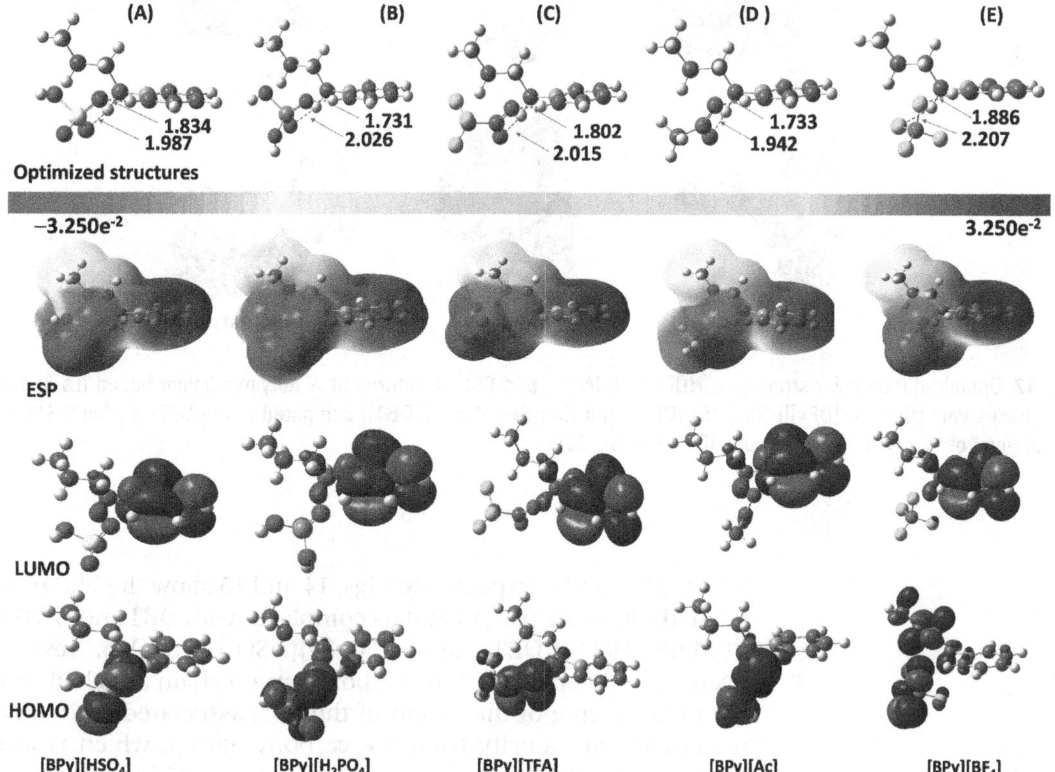

Fig. 11 Optimized molecular structures, HOMOs, LUMOs, and ESP structures of *N*-carboxymethylpyridinium based ILs: (A) [BPy][HSO_4]; (B) [BPy][H_2PO_4]; (C) [BPy][TFA]; (D) [BPy][Ac]; and (E) [BPy][BF_4] computed at the DFT ωb97xd/6-311++g(d,2p) level of theory in the gas phase (Frisch et al., 2019).

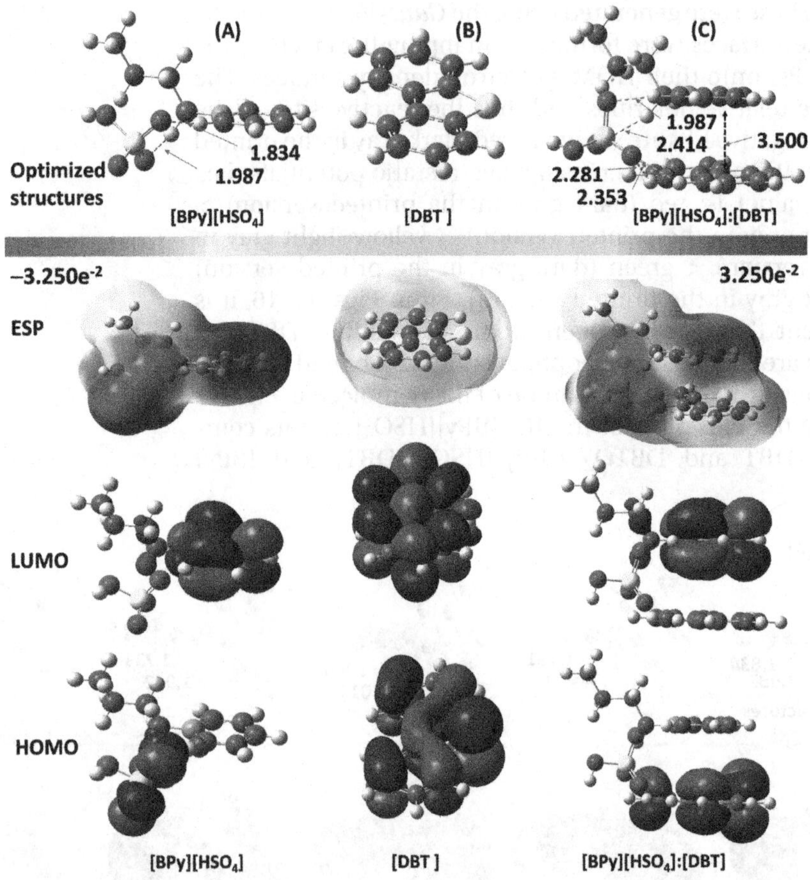

Fig. 12 Optimized molecular structures, HOMOs, LUMOs, and ESP structures of *N*-butylpyridinium based ILs and it complexes with DBT: (A) [BPy][HSO_4]; (B) [DBT]; and (C) ([BPy][HSO_4]:[DBT]) computed at the DFT ωb97xd/6-311++g (d,2p) level of theory in the gas phase (Frisch et al., 2019).

[HSO_4]:[$DBTO_2$]), respectively. Figs. 14 and 15 show the ESP plots of the ILs [CMePy][HSO_4] and its complexes with DBT and $DBTO_2$ ([CMePy][HSO_4]:[DBT] and [CMePy][HSO_4]:[$DBTO_2$]), respectively. Fig. 13 depicts that the carboxy carbon chain attached with the pyridine ring of the cation of the IL is associated with additional electron density from the carbonyl group, which is also electron-donating to the cationic charge of the pyridine ring. Consequently, the electrophilic HSO_4^- of the ionic liquid molecule is more reactive with the hydrogen atoms in both the DBT and $DBTO_2$ rings. On the other hand, in Figs. 11A–E and Figs. 12A

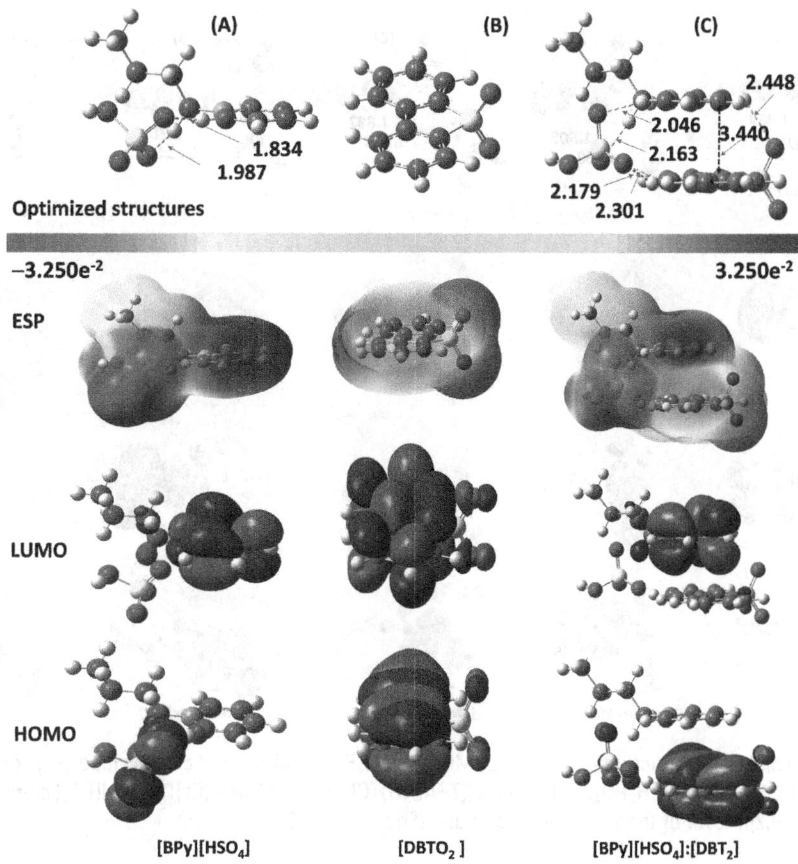

Fig. 13 Optimized molecular structures, HOMOs, LUMOs, and ESP structures of *N*-butylpyridinium based ILs and it complexes with $DBTO_2$: (A) [BPy][HSO_4]; (B) [$DBTO_2$]; and (C) ([BPy][HSO_4]:[DBT]) computed at the DFT ωb97xd/6-311++g(d,2p) level of theory in the gas phase (Frisch et al., 2019).

and 13A, it can be seen that the alkyl group attached to the nitrogen of the pyridine ring is associated with near-zero ESP value.

2.6 Interaction energies and thermodynamic calculations for the interactions with DBT and the ILs

Tables 4 and 5 summarize the computed interaction energies and the thermodynamic properties of the complexes formed between the two classes of the ILs, with DBT and with

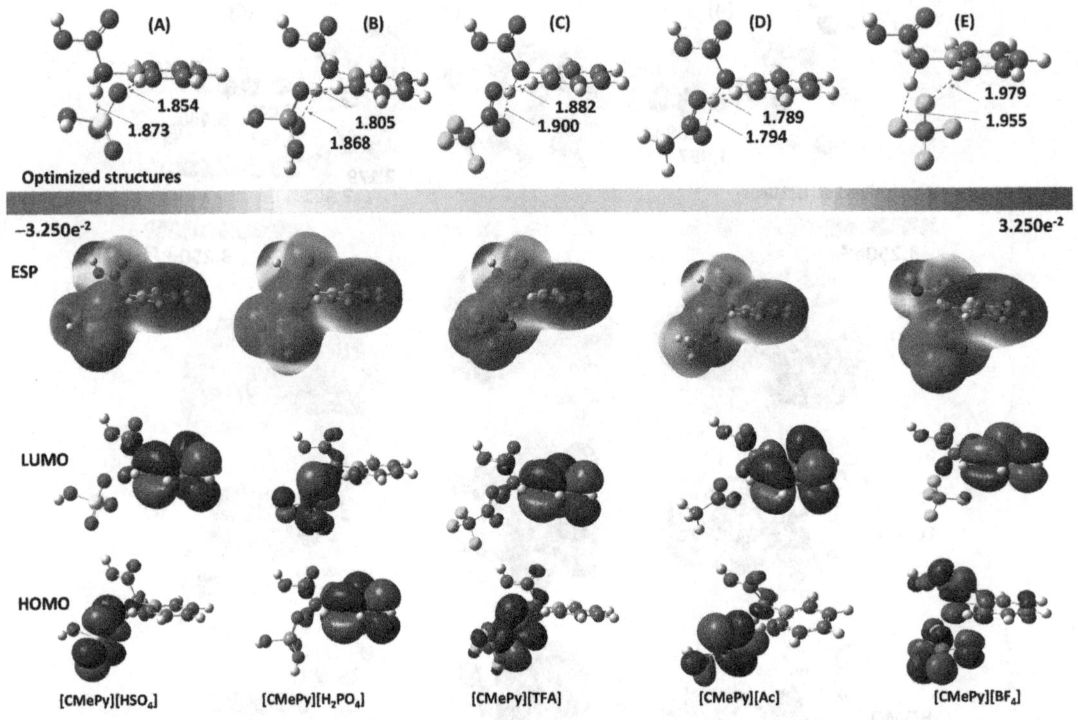

Fig. 14 Optimized molecular structures, HOMOs, LUMOs, and ESP structures of *N*-carboxymethylpyridinium based ILs: (A) [CMePy][HSO_4]; (B) [CMePy][H_2PO_4]; (C) [CMePy][TFA]; (D) [CMePy][Ac]; and (E) [CMePy][BF_4] computed at the DFT ωb97xd/6-311++g(d,2p) level of theory in the gas phase (Frisch et al., 2019).

$DBTO_2$. In all cases, as expected, the interaction energies and thermodynamic parameters favor the complexes formed between the ILs and the $DBTO_2$. The largest interaction energies (ΔIE) of all are seen for ([CMePy][SO_4]):[$DBTO_2$] ($-116.43\,kJ\,mol^{-1}$) and for ([OPy][HSO_4]):[$DBTO_2$] ($-113.18\,kJ\,mol^{-1}$). The corresponding free energy changes (ΔG) were found to be -53.35 and $-51.38\,kJ\,mol^{-1}$, respectively. For the complexes formed between the ILs and the DBT, the largest interaction energies (ΔIE), i.e., -106.81 and $-95.64\,kJ\,mol^{-1}$, respectively, are seen with the same two ILs, ([CMePy][HSO_4]):[DBT] and ([OPy][HSO_4]):[DBT]. The corresponding free energy changes (ΔG) were found to be -46.41 and $-37.47\,kJ\,mol^{-1}$, respectively. Interestingly, when the interaction energies and free energy changes are compared within each series of ILs, large variations are observed. Figs. 17 and 18 illustrate these variations from the data in Tables 4 and 5.

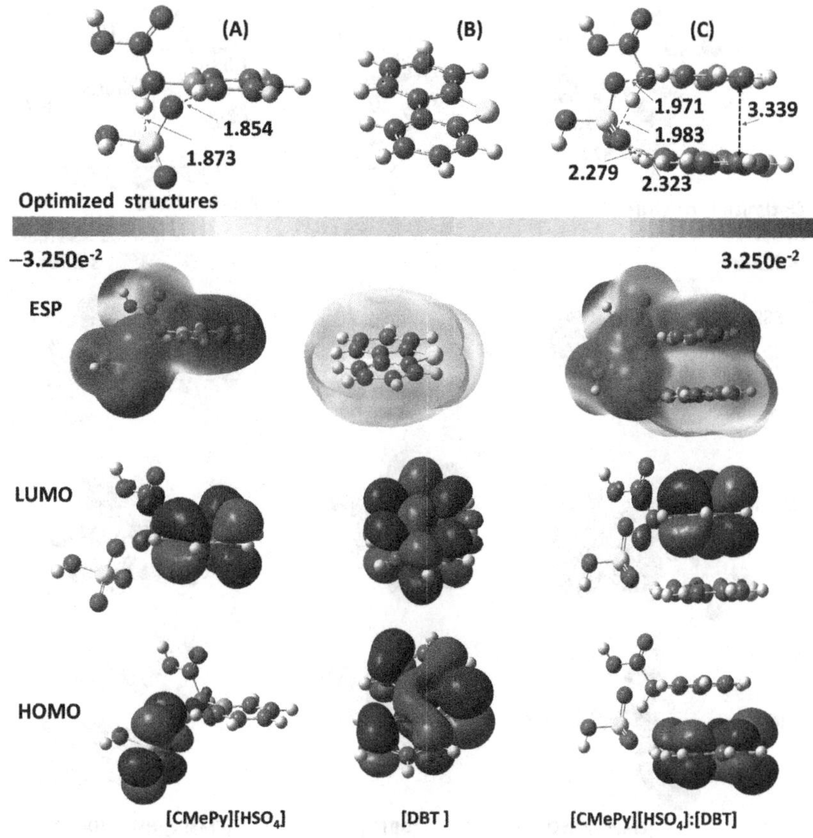

Fig. 15 Optimized molecular structures, HOMOs, LUMOs, and ESP structures of *N*-carboxymethylpyridinium based ILs and it complexes with DBT: (A) [CMePy][HSO_4]; (B) [DBT]; and (C) ([CMePy][HSO_4]:[DBT]) computed at the DFT ωb97xd/6-311++g(d,2p) level of theory in the gas phase (Frisch et al., 2019).

2.7 *N*-Alkylpyridinium ILs with HSO_4^- anions

With all of the ILs examined, clearly both their interaction energies (Δ*IE*) and their thermodynamic properties favor stronger binding or association with $DBTO_2$ over DBT (Figs. 17 and 18). In this respect, the data here are consistent with the findings of others cited previously. As indicated above, the largest Δ*IE* and Δ*G* values of −113.18 and −95.64 kJ mol^{-1} are noted for [OPy] [HSO_4] with $DBTO_2$ and DBT, respectively. However, if the free energy changes for this group of ILs are compared for each of the individual ILs, the order of the observed potential reactivities is reversed. Thus, for the *N*-ethylpyridium:HSO_4 IL, the difference in free energy changes for its interaction with $DBTO_2$ compared with that for DBT, given by $\Delta\Delta G = \Delta G_{[EPy][HSO4]:[DBTO2]} - \Delta G_{[EPy][HSO4]:[DBT]}$, is −14.5 kJ mol^{-1}, whereas the equivalent

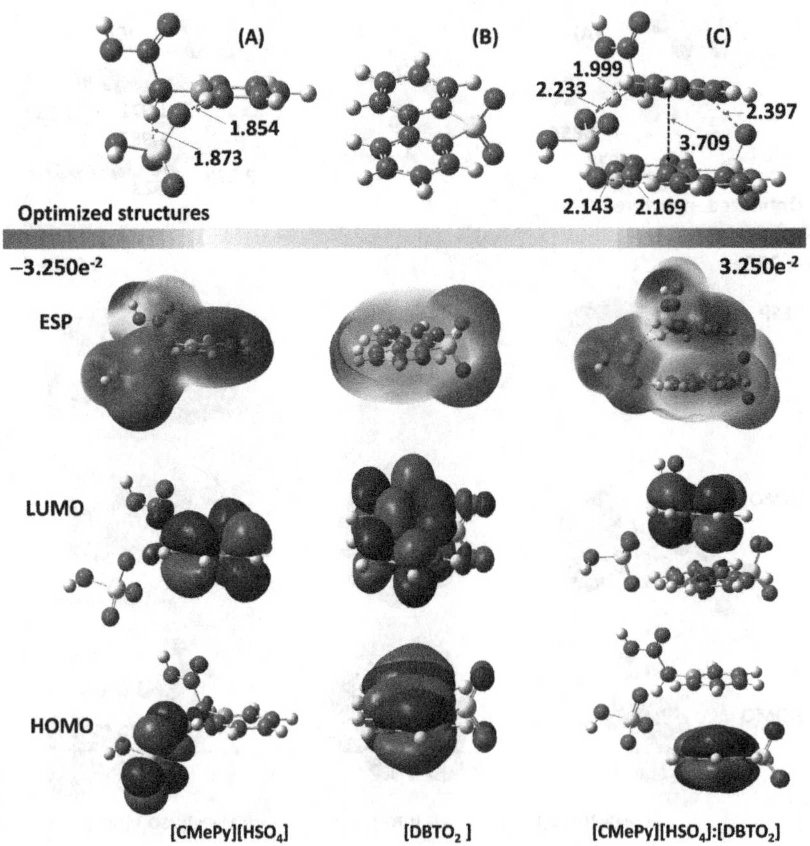

Fig. 16 Optimized molecular structures, HOMOs, LUMOs, and ESP structures of *N*-carboxymethylpyridinium based ILs and it complexes with $DBTO_2$: (A) [CMePy][HSO_4]; (B) [$DBTO_2$]; and (C) ([CMePy][HSO_4]:[DBT]) computed at the DFT ωb97xd/6-311++g(d,2p) level of theory, in the gas phase (Frisch et al., 2019).

$\Delta\Delta G$ for the *N*-octylpyridium:HSO_4 IL is only $-13.92\,kJ\,mol^{-1}$. This suggests that the more favorable free energy change overall would be with shorter alkyl chain ILs (see Table 4). None of the other quantum quantities offer much more insight into the potential significant benefits of one over the other members of this series.

2.8 *N*-Alkylpyridinium ILs with $H_2PO_4^-$ and Ac^- anions

Analysis of the corresponding data for the $H_2PO_4^-$ and also the Ac^- salts of the *N*-alkylpyridinium ILs yields similar conclusions as with the HSO_4^- ILs (see Table 4).

Table 4 Calculated interaction energies (ΔIE kJ mol^{-1}), enthalpy (ΔH), free energy (ΔG) changes and relative free energy changes ($\Delta\Delta G^*$) for the *N*-alkylpyridinium based ILs with DBT and $DBTO_2$ using the ωB97XD/6-311++G(d,2p) basis set in the gas phase (Frisch et al., 2019).

Complexes	ΔIE (kJ mol^{-1})	ΔH (kJ mol^{-1})	ΔS (kJ mol^{-1})	ΔG (kJ mol^{-1})	$\Delta\Delta G^*$ (kJ mol^{-1})
([EPy][HSO_4]):[DBT]	−92.42	−85.59	−0.1659	−36.14	
([EPy][HSO_4]):[$DBTO_2$]	−110.21	−103.16	−0.1762	−50.64	−14.50
([BPy][HSO_4]):[DBT]	−93.86	−87.22	−0.1694	−36.72	
([BPy][HSO_4]):[$DBTO_2$]	−111.55	−104.15	−0.1783	−51.03	−14.31
([HPy][HSO_4]):[DBT]	−95.19	−88.66	−0.1723	−37.31	
([HPy][HSO_4]):[$DBTO_2$]	−112.86	−105.16	−0.1806	−51.33	−14.03
([OPy][HSO_4]):[DBT]	−95.64	−88.99	−0.1729	−37.47	
([OPy][HSO_4]):[$DBTO_2$]	−113.18	−105.25	−0.1808	−51.38	−13.92
([EPy][H_2PO_4]):[DBT]	−92.23	−85.04	−0.1644	−36.04	
([EPy][H_2PO_4]):[$DBTO_2$]	−108.49	−102.59	−0.1753	−50.35	−14.31
([BPy][H_2PO_4]):[DBT]	−92.96	−86.39	−0.1668	−36.68	
([BPy][H_2PO_4]):[$DBTO_2$]	−108.70	−103.61	−0.1779	−50.59	−13.91
([HPy][H_2PO_4]):[DBT]	−94.41	−88.32	−0.1710	−37.36	
([HPy][H_2PO_4]):[$DBTO_2$]	−109.04	−104.81	−0.1803	−51.07	−13.71
([OPy][H_2PO_4]):[DBT]	−95.26	−89.37	−0.1732	−37.75	
([OPy][H_2PO_4]):[$DBTO_2$]	−109.31	−105.69	−0.1825	−51.30	−13.54
([EPy][TFA]):[DBT]	−91.27	−84.65	−0.1654	−35.35	
([EPy][TFA]):[$DBTO_2$]	−105.03	−95.41	−0.1802	−41.72	−6.36
([BPy][TFA]):[DBT]	−92.33	−85.31	−0.1662	−35.79	
([BPy][TFA]):[$DBTO_2$]	−106.45	−98.48	−0.1807	−44.63	−8.84
([HPy][TFA]):[DBT]	−93.18	−85.92	−0.1667	−36.24	
([HPy][TFA]):[$DBTO_2$]	−107.62	−101.03	−0.1814	−46.97	−10.73
([OPy][TFA]):[DBT]	−94.00	−86.43	−0.1669	−36.68	
([OPy][TFA]):[$DBTO_2$]	−108.44	−103.18	−0.1815	−49.09	−12.41

Continued

Table 4 Calculated interaction energies (ΔIE kJ mol^{-1}), enthalpy (ΔH), free energy (ΔG) changes and relative free energy changes ($\Delta\Delta G^*$) for the *N*-alkylpyridinium based ILs with DBT and $DBTO_2$ using the ωB97XD/6-311++G(d,2p) basis set in the gas phase—cont'd

Complexes	ΔIE (kJ mol^{-1})	ΔH (kJ mol^{-1})	ΔS (kJ mol^{-1})	ΔG (kJ mol^{-1})	$\Delta\Delta G^*$ (kJ mol^{-1})
([EPy][Ac]):[DBT]	−90.01	−82.35	−0.1851	−27.18	
([EPy][Ac]):[$DBTO_2$]	−104.38	−95.60	−0.1875	−39.72	−12.54
([BPy][Ac]):[DBT]	−91.50	−87.61	−0.1868	−31.94	
([BPy][Ac]):[$DBTO_2$]	−105.66	−98.20	−0.1877	−42.27	−10.33
([HPy][Ac]):[DBT]	−92.40	−91.86	−0.1888	−35.59	
([HPy][Ac]):[$DBTO_2$]	−106.15	−100.65	−0.1908	−43.79	−8.20
([OPy][Ac]):[DBT]	−93.73	−99.97	−0.1930	−42.45	
([OPy][Ac]):[$DBTO_2$]	−107.17	−104.32	−0.1941	−46.48	−4.02
([EPy][BF_4]):[DBT]	−90.02	−79.83	−0.1575	−32.91	
([EPy][BF_4]):[$DBTO_2$]	−104.03	−96.89	−0.1773	−44.05	−11.13
([BPy][BF_4]):[DBT]	−90.80	−82.40	−0.1625	−33.97	
([BPy][BF_4]):[$DBTO_2$]	−105.02	−98.84	−0.1783	−45.72	−11.75
([HPy][BF_4]):[DBT]	−91.56	−84.64	−0.1669	−34.92	
([HPy][BF_4]):[$DBTO_2$]	−106.23	−100.14	−0.1785	−46.95	−12.04
([OPy][BF_4]):[DBT]	−92.18	−87.71	−0.1738	−35.92	
([OPy][BF_4]):[$DBTO_2$]	−107.16	−102.99	−0.1818	−48.81	−12.88

Table 5 Calculated interaction energies (ΔIE kJ mol^{-1}), enthalpy (ΔH), free energy (ΔG) changes and relative free energy changes ($\Delta\Delta G^*$) for the *N*-carboxyalkylpyridinium-based ILs with DBT and $DBTO_2$ using ωB97XD/6-311++G(d,2p) basis set in the gas phase (Frisch et al., 2019).

Complexes	ΔIE (kJ mol^{-1})	ΔH (kJ mol^{-1})	ΔS (kJ mol^{-1})	ΔG (kJ mol^{-1})	$\Delta\Delta G^*$ (kJ mol^{-1})
([CMePy][HSO_4]):[DBT]	−106.81	−94.56	−0.1616	−46.41	
([CMePy][HSO_4]):[$DBTO_2$]	−116.43	−103.00	−0.1666	−53.35	−6.93
([CEtPy][HSO_4]):[DBT]	−99.06	−90.47	−0.1520	−45.18	
([CEtPy][HSO_4]):[$DBTO_2$]	−109.56	−94.02	−0.1564	−47.42	−2.24
([CPrPy][HSO_4]):[DBT]	−91.26	−105.16	−0.1434	−51.33	
([CPrPy][HSO_4]):[$DBTO_2$]	−104.00	−87.75	−0.1444	−44.72	−0.30
([CMePy][H_2PO_4]):[DBT]	−100.09	−92.54	−0.1605	−44.71	
([CMePy][H_2PO_4]):[$DBTO_2$]	−112.12	−99.56	−0.1647	−50.48	−5.76
([CEtPy][H_2PO_4]):[DBT]	−94.46	−89.83	−0.1582	−42.69	
([CEtPy][H_2PO_4]):[$DBTO_2$]	−107.41	−93.40	−0.1586	−46.13	−3.43
([CPrPy][H_2PO_4]):[DBT]	−88.58	−84.80	−0.1473	−40.91	
([CPrPy][H_2PO_4]):[$DBTO_2$]	−103.56	−88.08	−0.1551	−41.85	−0.95
([CMePy][TFA]):[DBT]	−94.04	−85.64	−0.1462	−42.08	
([CMePy][TFA]):[$DBTO_2$]	−105.60	−100.18	−0.1616	−52.02	−9.94
([CEtPy][TFA]):[DBT]	−91.42	−84.14	−0.1445	−41.09	
([CEtPy][TFA]):[$DBTO_2$]	−104.19	−97.98	−0.1568	−51.27	−10.18
([CPrPy][TFA]):[DBT]	−87.01	−81.88	−0.1421	−39.52	
([CPrPy][TFA]):[$DBTO_2$]	−102.41	−93.88	−0.1474	−49.96	−10.44
([CMePy][Ac]):[DBT]	−93.09	−84.73	−0.1495	−40.18	
([CMePy][Ac]):[$DBTO_2$]	−105.10	−100.11	−0.1706	−49.26	−9.08
([CEtPy][Ac]):[DBT]	−90.48	−82.70	−0.1481	−38.58	
([CEtPy][Ac]):[$DBTO_2$]	−103.56	−96.72	−0.1665	−47.09	−8.51
([CPrPy][Ac]):[DBT]	−86.83	−79.69	−0.1448	−36.55	
([CPrPy][Ac]):[$DBTO_2$]	−102.10	−92.96	−0.1655	−43.65	−7.10
([CMePy][BF_4]):[DBT]	−90.81	−84.13	−0.1684	−33.93	
([CMePy][BF_4]):[$DBTO_2$]	−103.25	−97.93	−0.1693	−47.46	−13.53
([CEtPy][BF_4]):[DBT]	−89.52	−83.39	−0.1672	−33.56	
([CEtPy][BF_4]):[$DBTO_2$]	−100.46	−94.17	−0.1607	−46.29	−12.74
([CPrPy][BF_4]):[DBT]	−88.16	−82.35	−0.1657	−32.97	
([CPrPy][BF_4]):[$DBTO_2$]	−95.99	−89.30	−0.1494	−44.78	−11.81

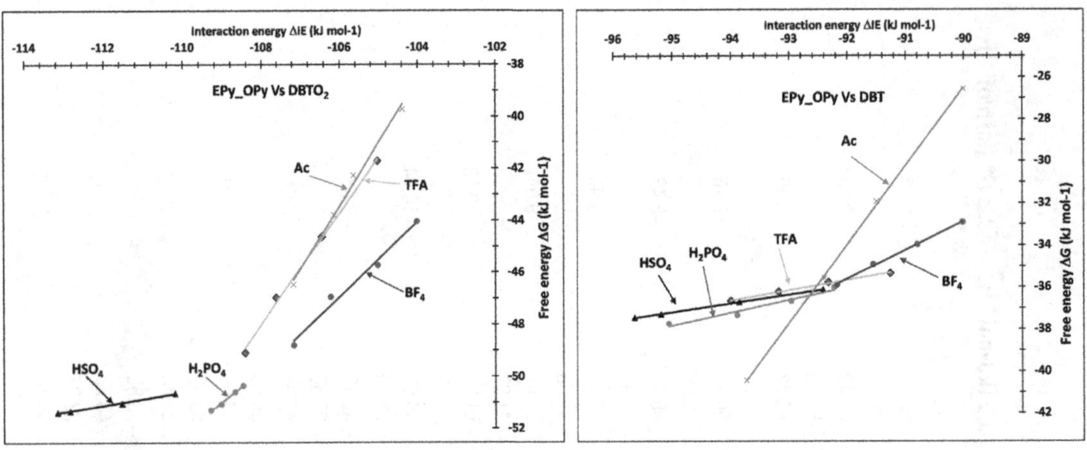

Fig. 17 Variations observed between the ΔE and ΔG values for the anion components of the *N*-alkylpyridinium based ILs for *left*: $DBTO_2$; *right*: DBT.

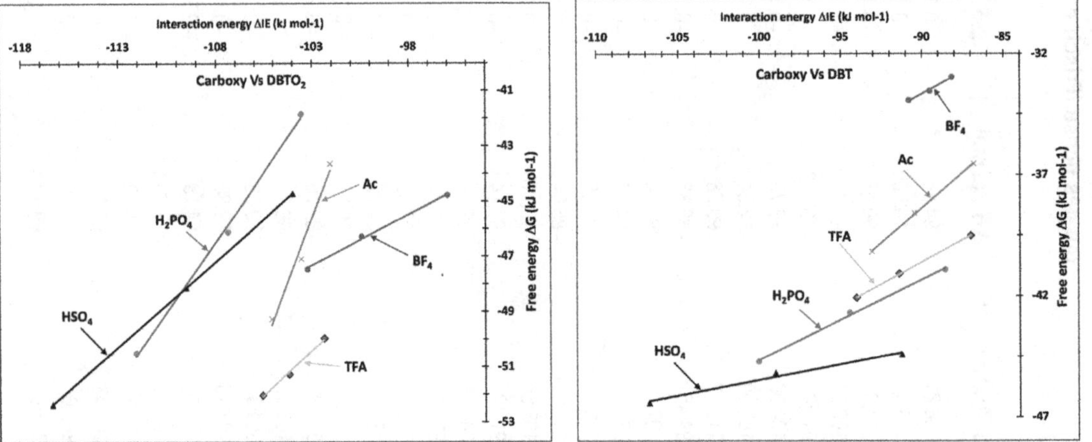

Fig. 18 Variations observed between the ΔIE and ΔG values for the anion components of the *N*-carboxyalkylpyridinium based ILs for *left*: $DBTO_2$; *right*: DBT.

2.9 *N*-Alkylpyridinium ILs with TFA^- and BF_4^- anions

On the other hand, for the ILs with TFA^- and BF_4^- anions, both the Interaction energies (ΔIE) and the thermodynamic properties favor the opposite order, which was seen with the HSO_4^-, $H_2PO_4^-$, and Ac^- anions, i.e., the [OPy]-based ILs have both the highest ΔIEs and thermodynamic properties. The presence of the fluorine atoms perhaps has a significant role in this regard due to the additional hydrogen-fluorine interactions, which are possible with these two anions and the longer alky chains. Interestingly, Gao and coworkers (Gao et al., 2008) reported a similar finding from their experimental study on the effect of the cation chain lengths of the three *N*-alkylpyridinium ILs, [BPy][BF_4], [HPy][BF_4], and [OPy][BF_4], on their extractability of TS, BT, and DBT. They found that these BF_4-containing ILs with the longer alkyl chain favored the extraction of these aromatic sulfur compounds from their model diesel oils in the order [OPy][BF_4] > [HPy][BF_4] > [BPy][BF_4] and that the removal of the sulfur compounds followed the order dibenzothiophene (DBT) > benzothiophene (BT) > thiophene (TS) under the same conditions. Their results are consistent with our gas-phase computational results.

2.10 *N*-Carboxyalkylpyridinium ILs with HSO_4^- anions

With these *N*-carboxyalkylpyridinium ILs, it is clear that both their Interaction energies (ΔIE) and their thermodynamic properties favor stronger binding or association with $DBTO_2$ over DBT (Table 5 and Fig. 18). In this respect, the data here are also consistent with all of the previous findings. The largest ΔIE and ΔG values of -116.43 and $-53.35\,kJ\,mol^{-1}$ are noted for [CMePy][HSO_4] with $DBTO_2$ and DBT, respectively. However, when both the ΔIE and ΔG values for this group of ILs are compared for each of these three individual ILs, in contrast to the corresponding *N*-carboxyalkylpyridinium ILs, the order of the observed potential reactivities favors the shortest chain [CMePy][HSO_4] IL (see Table 5).

2.11 *N*-Carboxyalkylpyridinium ILs with $H_2PO_4^-$ and Ac^- anions

Analysis of the corresponding data for the $H_2PO_4^-$ and Ac^- salts of the *N*-carboxyalkylpyridinium ILs yields similar conclusions as were noted for the corresponding data for the *N*-alkylpyridinium ILs (see Tables 4 and 5 and Fig. 18).

2.12 *N*-Carboxyalkylpyridinium ILs with TFA^- and BF_4^- anions

For the ILs with TFA^- anion, both the interaction energies (ΔIE) and the thermodynamic properties are different from those as seen with the corresponding *N*-alkylpyridinium TFA^- ILs; the [CMePy] IL has the highest ΔIEs, but its ΔG is lowest of the three *N*-carboxyalkylpyridinium ILs. These results suggest that the [CPrPy] IL would be the more favored IL. On the other hand, for the three ILs with BF_4^-, the trend is the same as seen for the corresponding HSO_4^- *N*-carboxyalkylpyridinium ILs. It is difficult to account fully for the differences seen here with those seen in the case of *N*-alkylpyridinium TFA^- and BF_4^- ILs. Finally, Fig. 18 shows the larger relative variations between the ΔE and ΔG values for the anion components compared with those in Fig. 17 can.

3 Conclusion and perspective for future developments

As this chapter reveals, the use of the DFT computational methodology for the study of possible ILs that could be employed for the desulfurization of fuel oils can yield a wealth of data. Much of the data can be used to guide experimental approaches. Two main additional conclusions, which could be drawn from the above study and those that have been reported, are that the selection of the appropriate anion with the optimum *N*-alkylpyridinium or *N*-carboxyalkylpyridinium cations could be targeted for further experimental confirmation. However, the DFT computational study can be also extended to account for the experimental observations. The evaluation of chemical reactivity and stability of ionic liquids can also be extended to and augmented by including the use of the COSMO-RS approach, which combines quantum chemical calculations with statistical thermodynamics (Vijayalakshmi et al., 2020; Zhao et al., 2018).

4 Summary

Quantum chemical density functional theory (DFT) computations can be used to investigate the interactions between ionic liquids (ILs) and heteroaromatic sulfur-containing compounds. An overview of the current state-of-the art is presented along with some preliminary results using DFT calculations which we have conducted with 35 ILs with dibenzothiophene (DBT) and dibenzothiophene sulfone ($DBTO_2$) as model heteroaromatic sulfur-containing compounds. *Gaussian 16* using the 6-311++G(d,2p)/wB97X-D basis set was used for the DFT calculations which can produce a wealth of information for the aim of optimizing oxidative desulfurization of fuel oils in nonvolatile, recyclable ILs which can ideally also be used to extract the oxidized sulfur compounds.

Acknowledgments

The authors extend their appreciation to the Deanship of Scientific Research, King Saudi University for funding through the Vice Deanship of Scientific Research Chairs; Research Chair for Tribology, Surface, and Interface Sciences. Compute Canada is also thanked for providing ongoing support for the computational facilities.

References

Bader, R. F. (1991). A quantum theory of molecular structure and its applications. *Chemical Reviews, 91*(5), 893–928.

Bhutto, A. W., Abro, R., Gao, S., Abbas, T., Chen, X., & Yu, G. (2016). Oxidative desulfurization of fuel oils using ionic liquids: A review. *Journal of the Taiwan Institute of Chemical Engineers, 62*, 84–97.

Bösmann, A., Datsevich, L., Jess, A., Lauter, A., Schmitz, C., & Wasserscheid, P. (2001). Deep desulfurization of diesel fuel by extraction with ionic liquids. *Chemical Communications*, 2494–2495.

Chai, J. D., & Head-Gordon, M. (2008). Long-range corrected hybrid density functionals with damped atom–atom dispersion corrections. *Physical Chemistry Chemical Physics, 10*(44), 6615–6620.

Dai, C., Zhang, J., Huang, C., & Lei, Z. (2017). Ionic liquids in selective oxidation: Catalysts and solvents. *Chemical Reviews, 117*(10), 6929–6983.

Dapprich, S., & Frenking, G. (1995). Investigation of donor-acceptor interactions: A charge decomposition analysis using fragment molecular orbitals. *The Journal of Physical Chemistry, 99*(23), 9352–9362.

Fang, D., Wang, Q., Liu, Y., Xia, L., & Zang, S. (2014). High-efficient oxidation–extraction desulfurization process by ionic liquid 1-butyl-3-methyl-imidazolium trifluoroacetic acid. *Energy & Fuels, 28*(10), 6677–6682.

Freemantle, M. (2010). *An introduction to ionic liquids*. Royal Society of Chemistry.

Frisch, M. J., et al. (2019). *Gaussian 16, Revision C.01*. Wallingford, CT: Gaussian, Inc.

Fukui, K., Yonezawa, T., & Shingu, H. (1952). A molecular orbital theory of reactivity in aromatic hydrocarbons. *The Journal of Chemical Physics, 20*(4), 722–725.

Gao, H., Luo, M., Xing, J., Wu, Y., Li, Y., Li, W., Liu, Q., & Liu, H. (2008). Desulfurization of fuel by extraction with pyridinium-based ionic liquids. *Industrial and Engineering Chemistry Research, 47*(21), 8384–8388.

Greer, A. J., Jacquemin, J., & Hardacre, C. (2020). Industrial applications of ionic liquids. *Molecules, 25*(21), 5207–5208.

Gu, P., Lü, R., Liu, D., Lu, Y., & Wang, S. (2014). Exploring the nature of interactions among thiophene, thiophene sulfone, dibenzothiophene, dibenzothiophene sulfone and a pyridinium-based ionic liquid. *Physical Chemistry Chemical Physics, 16*(22), 10531–10538.

Gui, J., Liu, D., Sun, Z., Liu, D., Min, D., Song, B., & Peng, X. (2010). Deep oxidative desulfurization with task-specific ionic liquids: An experimental and computational study. *Journal of Molecular Catalysis A: Chemical, 331*(1–2), 64–70.

Handy, S. (Ed.). (2011). *Ionic liquids: Classes and properties* BoD–Books on Demand.

Hizaddin, H. F., Anantharaj, R., & Hashim, M. A. (2014). A quantum chemical study on the molecular interaction between pyrrole and ionic liquids. *Journal of Molecular Liquids, 194*, 20–29.

Johnson, E. R., Keinan, S., Mori-Sánchez, P., Contreras-García, J., Cohen, A. J., & Yang, W. (2010). Revealing noncovalent interactions. *Journal of the American Chemical Society, 132*(18), 6498–6506.

Koopmans, T. (1933). Ordering of wave functions and eigenvalues to the individual electrons of an atom. *Physica, 1*, 104–113.

(a) Lehn, J. M., Hunter, C. A., & Sanders, J. K. (2007). From supramolecular chemistry towards constitutional dynamic chemistry and adaptive chemistry. *Chemical Society Reviews, 36*(2), 151–160 (b). The nature of. pi.-. pi. interactions. *Journal of the American Chemical Society, 112*(14), 5525–5534 (1990).

Lei, Z., Chen, B., Koo, Y. M., & MacFarlane, D. R. (2017). Introduction: Ionic liquids. *Chemical Reviews, 117*(10), 6633–6635.

Lin, J., Lü, R., Wu, C., Xiao, Y., Liang, F., & Famakinwa, T. (2017). A density functional theory study on the interactions between dibenzothiophene and tetrafluoroborate-based ionic liquids. *Journal of Molecular Modeling, 23*(4), 145.

(a) Lu, T., & Chen, F. (2012). Quantitative analysis of molecular surface based on improved Marching Tetrahedra algorithm. *Journal of Molecular Graphics & Modelling, 38*, 314–323. (b) Lu, T., & Chen, F. (2012). Multiwfn: A multifunctional wavefunction analyzer. *Journal of Computational Chemistry, 33*(5), 580–592.

For a recent study confirming the validity of Koopman's Theorem, see Luo, J., Xue, Z. Q., Liu, W. M., Wu, J. L., & Yang, Z. Q. (2006). Koopmans' theorem for large molecular systems within density functional theory. *The Journal of Physical Chemistry. A, 110*(43), 12005–12009.

Mohumed, H., Rahman, S., Imtiaz, S. A., & Zhang, Y. (2020). Oxidative-extractive desulfurization of model fuels using a pyridinium ionic liquid. *ACS Omega, 5*(14), 8023–8031.

Murray, J. S., & Politzer, P. (2011). The electrostatic potential: An overview. *Wiley Interdisciplinary Reviews: Computational Molecular Science, 1*(2), 153–163.

Parr, R. G., Donnelly, R. A., Levy, M., & Palke, W. E. (1978). Electronegativity: The density functional viewpoint. *The Journal of Chemical Physics, 68*(8), 3801–3807.

Parr, R. G., & Pearson, R. G. (1983). Absolute hardness: Companion parameter to absolute electronegativity. *Journal of the American Chemical Society, 105*(26), 7512–7516.

Pearson, R. G. (1963). Hard and soft acids and bases. *Journal of the American Chemical Society, 85*(22), 3533–3539.

Reed, A. E., Curtiss, L. A., & Weinhold, F. (1988). Intermolecular interactions from a natural bond orbital, donor-acceptor viewpoint. *Chemical Reviews, 88*(6), 899–926.

Savin, A., Umrigar, C. J., & Gonze, X. (1998). Relationship of Kohn–Sham eigenvalues to excitation energies. *Chemical Physics Letters, 288*(2–4), 391–395.

Vijayalakshmi, R., Anantharaj, R., & Brinda, L. A. (2020). Evaluation of chemical reactivity and stability of ionic liquids using ab initio and COSMO-RS model. *Journal of Computational Chemistry, 41*(9), 885–912.

Zhang, C., Pan, X., Wang, F., & Liu, X. (2012). Extraction–oxidation desulfurization by pyridinium-based task-specific ionic liquids. *Fuel, 102*, 580–584.

Zhao, D., Wang, Y., Duan, E., & Zhang, J. (2010). Oxidation desulfurization of fuel using pyridinium-based ionic liquids as phase-transfer catalysts. *Fuel Processing Technology, 91*(12), 1803–1806.

Zhao, X., Wu, H., Duan, M., Hao, X., Yang, Q., Zhang, Q., & Huang, X. (2018). Liquid-liquid extraction of lithium from aqueous solution using novel ionic liquid extractants via COSMO-RS and experiments. *Fluid Phase Equilibria, 459*, 129–137.

4

Molecular dynamics simulation in energy and chemical systems

Sohrab Zendehboudi
Department of Process Engineering, Memorial University, St. John's, NL, Canada

1 Introduction

A broadly used simulation approach to monitor the microscopic behavior of a system/phenomenon is molecular dynamics (MD) simulation. This method offers a bridge between experimental trials and conventional physical chemistry/thermodynamic models (Zhang et al., 2002). MD simulation is an advanced computational chemistry tool to study the mechanisms of various phenomena, such as interfacial tension (IFT), solubility, adsorption, and chemical aggregation. Also, MD is applied to estimate physical properties, including equilibrium concentrations and transport properties. MD can be combined with molecular/quantum mechanics to model chemical reactions along with dynamic simulation. By applying the statistical dynamics, properties of the system, such as IFT, can be acquired through analyzing the potential energy according to the position of the atoms (Zhang et al., 2002). The MD approach is attributed to the difficulty of classical modeling approach systems such as activity coefficient models or equations of state (EOSs) without experimental data. MD is a powerful method to disclose the physical interaction mechanisms such as adsorption, desorption, and deposition by considering the chemical structure of materials and suitable force-fields. MD simulation is one of the fundamental methods for studying chemical engineering systems from molecular to industrial scale. Chemical engineering systems involve multiple scales, including material levels of atoms, molecules, and particles to reaction levels of particles and clusters and process system levels. Computer simulation is becoming a more important approach in chemical engineering for quantifying multiscale systems with high accuracy. Using MD simulation, many force-fields have been

Modeling of Chemical Process Systems. https://doi.org/10.1016/B978-0-12-823869-1.00002-8

well-established to formulate the interactions between the elements and atoms for studying the system at the molecular level. Although it is computationally challenging to predict macroscale behaviors from an atomic scale, it can be applied to understand the entire mechanisms. MD is more appropriate when rigorous estimates are required for dynamical trajectories of all the atoms.

A practical method for fully understanding the structures and physicochemical properties at the molecular scale is MD simulation. MD simulation is capable of offering extensive molecular dynamic information, including molecular interactions. An interesting benefit of MD simulation is low-cost evaluation of a system, compared to the laboratory works. Additionally, MD simulation creates an adjustable platform for changing potentials to readily alter the potential for further understanding of the molecular behaviors (Karplus & McCammon, 2002).

2 Fundamentals of MD technique

In the molecular dynamics (MD) approach, the equation of motion is solved for every atom in the system. The position and velocity of molecules are determined along the simulation time based on the inter/intra-molecular potentials (Mcquarrie, 1965). The potentials between atoms and molecules are assigned based on the selected force-field. Force-fields are the empirical potential energy functions and parameters that are developed by fitting to experimental data. MD has three main applications. First, it studies the mechanisms of phenomena by both qualitative and quantitative evaluation. Second, it can predict the static and dynamic characteristics of systems. Third, it can be coupled with the equations of state (EOSs) to estimate the adjustable factors, tune them, and predict experimental results more accurately. MD provides an accurate prediction of the system behaviors through precise estimation of interactions between the molecules. This modeling technique has been widely employed due to several benefits. For example, it appears to be an exceptional approach to obtain the thermodynamic properties of a system in addition to the physical and chemical property estimation through statistical and fundamental approaches (Mcquarrie, 1965).

In the MD simulations, the empirical potential functions and Newton's second law are used to calculate the forces and interaction between all atoms in different molecules to describe the velocity and movement of each atom in the simulation box. In MD simulation, as the first step, the force of different molecules is calculated by employing the summation of all potential energies. Then, it is used to obtain the new positions for atoms by employing a proper algorithm to solve Newton's equation after a time step (Gear, 1971; Swope et al., 1982; Verlet, 1967).

MD simulation is governed by force-field models, which provide a set of functional forms. Based on the particle position and selected force-field, atoms interact along the simulation time, and the potential energy is calculated for the system. There are many research investigations in the literature that include theory, background, and practical aspects of force-fields (Geng et al., 2010; Kondori et al., 2017; Li et al., 2018; Sakemoto et al., 2010; Smirnov & Stegailov, 2012; Wan et al., 2009; Zhang & Pan, 2011). Various force-fields are used in hydrocarbon and nonhydrocarbon studies, such as COMPASS force-field (Amjad-Iranagh et al., 2015; Fang et al., 2018; Pacheco-Sánchez et al., 2003; Pacheco-Sánchez et al., 2004; Tirjoo et al., 2019), GROMOS96 force-field (Gao et al., 2014; Jian et al., 2013; Jian et al., 2018; Kuznicki et al., 2008; Lv et al., 2017; Mehana et al., 2019; Silva et al., 2016; Sodero et al., 2016; Song et al., 2018), and OPLS-AA force-field (Goual et al., 2014; Goual & Sedghi, 2015; Headen et al., 2009; Headen et al., 2017; Headen & Boek, 2011a; Headen & Boek, 2011b; Headen & Hoepfner, 2019; Khalaf & Mansoori, 2018; Khalaf & Mansoori, 2019; Lowry et al., 2016; Lowry et al., 2017; Mikami et al., 2013; Mohammed & Gadikota, 2019a; Mohammed & Gadikota, 2019b; Mohammed & Mansoori, 2018a; Mohammed & Mansoori, 2018b; Sedghi et al., 2013; Sedghi & Goual, 2016; Yaseen & Mansoori, 2017; Yaseen & Mansoori, 2018). Two latter ones are more popular to use. Based on a study conducted by Fu and Tian (Fu & Tian, 2011), OPLS-AA is even more accurate and reliable in predicting the experimentally available thermodynamic properties. Regarding force-fields, the total potential energy function of the system is the summation of bonded and nonbonded interactions between atoms (Seyyedattar et al., 2019). The bonded potential is related to the interaction between neighboring atoms, which are directly bonded. In contrast, the non-bonded potential is defined between atoms that are not directly bonded, either in the same or different molecules. The bonded potential includes three intramolecular potentials named bond stretching, angle bending, and dihedral angle torsion.

3 Emerge of MD technique

Since the 1950s, MD simulation methods have been developed to simulate processes at the atomic level and provide accurate information for prediction/validation of experimental outcomes

(Ebro et al., 2013; Meller, 2001; Zendehboudi et al., 2018). After that, significant improvements and advances in computational and computer technology have caused considerable interest and applicability in the MD approach as a strong simulation tool in various branches of science and engineering, particularly for research investigations on micromolecules and macromolecules (Meller, 2001). In addition, the MD technique has high potential to provide quantitative/qualitative information about the system by integrating concepts from various branches of science, including computer science, mathematics, chemistry, and physics, and fill out the knowledge gaps in experimental methods and modeling approaches (Ebro et al., 2013; Goodarzi & Zendehboudi, 2019; Zendehboudi et al., 2018).

In fact, the molecular-scale active phenomena in different disciplines, such as chemistry, chemical engineering, biochemical, and medicine, are the primary motivation of emerging MD methods (Ebro et al., 2013; Meller, 2001; Zendehboudi et al., 2018). For instance, GROMACS is one of the powerful and universal MD software (Ghamartale et al., 2021; Zendehboudi et al., 2018). It is used in material science, biochemistry, biophysics, and recently in engineering. GROMACS is open-source software, which is of great importance in various science and engineering subjects. It might not be very user-friendly initially since users need to write a code to perform their simulation; however, it gives the user the opportunity to manipulate and change the background of the software in a way that is compatible and useful for targeted research studies. Sometimes, users can add the types of atoms that are not available in the software library; this improves the reliability and capability of the software. GROMACS simulation is good to be run in both CPU and GPU, which allows the simulation to be implemented faster and more efficiently. GROMACS comes with a set of general force-fields, including AMBER, CHARMM, GROMOS, and OPLS (Ghamartale et al., 2021; Zendehboudi et al., 2018). Although GROMACS is very powerful to provide effective calculation, it is relatively basic, and some additional software packages need to be involved before and after simulation. For instance, a software is needed to sketch the molecules, and another software is needed to visualize the molecules and systems. The outputs of the MD simulation include atom position, velocity, and force, but they are not adequately meaningful and understandable at their initial forms. Thus, it needs further processing and analysis to obtain useful results. GROMACS involves powerful built-in tools to conduct post-processing. However, sometimes the users might like to conduct specific analysis, which is missing in the built-in tools (Ghamartale et al., 2021; Zendehboudi et al., 2018). The outputs of GROMACS are designed to be easily incorporated in Python, which enables the users to script a code for their specific analysis. Some libraries already exist in the Python with basic analysis such as MD Analysis (Ghamartale et al., 2021; Zendehboudi et al., 2018). The users can use the prepared code directly, develop their code from scratch, or improve the existing code effectively. The application of MD simulation can be found in several research areas. For example, Kondori et al. (Kondori et al., 2017) conducted a comprehensive literature review on the different potential functions and applications of MD simulations to describe gas hydrate decomposition. Methane gas decomposition in a system with methanol and ethanol (as inhibitors) was simulated with MD simulation by Wan et al. (Wan et al., 2009). Sakemoto et al. (2010) compared the mechanical and thermodynamic parameters of methane hydrate structures obtained from MD simulation runs and previous experimental studies. Also, Zhang and Pan (Zhang & Pan, 2011) calculated the structural and dynamic features of the methane hydrates, such as potential energy, radial distribution function (RDF), mean square displacement (MSD), and density profile, by employing the MD technique. Analysis of the methane hydrate dissociation was conducted by Smirnov and Stegailov (2012) using various potential functions for pressure ranges under 5 GPa. They showed that the methane hydrate decomposition occurs upon reduction in cage occupancy. Wan et al. (Geng et al., 2010) investigated the stability and decomposition of methane hydrate after adding ethylene glycol as an inhibitor. It was found that the clathrate hydrate is dissociated at a higher temperature after adding inhibitor molecules. In a study conducted by Wen et al. (Wan et al., 2009), it was shown that the stability of methane hydrate would increase after injection of tetrahydrofuran molecules as a promotor in large cavities of hydrogen hydrate (structure II). Li et al. (Li et al., 2018) investigated the effect of methanol injecting to the propane structure II hydrate. It was concluded that methanol molecules accelerate propane hydrate dissociation. Ghamartale et al. (Ghamartale et al., 2021) employed the MD simulation approach to investigate the effects of thermodynamic conditions and inhibitor concentration on asphaltene aggregation. Their research study could help researchers and engineers design an optimal chemical inhibitor for asphaltene precipitation (Ghamartale et al., 2021). Aghaie et al. (2020a) introduced a new effective strategy to determine the molecular and transport phenomena behaviors of [Bmim][Ac]/Water and

[Bmim] [Ac] /Water/CO_2 systems in the presence of water through employing MD simulation. The important absorption mechanisms were explored during carbon capture with ionic liquids (ILs). In addition, there was a good match between the real data and modeling results. In general, their research work offers useful tips to further understand the solubility and diffusivity behaviors of ILs, excess volume and enthalpy upon CO_2 absorption, and dynamic behaviors of CO_2-ILs-water. Such data/info are crucial in designing appropriate ILs for various energy and environment applications at broad process and thermodynamic conditions (Aghaie et al., 2020a).

4 Architecture of MD technique

There are some useful computational engines such as GROMACS and Materials Studio for energy minimization and dynamic simulation (Aghaie et al., 2020a; Ghamartale et al., 2021; Kondori et al., 2019a; Zendehboudi et al., 2018). Modeling at the molecular scale is addressed by the assessment of complicated chemical systems considering a proven atomic model. Indeed, the main purpose of employing molecular simulation is to fully understand and obtain the macroscopic characteristics at atomic scale. Macroscopic physical properties are determined based on static equilibrium specifications, e.g., binding constant of inhibitor/enzyme, system's average potential, and liquid radial distribution function (Aghaie et al., 2020a; Meller, 2001; Zendehboudi et al., 2018). Also, they can be identified with dynamic/nonequilibrium characteristics, such as fluid viscosity, gas/liquid diffusion, and reaction kinetics.

MD simulation investigations are based on solving Newton's equations of motion for N interacting atoms in a system (Goodarzi et al., 2019; Kondori et al., 2019a; Kondori et al., 2020). The MD simulation considers the negative derivatives of the potential function (as the forces) and solves the equations simultaneously in predefined small time-steps. Also, it should be ensured that pressure and temperature are correctly set and the coordinates are transferred to the output file at the desired intervals (Goodarzi et al., 2019; Kondori et al., 2019a; Kondori et al., 2020; Zendehboudi et al., 2018). Indeed, the coordinates show the trajectory of the system as a function of time. Typically, the system will reach an equilibrium state after initial changes. Then, several macroscopic characteristics can be obtained toward averaging over the equilibrium trajectory. In order to run the MD simulation, a set of initial coordinates and velocities of particles should be defined (Aghaie et al., 2020a; Kondori et al., 2019a; Kondori et al., 2020; Zendehboudi et al., 2018).

The global MD algorithm consists of four steps: (1) input initial conditions, potential interaction as a function of atom positions, positions *r*, and velocities *v* of all atoms in the system; (2) compute forces, the forces due to bonded interactions and nonbonded interactions; (3) update configurations, the movement of all atoms is simulated by numerically solving Newton's equations of motion; and (4) output step (if required), write positions, velocities, energies, temperatures, and pressures (Meller, 2001; Zendehboudi et al., 2018).

5 Theoretical frameworks of MD technique

The MD method considers molecule interactions in the target system at the microscopic scale. MD also provides macroscopic properties through statistical measurements. Indeed, MD employs statistical mechanics to bridge microscopic and macroscopic properties (Goodarzi et al., 2019;Seyyedattar et al., 2019 ; Zendehboudi et al., 2018). Computation at a molecular scale involves classical molecular mechanics, quantum mechanics, and hybridized classical and quantum approaches. Quantum mechanics mainly operates through the electronic structure and molecular motion and considers wave functions of the subatomic particles. Quantum mechanics is a reliable technique to explore atomic features; however, it suffers from computational costs (Seyyedattar et al., 2019; Zendehboudi et al., 2018). On the other hand, classical molecular mechanics utilizing Newtonian forces/equations are faster but less accurate. Therefore, combining these two methods can result in faster computations, lower costs, and reliable outputs for 104–105 molecules per 10–100 ps (Cuendet, 2008; Springer, 2006).

5.1 Atomic interactions and forces

MD simulation is developed to study the interaction between atoms based on interatomic forces introduced to the software by atomic force-field models. The selected force-field governs the potential energy between each atom, and the system potential energy is calculated based on the relative positions of the atoms (Ebrahimi, 2014;Seyyedattar et al., 2019 ; Springer, 2006). The total potential energy, *U*, for a system with *N* interacting atoms, is a function of the atomic positions, $r^{(N)}$, indeed the coordinates of all the atoms, as given below (Ebrahimi, 2014; Zendehboudi et al., 2018):

$$r^{(N)} = (r_1, r_2, \ldots, r_N) \tag{1}$$

$$r_i = (x_i, y_i, z_i) \tag{2}$$

In Eqs. (1) and (2), r_i denotes the atom i position, which is defined by the Cartesian coordination of x, y, and z in three directions. This means a system with N atoms has 3 N atomic coordinates at each time step during the simulation. The total potential energy, U, refers to the amount of potential energy of the system at each time step. The force on each atom (F_i) is defined by the gradient of the potential energy function with respect to the atomic displacement, as follows (Seyyedattar et al., 2019; Springer, 2006):

$$F_i = \nabla_{r_i} U(r_1, r_2, \ldots, r_N) \tag{3}$$

Meanwhile, the laws of classical mechanics define the force and the motion of the atoms integrating Newton's equations. Therefore, based on Newton's second law, the atomic force is expressed, as follows (Seyyedattar et al., 2019; Springer, 2006):

$$F_i = m_i \frac{d^2 r_i}{dt^2}, \tag{4}$$

where m_i represents the mass of atom i and t introduces the time. Combining Eqs. (3) and (4) results in (Seyyedattar et al., 2019; Springer, 2006)

$$\nabla_{r_i} U(r_1, r_2, \ldots r_N) = m_i \frac{d^2 r_i}{dt^2}. \tag{5}$$

After simplification, it is concluded that the potential energy function (U) empirically includes bonded and non-bonded atomic interactions, as shown below (Meller, 2001; Zendehboudi et al., 2018):

$$U(r_1, r_2, \ldots r_N) = U_{\text{nonbonded}} + U_{\text{bonded}} \tag{6}$$

The bonded potential (U_{bonded}) represents the interaction between the directly bonded atoms. However, the non-bonded potential ($U_{\text{nonbonded}}$) indicates the potentials between atoms that are not directly connected with covalent bonds. They can be either atoms from different molecules or atoms from the same molecules. Fig. 1 depicts examples of nonbonded and bonded potentials (Meller, 2001; Zendehboudi et al., 2018).

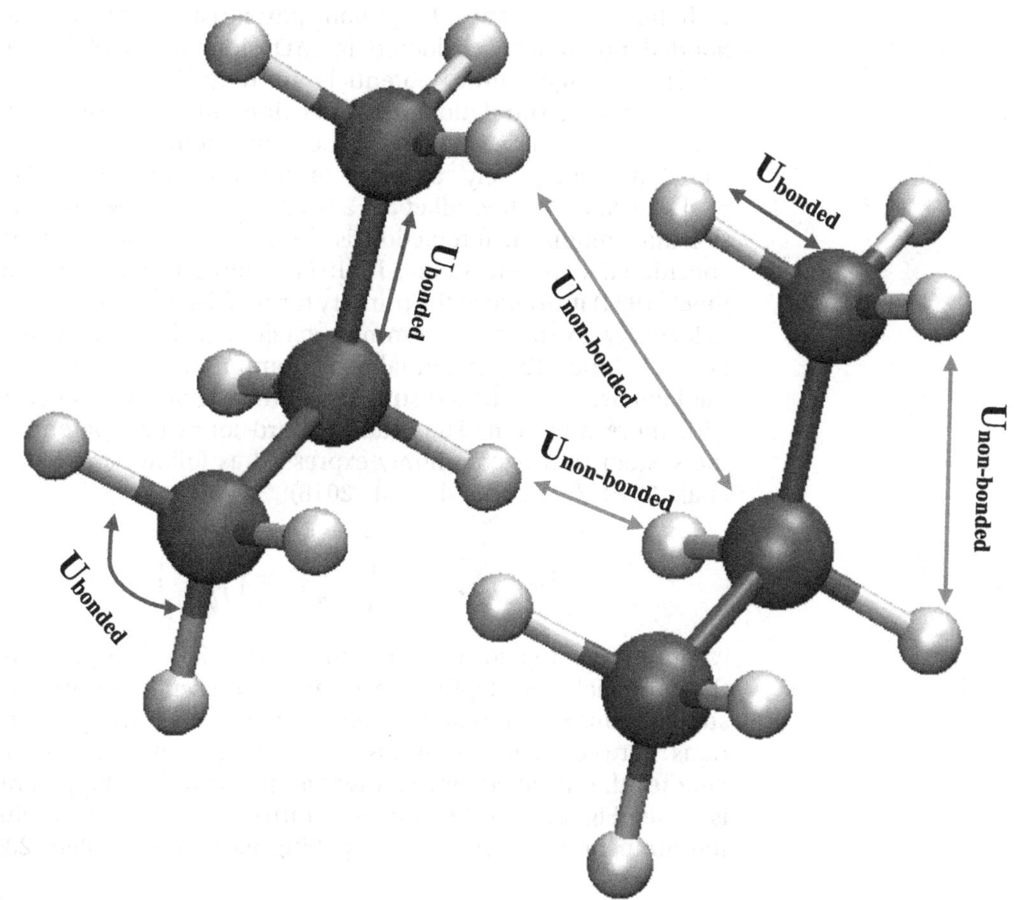

Fig. 1 Simple schematic of bonded and nonbonded and potentials shown by arrows. Modified after Max Born Institute, "Dynamics of condensed phase molecular systems," [Online]. Available: https://www.mbi-berlin.de/index_en.html. [Accessed 22 November 2021].

The nonbonded potentials are usually described in the following form (Allen, 2004; Zendehboudi et al., 2018):

$$U_{\text{nonbonded}} = \sum_{i} u(r_i) + \sum_{i} \sum_{j>i} v(|r_i - r_j|) + \ldots, \tag{7}$$

where $u(r_i)$ represents the 1-body potential which either originates from an external potential field or indicates the effect of walls. In the fully periodic boundary simulation, 1-body can be neglected (Bondor et al., 2005; Seyyedattar et al., 2019; Zendehboudi et al., 2018). $v(|r_i - r_j|)$ or 2-body term introduces the pair potentials between atoms that are not bonded directly with the $|r_i - r_j|$ distance. The 2-body part is usually the main nonbonded potentials considered in MD simulations of complex fluids. The higher-order potentials are only significant in condensed matter simulation scenarios (Ippoliti, 2011; Seyyedattar et al., 2019); otherwise, they can be conveniently neglected and still have adequately accurate approximations (Seyyedattar et al., 2019; Zendehboudi et al., 2018). The van der Waals dispersion and repulsion interaction is the pairwise potential that is considered between atoms. In 1924, Lennard-Jones (Lennard-Jones, 1924) introduced the primary form of this developed potential energy to the most common version, which is named the Lennard-Jones 12-6 potential. The computational simplicity of the Lennard-Jones 12-6 distinguishes that from other potentials with more precision. The total Lennard-Jones pair potential of the system (v_{LJ}) is commonly expressed as follows (Seyyedattar et al., 2019; Zendehboudi et al., 2018):

$$v_{LJ} = \sum_{\substack{i,j \\ i<j}} 4\varepsilon_{ij} \left(\left(\frac{\sigma_{ij}}{r_{ij}} \right)^{12} - \left(\frac{\sigma_{ij}}{r_{ij}} \right)^{6} \right), \tag{8}$$

where ε refers to a local minimum of potential energy, known as the potential well depth. The ε value represents the attraction strength; this means that the higher ε values show stronger attractions between the two atoms or molecules and vice versa. σ denotes the distance between two atoms at which the potential is 0 (see Fig. 2). In other words, σ introduces a radius at which the atom or molecule starts repelling each other (Allen, 2004;

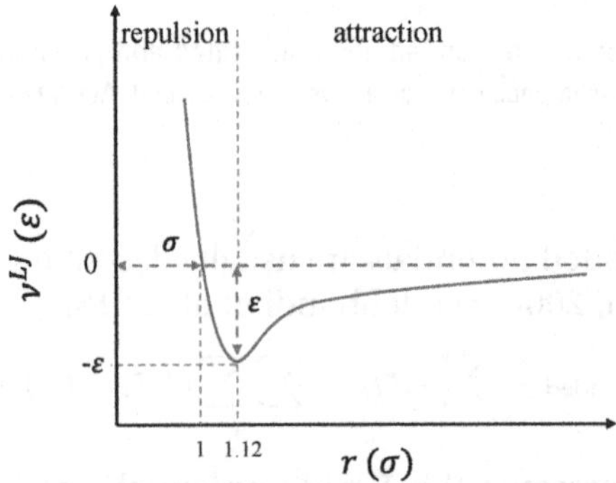

Fig. 2 Simple schematic of Lennard-Jones potential function to depict repulsion and attraction zones. Modified after "Molecular dynamics (MD)," Atoms In Motion, [Online]. Available: http://atomsinmotion.com/. [Accessed 05 December 2021]; R. Naeem, "Lennard-Jones potential," The LibreTexts, 2017. [Online]. Available: https://chem.libretexts.org/. [Accessed 05 December 2021].

Naeem, 2017; Seyyedattar et al., 2019). r signifies the distance between the centers of two atoms.

Based on the Lennard-Jones potential function, the two particles are not interacting at an infinite distance, which implies they are at zero potential (Fig. 2). The large dipole-dipole attraction (van der Waals dispersion forces), represented by the term $(\sigma_{ij}/r_{ij})^6$ in the Lennard-Jones potential function, can initiate moving the two particles and bringing them closer to each other. At a closer distance, where the Lennard-Jones potential reaches a minimum magnitude of $-\varepsilon$ at around 1.122σ, two particles become bound. At $r_{ij}=\sigma$, the potential energy is 0 between atoms, showing an equilibrium state. For $r_{ij}<\sigma$, a strong repulsion develops between the particles as the atomic electron clouds overlap. Thus, the repulsion force, accounted for by the term $(\sigma_{ij}/r_{ij})^{12}$ in the Lennard-Jones potential, experiences an appreciable increase in the system potential energy (Naeem, 2017; Seyyedattar et al., 2019).

If the Lennard-Jones potential energy is taken into account at an infinite range, large computational resources are required to implement MD simulations (Seyyedattar et al., 2019; Springer, 2006). Therefore, a cutoff radius (R_c) is usually determined beyond which the van der Waals atomic interactions for any two atoms are ignored (Seyyedattar et al., 2019; Springer, 2006). In MD simulations, the cutoff radii of 2.5σ and 3.2σ are generally utilized to speed up the computations and still have accurate findings (Seyyedattar et al., 2019; Springer, 2006).

Simple truncation, shift, and switch cutoff methods are three approaches to apply termination of the atomic interactions beyond the cutoff radius (Seyyedattar et al., 2019; Stote et al., 1999). In the simple truncation technique, the Lennard-Jones potential energy is calculated for $r_{ij}<R_c$ and truncated by setting the interactions to zero for $r_{ij}>R_c$. However, a small fluctuation occurs at potential energy, which can spoil the energy conservation when their number increases along with the simulation. To avoid this issue, the shift method is introduced, and it shifts the entire Lennard-Jones potential energy function to zero potential at interatomic separation distances $r_{ij}=R_c$; introduced by the following expression (Seyyedattar et al., 2019; Springer, 2006):

$$v(r_{ij}) = \begin{cases} v_{LJ}(r_{ij}) - v_{LJ}(R_c) \text{ for } r_{ij} \leq R_c \\ 0 \text{ for } r_{ij} > R_c \end{cases}. \qquad (9)$$

The drawback of the shift strategy causes a slight decrease in the equilibrium distances (Seyyedattar et al., 2019; Stote et al., 1999). The switch method introduces a smaller cutoff radius besides the main cutoff distance. The potential energy values are gradually tapered in this method, starting from the smaller cutoff radius and reaching zero at the main cutoff distance. The strong forces in the distance range between the two cutoff radii result in a slight perturbation of the equilibrium structure, which is the main drawback of the switch cutoff approach. This method has a large error for the short cutoff regions, and it is not recommended in most cases (Seyyedattar et al., 2019; Stote et al., 1999).

Coulomb's electrostatic potential is the second pairwise potential for non-bonded particles in MD simulations. This potential is based on the atoms' charges, which implies the atoms with opposite charges attract each other, while the atoms with like charges repel each other. The Coulomb's electrostatic potential (v_C) for the system is expressed as follows (Seyyedattar et al., 2019; Zendehboudi et al., 2018):

$$v_C = \sum_{\substack{i,j \\ i<j}} \frac{1}{4\pi\varepsilon_0} \cdot \frac{q_i q_j}{r_{ij}}, \tag{10}$$

where the particles i and j carry the q_i and q_j electrostatic point charges, respectively; r_{ij} represents the distance between the two particles; and ε_0 stands for the permittivity of the free space. Although there are long-range electrostatic interactions besides short-range interactions, they cannot be included in MD computations as the computational time and cost will increase dramatically. Nevertheless, several models have been developed to include the long-range electrostatic potentials for periodic systems (Allen & Tildesley, 2017; Brooks et al., 1983; York et al., 1993). However, the electrostatic interactions are split into the short-range and long-range components in most models for non-periodic systems (Brooks et al., 1983; Shimada et al., 1994; Stote et al., 1991). In these models, the short-range component is dealt with the pairwise interactions, while the long-range interactions are estimated by the multipole approximations (Seyyedattar et al., 2019; Stote et al., 1999).

Hence, Eq. (11) shows the general form of the total nonbonded pairwise potential, which is the summation of Lennard-Jones and Coulomb's electrostatic potentials, as given below (Seyyedattar et al., 2019; Zendehboudi et al., 2018):

$$\begin{aligned} U_{\text{nonbonded}} &= v_{LJ} + v_C \\ &= \sum_{\substack{i,j \\ i<j}} 4\varepsilon_{ij} \left(\left(\frac{\sigma_{ij}}{r_{ij}} \right)^{12} - \left(\frac{\sigma_{ij}}{r_{ij}} \right)^{6} \right) + \sum_{\substack{i,j \\ i<j}} \frac{1}{4\pi\varepsilon_0} \cdot \frac{q_i q_j}{r_{ij}}. \end{aligned} \tag{11}$$

The second portion of the total potential energy is U_{bonded}, which originates from the intramolecular bonding interactions (Eq. (6)). The covalent bonds should have specific lengths, bend angles, and torsion angles to have a stable molecule. Fig. 3 illustrates the three components of the bonded potential energy. The bonded potential energy can be written in the following form (Seyyedattar et al., 2019; Zendehboudi et al., 2018):

$$U_{\text{bonded}} = \upsilon_{\text{bond length}} + \upsilon_{\text{bend angle}} + \upsilon_{\text{torsion}}. \tag{12}$$

The term $\upsilon_{\text{bond length}}$ refers to the 2-body potential energy that results in the deformation of the bond length. $\upsilon_{\text{bond length}}$ is a harmonic vibrational motion between the two covalently bonded atoms, which can be defined by the following expression (Seyyedattar et al., 2019; Zendehboudi et al., 2018):

$$\upsilon_{\text{bond length}} = \frac{1}{2}\sum_{i,j} a_{ij}\left(r_{ij} - r_{\text{eq}}\right)^2, \tag{13}$$

where r_{ij} refers to the separation distance; r_{eq} stands for the equilibrium separation distance between the two covalently bonded atoms; and a_{ij} introduces a force constant (spring constant) that depends on the type of atoms. The term $\upsilon_{\text{bend angle}}$ reflects the 3-body potential energy resulting from the angular vibrational motion occurring between three consecutively connected atoms. The atoms should be bonded with the covalence bonds, which means the middle atom is bonded to the two other atoms. The bend angle potential ($\upsilon_{\text{bend angle}}$) can be defined as follows (Seyyedattar et al., 2019; Zendehboudi et al., 2018):

$$\upsilon_{\text{bend angle}} = \frac{1}{2}\sum_{i,j,k} b_{ijk}\left(\theta_{ijk} - \theta_{\text{eq}}\right)^2, \tag{14}$$

where θ_{ijk} introduces the angle between the ij and jk atom pair bonds; θ_{eq} denotes the equilibrium angle; and b_{ijk} stands for a force constant that depends on the type of atoms. The last part of the bonded potential energy is 4-body potential energy, named $\upsilon_{\text{torsion}}$. This potential is related to the rotations around a chemical bond. It models the steric barriers between atoms separated by three consecutive covalent bonds. The torsion angle (dihedral angle) potential has a periodic nature that is simply defined in the following form (Seyyedattar et al., 2019; Springer, 2006):

$$\upsilon_{\text{torsion}} = \frac{1}{2}\sum_{i,j,k,l} c_{ijkl}(1 + \cos(n\phi)), \tag{15}$$

where ϕ denotes the angle between the ijk plane and the jkl plane in the $ijkl$-quadruple of covalently bonded atoms; n is a constant

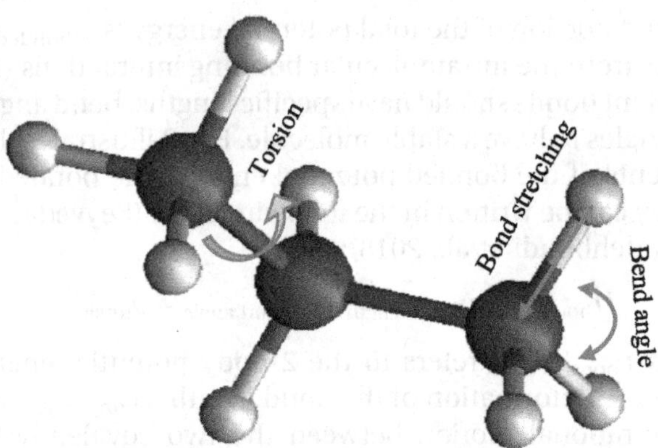

Fig. 3 A simple cartoon showing the key three elements of bonded potential energy including torsion bend angle and stretching of bond length (known as bond stretching). Modified after Wikipedia, "Molecular mechanics," Wikimedia Foundation, Inc., [Online]. Available: https://en.wikipedia.org/wiki/Molecular_mechanics. [Accessed 20 November 2021].

integer indicating the periodicity; and c_{ijkl} symbolizes the height of the rotational barrier. Therefore, the extended version of the total bonded potential energy can be expressed by the following relationship (Seyyedattar et al., 2019; Springer, 2006):

$$U_{\text{bonded}} = \frac{1}{2}\left(\sum_{i,j} a_{ij}\left(r_{ij} - r_{\text{eq}}\right)^2 + \sum_{i,j,k} b_{ijk}\left(\theta_{ijk} - \theta_{\text{eq}}\right)^2 + \sum_{i,j,k,l} c_{ijkl}(1 + \cos(n\phi))\right). \tag{16}$$

Therefore, the total potential for the system can be determined by the following equation (Seyyedattar et al., 2019; Springer, 2006):

$$U_{\text{total}} = \frac{1}{2}\sum_{i,j} a_{ij}\left(r_{ij} - r_{\text{eq}}\right)^2 + \frac{1}{2}\sum_{i,j,k} b_{ijk}\left(\theta_{ijk} - \theta_{\text{eq}}\right)^2 + \frac{1}{2}\sum_{i,j,k,l} c_{ijkl}(1 + \cos(n\phi))^2 + \sum_{\substack{i,j \\ i<j}} 4\varepsilon_{ij}\left[\left(\frac{\sigma_{ij}}{r_{ij}}\right)^{12} - \left(\frac{\sigma_{ij}}{r_{ij}}\right)^{6}\right] + \sum_{\substack{i,j \\ i<j}} \frac{1}{4\pi\varepsilon_0} \times \frac{q_i q_j}{r_{ij}}. \tag{17}$$

After the calculation of system potential energy in each configuration, the acting force on each atom is obtained as the derivation of potential to the atom displacement. Then, the motion of atoms in the system is measured through the integration of Newton's equations. The position and velocity vector for each atom are determined to solve Newton's equations of motion, which can be used to calculate other parameters (Seyyedattar et al., 2019; Springer, 2006).

The MD equation system can be solved in four main statistical ensembles; this represents an ideal system with numerous possible states and compatible with a set of constraints. Statistical ensembles are the microcanonical ensemble (*NVE*), canonical ensemble (*NVT*), grand canonical ensemble (μVT), and isobaric-isothermal ensemble (*NPT*). *NVE* is an isolated system, in which the number of molecules, total volume, and total energy are constant. *NVT* is a system with a constant number of molecules, total volume, and temperature. μVT is a system with constant chemical potential, total volume, and temperature. *NPT* introduces a system in which the number of molecules, pressure, and temperature are specified and kept constant during the simulation (Aghaie et al., 2020b; Kondori et al., 2019b; Tazikeh et al., 2021).

5.2 Periodic boundary conditions

An essential step in MD simulation studies is defining the size of the system. The size of the system could differ, depending on the physical/chemical properties of the substance and the physical phenomenon occurring within the system. Generally, two concepts are taken into account for the system's size in MD simulations: (1) finite system and (2) infinite system (Seyyedattar et al., 2019; Springer, 2006). When only a part of the system containing a finite number of molecules or atoms is selected for simulation, the system is called a finite system. In this situation, the selected part of the system is considered to be covered by a virtual box, a cube, for example. In other words, the system's volume is limited by the surfaces/boundaries of the box, which are the system's boundaries. Compared to the particles closer to the center of the box, a small number of particles exists next to the particles near the surfaces. This approach seems unrealistic, mainly when a bulk property of the macroscopic system is planned to be calculated by MD simulation. There might be over 10^{23} molecules or atoms in a piece of a substance in macroscopic scale (Seyyedattar et al., 2019; Springer, 2006). For instance, around 3.3×10^{25} water molecules are present in 1 L of water at room temperature. Assuming the finite boundaries for the water (a cube of 10 cm length), the surface molecules could interact with up to 10 layers of water molecules, considering a spherical molecule of water having 2.8 Å diameter (Seyyedattar et al., 2019; Springer, 2006). Ten layers of water molecules include around 2×10^{19} molecules, where this number is substantially less than the total number of water molecules in the system (i.e., only 0.00006% of the total number of water molecules interact on the surface) (Roux & Petkov, n.d.; Seyyedattar et al., 2019). Since the current MD simulations can only model thousands of atoms, the wall boundary assumption results in exaggerated surface effects due to the larger ratio of surface atoms to the total number of atoms. In the MD simulation of macroscopic systems, periodic boundary conditions could address this so-called length problem (Seyyedattar et al., 2019; Springer, 2006).

The concept of periodic boundary conditions assumes a simulation box surrounded by infinite similar imaginary boxes in all directions of the three-dimensional space. The surfaces of the simulation box are open so that particles can leave and enter the box; when a particle leaves the system from one side/surface, a similar particle (or the image of the particle) enters the box from the opposite surface. Hence, the length problem is eliminated, and the molecules close to the box surfaces are exposed to a nano-size system to infinity (Seyyedattar et al., 2019; Zendehboudi et al., 2018). Thus, in addition to the interactions between each atom and other atoms in the box, there will be interactions with the images of atoms in the neighboring similar boxes. Fig. 4 demonstrates a simulated 2-D system using the periodic boundary conditions; the main simulation box is surrounded by its imaginary replications (Seyyedattar et al., 2019; Zendehboudi et al., 2018).

Suppose any atom j and its images in the periodic array, the atom i interrelates only with the nearest on the basis of minimum image convention. It should be noted that the cutoff distance is selected so that the atoms in the key simulation box do not face their own images in the adjacent imaginary boxes (Seyyedattar et al., 2019).

Although the periodic boundary conditions allow the definition of an infinite system, it is not able to provide the exact image of a large macroscopic system. The system is assumed to be symmetrical in the periodic boundary conditions, having translational invariance (Seyyedattar et al., 2019; Springer, 2006).

5.3 Numerical integration algorithms

The main purpose of MD is that the numerical integration of Newtonian equations of motion of the involved molecules/particles is conducted such that the time evolution of the target

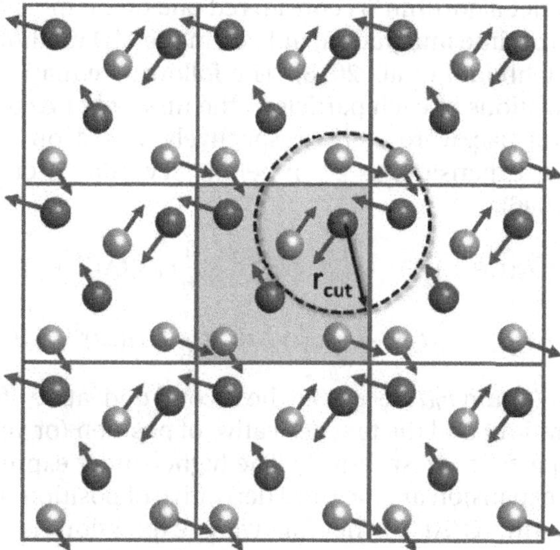

Fig. 4 The main simulation box bounded by the imaginary copies in periodic boundary conditions (Chapman, 2018).

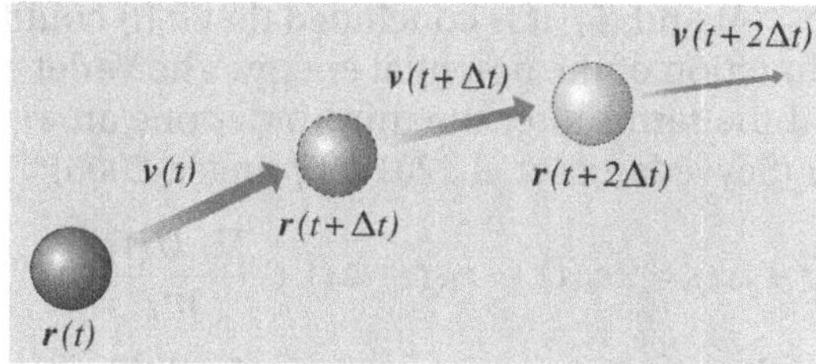

Fig. 5 A simple schematic of time integration stages during the MD simulation process (Springer, 2006).

process is obtained to describe the system in terms of velocity and position vectors (Seyyedattar et al., 2019). The second law of Newton (see Eq. (4)), which appears in the form of a second-order differential equation, should be solved through a numerical technique where the initial values of velocities and positions as well as the instantaneous forces exerted on the particles are given. All main integration steps to determine each particle's trajectory over the MD simulation strategy are reported in Fig. 5. The position of each particle at each time step is obtained from the particle's previous position using numerical integrators. This process is repeated for several time steps. For time integration, a finite difference strategy is employed. Δt is the symbol for the time step (Seyyedattar et al., 2019).

The "Verlet algorithm" is considered one of the most common algorithms of time integration, utilized in the MD method (Verlet, 1967; Zendehboudi et al., 2018). The following equations represent the positions of each particle at the times of $t+\Delta t$ and $t-\Delta t$, forward and backward steps, respectively, based on the third-order Taylor expansion, as given below (Seyyedattar et al., 2019; Springer, 2006):

$$r_i(t+\Delta t) = r_i(t) + v_i(t)\Delta t + \frac{1}{2}a_i(t)(\Delta t)^2 + \frac{1}{6}b_i(t)(\Delta t)^3 + O(\Delta t^4), \quad (18)$$

$$r_i(t-\Delta t) = r_i(t) - v_i(t)\Delta t + \frac{1}{2}a_i(t)(\Delta t)^2 - \frac{1}{6}b_i(t)(\Delta t)^3 + O(\Delta t^4), \quad (19)$$

in which, $a_i(t)$ and $v_i(t)$ resemble the second derivative of position (or acceleration) and the first derivative of position (or velocity) at time t for particle i, respectively. The higher-order expressions in the Taylor expansion are the third derivative of position ($b_i(t)$) and the last term, $O(\Delta t^4)$. The following expression is obtained through integrating Eqs. (18) and (19) (Seyyedattar et al., 2019; Springer, 2006):

$$r_i(t+\Delta t) = 2r_i(t) - r_i(t-\Delta t) + a_i(t)(\Delta t)^2 + O(\Delta t^4). \quad (20)$$

Based on Eqs. (4) and (5), it is concluded that $a_i(t)$ could be determined as a function of the potential energy. The Verlet expression is obtained if the terms after the third-order one are removed, as given below (Seyyedattar et al., 2019; Springer, 2006):

$$r_i(t+\Delta t) = 2r_i(t) - r_i(t-\Delta t) + \frac{\nabla_{r_i}U(t)}{m_i}(\Delta t)^2. \quad (21)$$

Considering the current computational facilities, it is recommended to use the Verlet algorithm in MD simulation runs, which is stable, simple, and precise (Seyyedattar et al., 2019; Zendehboudi et al., 2018). As just the two newest positions of each particle is required to be obtained over the computation process, the Verlet algorithm does not demand high computational memory. As is clear from Eq. (21), only the positions of each particle are included, and the velocities are not obtained directly. In order to compute the magnitude of the kinetic energy and examine the conservation of total energy during the MD run, the values of the velocities are needed. If the particle's positions are known at the last and next time steps, the velocity at the specific time (v_i) is calculated through the following simple way (Seyyedattar et al., 2019; Springer, 2006):

$$v_i(t) = \frac{r_i(t+\Delta t) - r_i(t-\Delta t)}{2\Delta t}. \quad (22)$$

There are some errors in the order of Δt^2 with Eq. (22) due to excluding some terms from the Taylor expansion, despite a minor error in the order of Δt^4 while calculating the positions of each particle, based on Eqs. (18)–(21) (Seyyedattar et al., 2019; Springer, 2006). The "Velocity Verlet algorithm" can be used as a robust strategy to obtain velocities with higher accuracy so that the values of accelerations and velocities of each particle are determined at $t+\Delta t$ when their extent is known at the time of t. Hence, the positions and velocities of each particle could be obtained at time t, as expressed below (Seyyedattar et al., 2019; Springer, 2006):

$$r_i(t+\Delta t) = r_i(t) + v_i(t)\Delta t + \frac{1}{2}a_i(t)(\Delta t)^2 + O(\Delta t^3). \tag{23}$$

$$v_i(t+\Delta t) = v_i(t) + a_i(t)(\Delta t) + O(\Delta t^2). \tag{24}$$

Considering the time intervals of $\Delta(t/2)$, the following equations for the acceleration and velocity are obtained if the velocity at the first-order term and the position at the second-order term are truncated (Seyyedattar et al., 2019; Springer, 2006):

$$v_i\left(t+\frac{\Delta t}{2}\right) = v_i(t) + a_i(t)\left(\frac{\Delta t}{2}\right). \tag{25}$$

$$a_i(t+\Delta t) = -\frac{1}{m_i}\nabla U(r_i(t+\Delta t)). \tag{26}$$

$$v_i(t+\Delta t) = v_i\left(t+\frac{\Delta t}{2}\right) + \frac{1}{2}a_i(t+\Delta t)(\Delta t). \tag{27}$$

Since the magnitudes of acceleration, velocity, and position for each particle do not require to be stored for various times at the same time, the velocity Verlet algorithm is considered a stable procedure and needs low memory (Seyyedattar et al., 2019; Springer, 2006). The "leap-frog algorithm" is recognized as another version of the Verlet algorithm case as the velocity and position of each particle are calculated at inserted time points. For instance, the values of velocities for each particle might be determined at a half time step once the positions are computed. The following equations can be used to obtain the velocity and position for particle i (Seyyedattar et al., 2019; Springer, 2006):

$$r_i(t+\Delta t) = r_i(t) + v_i\left(t+\frac{\Delta t}{2}\right)(\Delta t). \tag{28}$$

$$v_i\left(t+\frac{\Delta t}{2}\right) = v_i\left(t-\frac{\Delta t}{2}\right) + a_i(t)(\Delta t). \tag{29}$$

More advanced "predictor-corrector" approaches have been developed to determine the velocity of each particle when the MD simulation run requires high time steps. The suggested techniques are on the basis of the temporary computation of the position of the targeted particle at a forward step and their following modification utilizing the forces calculated at the new time step. However, the Verlet technique leads to more precise outputs based on the energy conservation law in some practical scenarios. Certainly, the available computational facilities, the goal of MD simulation, and type/nature of the simulated systems help choose the proper methods (Seyyedattar et al., 2019; Springer, 2006).

5.4 Statistical ensembles

According to the statistical mechanics, the physical properties of a macroscopic system are obtained by the average of microscopic states of the system whose distribution is based on the statistical ensemble (Seyyedattar et al., 2019; Zendehboudi et al., 2018). The term "statistical ensemble" denotes an idealization of the system comprising its various potential states which are well-suited with a series of constraints, including applied temperature or pressure (Seyyedattar et al., 2019; Zendehboudi et al., 2018). In other words, a statistical ensemble is defined as the probability distribution of the possible states of the system (http://www.quimica.urv.es/~bo/MOLMOD/General/Dynamics/Ensembles.html, n.d.; Gibbs, 1902). The main statistical ensembles are briefly explained here (http://www.quimica.urv.es/~bo/MOLMOD/General/Dynamics/Ensembles.html, n.d.; Gibbs, 1902; Seyyedattar et al., 2019; Springer, 2006).

Microcanonical ensemble (NVE). The microcanonical ensemble considers an entirely isolated system with a constant total number of particles (N), total volume (V), and total energy (E). To be in statistical equilibrium, it is necessary to have complete isolation of the system. Temperature and pressure and conjugate intensive variables of internal energy and volume could vary during the MD simulation run. Thus, the Newtonian equations of motion will be solved in MD simulations with a microcanonical ensemble without any control on temperature and/or pressure. This ensemble is suitable for exploring the transport properties of the system. Due to the difficulty of maintaining the system at a constant energy state, this ensemble is not feasible to be utilized (http://www.quimica.urv.es/~bo/MOLMOD/General/Dynamics/Ensembles.html, n.d.; Seyyedattar et al., 2019; Springer, 2006).

Canonical ensemble (NVT). In the canonical ensemble, the total number of particles (N), total volume (V), and system temperature (T) are fixed, while the energy of the system is not precisely known. Although this ensemble works for an isolated system in weak thermal contact with a heat reservoir, it cannot exchange system energy with any other object. The system can have different values of total energy, depending on the system's state. This ensemble also evaluates the phase properties (http://www.quimica.urv.es/~bo/MOLMOD/General/Dynamics/Ensembles.html, n.d.; Gibbs, 1902; Seyyedattar et al., 2019; Zendehboudi et al., 2018).

Grand canonical ensemble (μVT). The grand canonical ensemble operates at specified chemical potential (μ), total volume (V), and temperature (T). Since the chemical potential is constant, the system is in thermodynamic equilibrium with a reservoir. The system is allowed to exchange energy and particle with the reservoir, which makes this ensemble suitable to describe an open system. This ensemble can be treated as a canonical ensemble for the value of N. The μVT approach is appropriate for investigating vapor-liquid equilibria, adsorption isotherms, and selectivity (http://www.quimica.urv.es/~bo/MOLMOD/General/Dynamics/Ensembles.html, n.d.; Gibbs, 1902; Seyyedattar et al., 2019; Springer, 2006).

Isobaric-isothermal ensemble (NPT). The NPT ensemble assumes a constant total number of particles (N), pressure (P), and temperature (T) of the system. The volume of the system can be changed to keep the pressure constant. The volume change occurs using the flexible walls. Also, the system might exchange energy by flowing in or out of streams. This ensemble is very useful when correct values of pressure, volume, and densities are required. In addition, the phase properties of the system could be achieved using this ensemble (http://www.quimica.urv.es/~bo/MOLMOD/General/Dynamics/Ensembles.html, n.d.; Gibbs, 1902; Seyyedattar et al., 2019; Springer, 2006).

Gibbs ensemble. The Gibbs ensemble is used for most fluid phase equilibria calculations. It could be employed at imposed global volume of the two phases or at imposed global pressure. The temperature (T) and total number of particles (N) should be specified. The Gibbs ensemble is used for pure components to calculate equilibrium properties, including saturated vapor pressure and phase densities. Applying on binary and multicomponent mixtures, it could be used to evaluate the equilibrium phase compositions and densities at a constant volume or pressure. Gibbs ensemble is appropriate for studying near-critical behavior, adsorption isotherms, and selectivity (http://www.quimica.urv.es/~bo/MOLMOD/General/Dynamics/Ensembles.html, n.d.; Gibbs, 1902; Seyyedattar et al., 2019; Springer, 2006).

5.5 Property calculation

MD simulation shifts the microscopic variables into macroscopic specifications, including mole numbers, pressure, temperature, and energy content using statistical mechanics (Seyyedattar et al., 2019; Springer, 2006). MD parameters are defined based on the time average of physical properties observed toward the system trajectories when adequate system configurations are available (Seyyedattar et al., 2019; Springer, 2006; Ungerer et al., 2007). The MD parameters (e.g., physical features) are presented in terms of particles' velocities and coordinates. A time-dependent instantaneous property, $X(t)$, is defined as follows (Seyyedattar et al., 2019; Springer, 2006):

$$X(t) = f\left(r_1(t), r_2(t), \ldots, r_N(t), v_1(t), v_2(t), \ldots, v_N(t)\right). \quad (30)$$

The property X has an ensemble average value $\langle X \rangle$ over the MD simulation run, which is introduced as follows (Seyyedattar et al., 2019; Springer, 2006):

$$\langle X \rangle = \frac{1}{N_T} \sum_{t=1}^{N_T} X(t), \quad (31)$$

where t and N_T resemble the time step and the number of time steps, respectively. Note that N_T agrees with the points in the phase space. MD offers valuable information, such as kinetic and potential energy, pressure, temperature, diffusivity, viscosity, radial distribution function, and elastic moduli. Some of these properties (e.g., potential energy) are directly obtained through the particle trajectories and can help us determine other thermodynamic properties. As an example, the average kinetic energy can obtain the system's temperature as follows (Seyyedattar et al., 2019; Springer, 2006):

$$T = \frac{2}{3} \frac{\langle K \rangle}{N k_B}, \quad (32)$$

where N denotes the number of particles; k_B expresses the Boltzmann constant; and $\langle K \rangle$ represents the average kinetic energy obtained from the instantaneous kinetic energy $K(t)$, as follows (Seyyedattar et al., 2019; Springer, 2006):

$$K(t) = \frac{1}{2} \sum_{i=1}^{N} m_i . [v_i(t)]^2 \quad (33)$$

The instantaneous values can be updated at each time step as well as average values (see Eq. (2)) for MD simulation. Otherwise, a separate program can also be dedicated to a trajectory file containing particle positions and velocities (considering the system configuration) to be processed when the simulation is done. The former needs lower storage space and fits the simple properties, while the latter is more suitable for the properties facing complexity and requiring additional inputs (Seyyedattar et al., 2019; Springer, 2006).

6 Algorithms and simulation packages for MD technique

MD simulation investigates the concerning properties through atomic trajectories (GROMACS, n.d.; Haile, 1992). The typical steps of MD simulations include initialization, equilibrium, and production. A discussion on the MD workflow is provided in this section.

First of all, the desired properties and appropriate force field are chosen. Other steps of the initialization stage involve defining the number of atoms/molecules and their types, coordinates, system size, units, and boundary conditions. Also, the simulation details (e.g., particles' velocity and initial positions) should be set (Seyyedattar et al., 2019; Springer, 2006). The particles with close initial position experience a large inter-particle force that may crash the MD simulation. Accordingly, the MD simulation initiation should follow an energy minimization step to avoid any unsatisfactory simulation process (Seyyedattar et al., 2019; Springer, 2006) The equilibrium step mainly runs the MD simulation based on the Newtonian equations of motion to achieve a steady-state condition, considering the initial random structure. The steady-state is defined as the fluctuation of a variable around a stabilized average value. The main intention of the equilibrium stage is to achieve the stabilized values at their equilibrium conditions. Then, the system will take a phase space sample based on the probability distribution in accordance with a specified statistical ensemble (Seyyedattar et al., 2019; Springer, 2006). The larger systems with a greater phase space need a longer time for equilibrium. The energy information, pressure, and temperature data should be interpreted to obtain the systems' condition (Seyyedattar et al., 2019; Zendehboudi et al., 2018). The last step is the production stage; this stage also works based on the motion equations. Once the system stabilizes at the desired ensemble, the data can be collected. The data belong to a specific time with a predetermined time step. The system size and target properties impact the production run. The obtained trajectory data are employed for visualization, property calculation, and assessing phenomenon/interactions (Schäfer et al., 2000; Seyyedattar et al., 2019; Zendehboudi et al., 2018).

MD procedure and details. Typically, MD simulation is performed for each time step (e.g., 1 fs) and periodic boundary condition (Panizon et al., 2015; Sun & Zhou, 2013). The MD simulation employs the particle-mesh-Ewald (PME) method to calculate the electrostatic interactions with a real space cut-off of 1.4 nm (Chellehbari et al., 2021; Seyyedattar et al., 2019). Moreover, the Lennard-Jones equation is utilized to calculate the *v*dW interactions. The Nose-Hoover thermostat (Evans & Holian, 1985) and extended Lagrangian method (Liou, 1995) are employed to check the temperature and pressure, respectively. Fig. 6 illustrates the key steps of the MD simulation in the form of a flowchart.

MD simulation can be conducted using GROMACS software. GROMACS 2019 (Abraham et al., 2015; Lindahl et al., 2019) and OPLS-AA force-field are chosen to show the general algorithm of MD simulation. Table 1 describes the general algorithm of MD simulation for investigating molecular interactions. MD simulation includes three main steps: initialization, equilibration, and production. The initialization step identifies the molecule type, molecule number, molecule coordination, and system size. Before the equilibration step, the system's initial configuration should arrive at the minimum energy level to avoid having atoms overlapped or very close, which causes atoms to shoot out of the box at the beginning of the simulation (Seyyedattar et al., 2019; Zendehboudi et al., 2018). Usually, the steepest descent method is used to relax the system and reach the local minimum energy. The equilibration step is to let the system properties reach the average designed values. In this system, the temperature and pressure need to be fixed during the simulation. Hence, *NVT* and *NPT* simulations are run to reach the designed temperature and pressure, respectively. In the *NVT* step, the pressure changes, and the temperature reaches the designed temperature at constant volume (Seyyedattar et al., 2019; Springer, 2006). The volume changes in the *NPT* step, and the pressure reaches the designed pressure at constant temperature. The *NVT* step is conducted using a velocity rescaling thermostat (for 100 ps) (Bussi et al., 2007), and the *NPT* step is conducted using a velocity rescaling thermostat and Berendsen barostat (for 1 ns) (Berendsen et al., 1984). The production step is only for sampling from the system at the designed condition. Therefore, enough time (120 ns) with a suitable time step (2 fs) should be given to the system to evaluate the expected properties (Headen et al., 2009). In the production run, we use the Parrinello–Rahman barostat (Parrinello &

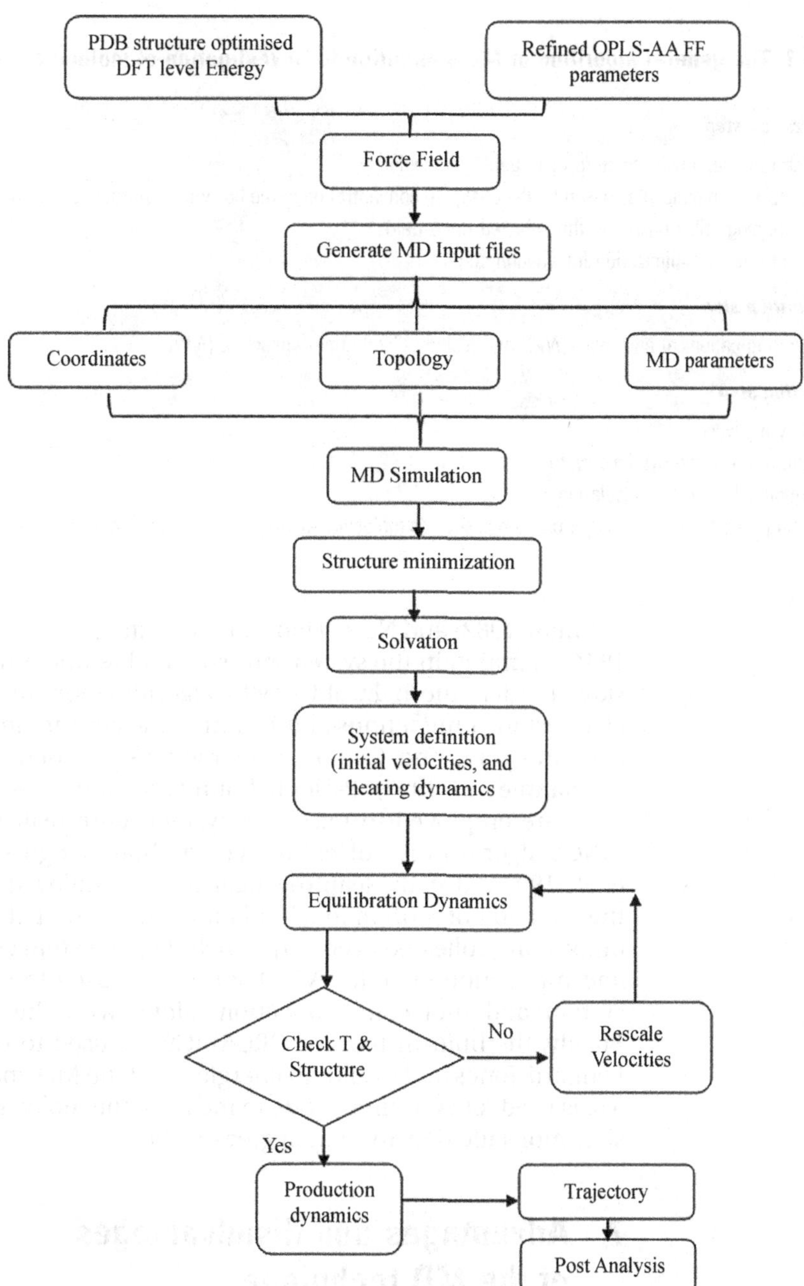

Fig. 6 A simple schematic of an algorithm to demonstrate the MD simulation's main parts.

Table 1 The general algorithm of MD simulation for investigation of molecular interactions.

Initialization step
1. Sketch representative structure of targeted molecules
2. Insert optimized molecules based on the designed concentration in the box with a suitable size (regular or random)
3. Build topology files based on the selected force-field
4. Conduct energy minimization for enough steps
Equilibration step
Implementing canonical ensemble (*NVT*) and isobaric–isothermal ensemble (*NPT*)
Production step
Trajectory analysis
1. Visualization of the dynamic system
2. Intermolecular energy calculation
3. Scripting code to measure target properties such as molecular gyration size, molecule density, and molecule shape

Rahman, 1982) and Nose–Hoover thermostat (Hoover, 1985; Nosé, 1984) to maintain the system pressure and temperature. We consider particle-mesh Ewald (PME) algorithm for the long-range electrostatic interactions and 1.2 nm as the cut-off radius for both Lennard-Jones and Coulomb interactions (Darden et al., 1993; Essmann et al., 1995). Periodic boundary conditions in all directions are employed to make the system more realistic, and the LINCS algorithm is utilized to keep all bond lengths rigid (Hess et al., 1997). Also, the leapfrog algorithm is employed to integrate the equation of motion for simulation (Hockney et al., 1974). The outputs are collected every 10 ps, including the atom coordination and interaction energies. VMD software is used to visualize the system and molecule interaction, along with the simulation. Finally, the built-in tools of GROMACS are used to evaluate the Lennard-Jones and Coulomb energies, and the MD analysis package is used for scripting a code to measure the molecular gyration size, molecule density, and aggregate shape.

7 Advantages and disadvantages of the MD technique

There are many different experimental, theoretical, and molecular methodologies to assess the stability and behaviors of various chemical processes and phenomena. Each has its strengths and weakness. The benefit of the MD approach is that it needs only atomic and intermolecular parameters for the simulation. MD simulation could be selected for analyzing systems for a wide range of temperature and pressure (Seyyedattar et al., 2019; Springer, 2006). Another advantage of the MD approach is the ability to assess the behavior of molecules in complex systems, such as polymer, protein, and enzyme molecules. Also, the MD technique is the best strategy to evaluate the structural, thermodynamic, and mechanical characteristics of rare and expensive molecules. However, there are certain problems and drawbacks associated with the use of MD simulation. The main disadvantages of the MD simulations are the number of molecules and simulation time. Also, a strong CPU is required for MD simulations to provide comprehensive results after simulation with all aspects and properties of the proposed mixture (Seyyedattar et al., 2019; Springer, 2006).

As mentioned earlier, although MD simulation is a powerful tool for research at the atomic and molecular levels to complement experiments and theory, it has some limitations that should be considered while employing this technique. The main limitations of MD simulation are briefly discussed in the following subsections.

Use of classical mechanics. Each molecule's velocity and position are calculated based on Newton's second law of motion in MD simulation tools/approaches, while the laws of quantum mechanics can more accurately describe the particle dynamics. In fact, Schrödinger's equation is directly solved in the quantum mechanics for system modeling. In addition, states of both atomic nuclei and electrons are taken into account in quantum mechanics (Seyyedattar et al., 2019; Zendehboudi et al., 2018). At relatively high temperature conditions, most atoms can be considered the point particles for which utilization of the classical mechanics can provide satisfactory outputs. However, classical mechanics is a relatively poor approximation to simulate and describe the dynamics of very light particles such as hydrogen (H_2), helium (He), and neon (Ne). In addition, the classical mechanics cannot capture the quantum effects during MD simulation, including chemical bonds (e.g., formation or breaking of the covalent bonds), existence of noncovalent intermediates, specific heat drop in crystals below a particular temperature, and thermal expansion coefficient anomalies. This limitation is mainly noticed at fairly low temperature conditions (Seyyedattar et al., 2019; Springer, 2006).

Force-field imperfections. All atoms/molecules move in the system because of the interaction force between them. The force between the atoms/molecules results in potential energy. Hence, the correct definition of potential energy and introduction of proper forces between atoms during MD simulation significantly affect outcomes' reliability (Seyyedattar et al., 2019; Zendehboudi et al., 2018). The force-field estimates forces between atoms/molecules and calculates the potential energy by considering functional forms and parameters, as mentioned in the previous sections. There are different force-fields that are used for MD simulation. However, their accuracy is debated because these force-fields are based on experimental data. Hence, the available force-fields present approximations of the real forces between atoms/molecules (Seyyedattar et al., 2019; Springer, 2006).

Despite the significant improvements in the development of effective force-fields over the recent years, there are still some limitations with them. For instance, the dielectric constant of vacuum is used to calculate potential energy in force-fields, while the system consists of molecules surrounded by aqueous solutions with much greater dielectric constants (Seyyedattar et al., 2019; Springer, 2006). On the other hand, these dielectric constants are determined at a macroscopic scale, and use of them to calculate the electrostatic forces at short interatomic distances (microscopic scale) increases the chance of error. Moreover, the accuracy of van der Waals forces depends on the interaction environment. The Lennard-Jones potential is used to account for short-range van der Waals interactions in MD simulations that work based on London theory and are valid at vacuum condition (Seyyedattar et al., 2019; Springer, 2006). Intermolecular hydrogen bonds are also not incorporated in most of force-fields. Some newly developed force-fields have implicitly included the bonds as the electrostatic interactions of the atomic point charges. This approximation may not be adequately accurate due to the hydrogen bonds' partially quantum mechanical and chemical nature (Seyyedattar et al., 2019; Springer, 2006).

Time and size scales: The feasibility of any MD simulation depends on the available computational resources. The total duration of the simulation, length of the time steps, and the number of particles are sources of errors/problems in MD simulation that can be solved using powerful computational resources (Seyyedattar et al., 2019; Springer, 2006).

Most of real reactions and experimental investigations happen at the macroscopic scale. Hence, the total time of MD simulation should be considered long to obtain reliable results. On the other side, the time step during calculation should be small to minimize the discretization errors and calculate the realistic numerical constant of the integrations. The long time step causes instability of MD simulation runs. Hence, it is necessary to choose an appropriate time step to enhance the precision of energy conversion and numerical stability (Seyyedattar et al., 2019; Springer, 2006). The maximum time step length for integrations in MD simulations is selected based on the fastest vibrational atomic motions, which can be in the order of several femtoseconds (10–15 s). This means that there might be a large number of iterations (millions) required to track the motions of a single atom for a total time period of a microsecond, while the simulated system normally is made of many atoms (e.g., tens or hundreds of thousand). The number of atoms/molecules directly affects the size (dimensions) of the simulation box. Therefore, it is required to consider several atoms/molecules when we have plan to simulate a system at a nanoscale, leading to a considerably large and complex system. This is a serious limitation for many studies that need a simulation box at the nanometer scale (Seyyedattar et al., 2019; Springer, 2006). Using more advanced computational resources and considering longer time steps and trajectories for more particles in MD simulations might overcome these problems in the future.

8 Applications/case studies of MD

This section presents four case studies, including gas hydrate formation, asphaltene precipitation, CO_2 capture using ionic liquids, and water/oil/salt, where the MD simulation approach is employed.

Hydrate dissociation. The decomposition process and stability of CH_4, CH_4+CO_2, $CH_4+C_3H_6$, and $CH_4+iC_4H_{10}$ mixtures are investigated for different pressures, temperatures, and compositions, in the absence and/or presence of inhibitors. In addition, the physical and structural properties, such as density, unit cell, and potential energy, are calculated for the studied mixtures (Kondori et al., 2017; Kondori et al., 2019a; Kondori et al., 2020).

The radial distribution functions (RDF) for the oxygen atoms of water molecules and carbon-carbon of methane and isobutane are shown in Fig. 7, which includes 700-ps *NPT* simulation results for structure II of $CH_4+iC_4H_{10}$ hydrate at various temperatures and constant pressure.

As observed in Fig. 8A, the peaks follow a decreasing trend when the temperarure increases from 280 to 320 K. The second and third peaks, represented by the distance of oxygens in hydrate cages (tetrahedral and hexahedral rings), gradually disappear due to decomposition of gas hydrate cages. In Fig. 8B, the trend of RDFs of methane is exactly similar to that of water molecules.

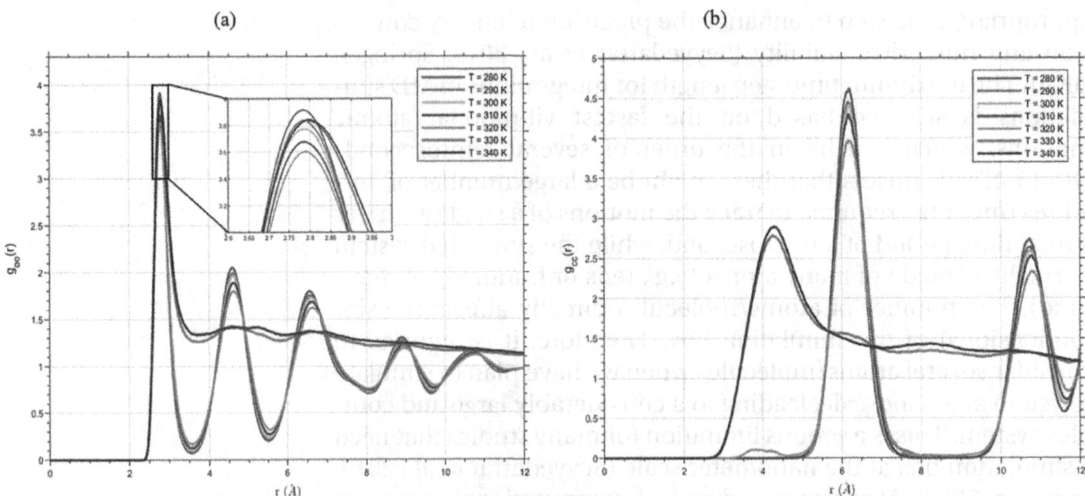

Fig. 7 RDFs for isobutane/methane hydrate case at 700 ps, $P=20$ MPa, and different temperatures: (A) oxygen atoms in water molecules and (B) carbon atoms in methane molecules. Modified after Kondori, J., Zendehboudi, S., & James, L. (2019b). Molecular dynamic simulations to evaluate dissociation of hydrate structure II in the presence of inhibitors: A mechanistic study. *Chemical Engineering Research and Design 149*, 81–94.

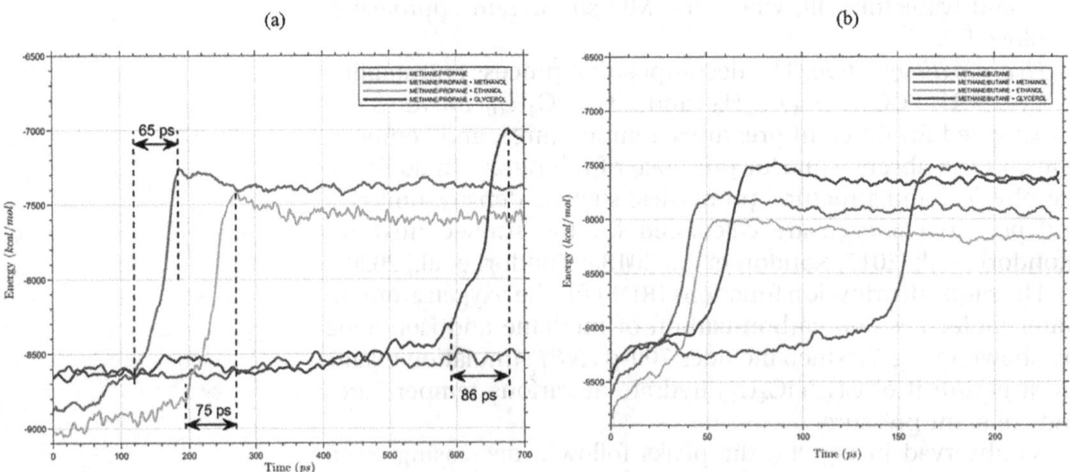

Fig. 8 Varaiation of potential energy with time at $P=20$ MP and $T=330$ K for (A) propane/methane and (B) isobutane/methane hydrate systems. Modified after Kondori, J., Zendehboudi, S., & James, L. (2019b). Molecular dynamic simulations to evaluate dissociation of hydrate structure II in the presence of inhibitors: A mechanistic study. *Chemical Engineering Research and Design 149*, 81–94.

The new peaks appear after decomposition ($T=340$) of hydrate clathrates addressed by the distance of methane molecules after creating a cluster. The potential energy changes for two different systems in the presence and absence of inhibitor molecules through the MD simulation are illustrated in Fig. 8 (Kondori et al., 2019b).

According to Fig. 8, the potential energy fluctuates around one value (equilibrium) before decomposition. Hence, there is no decomposition for the mixture of $CH_4+C_3H_6$. The potential energy of the mixture is dramatically increased after adding inhibitors (methanol, ethanol, and glycerol). It is noticed that the dissociation occurs for $CH_4+C_3H_6$ hydrate in the presence of methanol, ethanol, and glycerol as added inhibitors at almost 110, 195, and 595 ps, respectively (Fig. 8B). Also, the decomposition of $CH_4+C_3H_6$ hydrates requires about 65, 75, and 86 ps upon addition of methanol, ethanol, and glycerol, respectively. The hydrate dissociation for methane/isobutane happens at 140 ps without inhibitors, as depicted in Fig. 8B. The decomposition takes place before 140 ps for the $CH_4+iC_4H_{10}$ system in the presence of inhibitors. Hence, the influence of different inhibitors can be analyzed by following the decomposition time for other systems. The snapshots of the MD simulation box for methane hydrate at various temperature conditions at constant pressure (5 MPa) are displayed in Fig. 9 (Kondori et al., 2019a).

Panels A, B, and C of Fig. 9 reveal that the structure of gas hydrate cages is almost firm, and the methane gases vibrate and rotate at the center of cages for temperatures before 280 K (decomposition temperature). Then, the structure of methane hydrates will vanish by increasing temperature to 280 K. The methane hydrate structure will be decomposed by moving the methane molecules and creating gas clusters, as clearly observed in Fig. 9D (Kondori et al., 2019a).

One of the most important structural parameters of gas hydrate is the unit cell or lattice parameter. The gas hydrate structure I unit cell contains two dodecahedral (5^{12}) and six tetrakaidekahedral ($5^{12}6^2$) cages of water molecules. Fig. 10 demonstrates the unit cell for the CH_4+CO_2 structure I for different pressures and temperatures at the equilibrium condition calculated by MD simulation. There is acceptable agreement between the calculated lattice parameters and those reported in the previous simulation and experimental studies. In addition, it is found that the unit cell parameter increases by increasing the size of guest molecules and temperatures under constant pressure condition. Lower values for the unit cell parameter are obtained at lower pressures (Kondori et al., 2020).

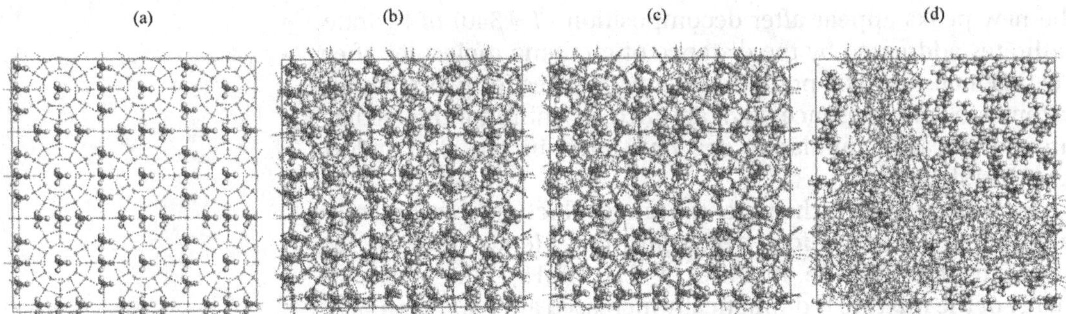

Fig. 9 Snapshots of MD simulation for methane hydrate case at $P=5$ MPa after 200 ps (A) Initial configuration, (B) $T=260$ K, (C) $T=270$ K, and (D) $T=280$ K. Modified after Kondori, J., Zendehboudi, S., & James, L. (2019a). New insights into methane hydrate dissociation: Utilization of molecular dynamics strategy. *Fuel 249*, 264–276.

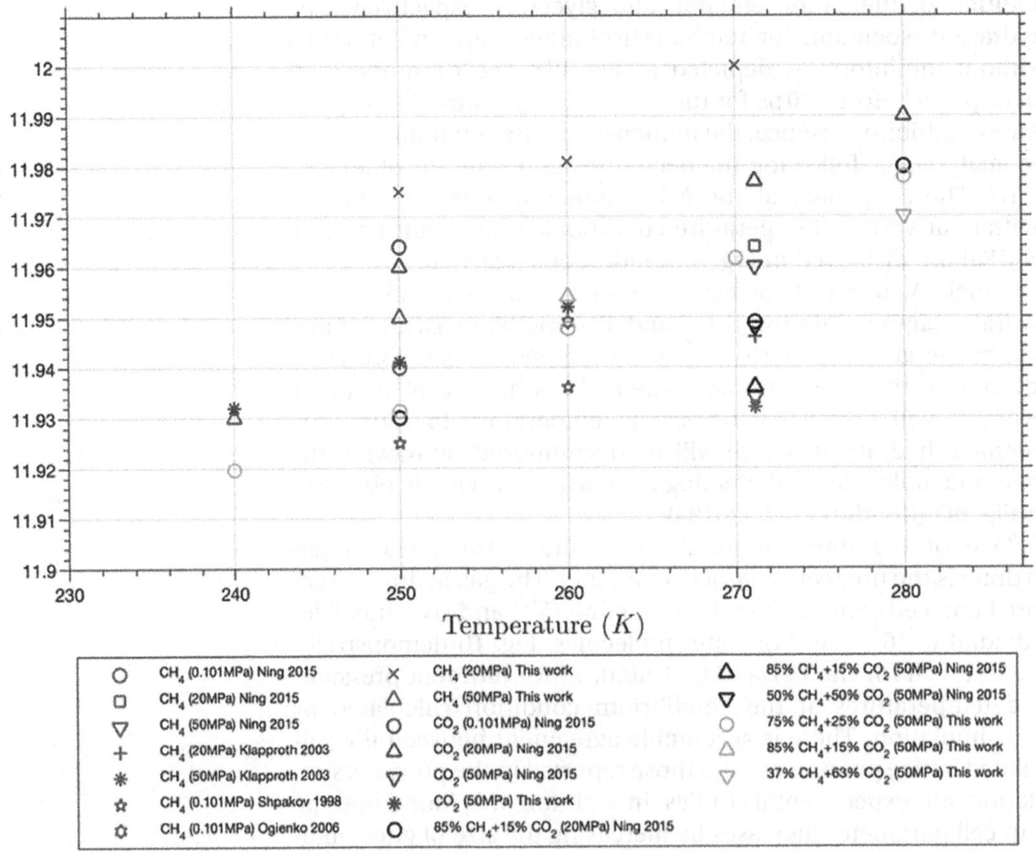

Lattice parameter (Å)
12
11.99
11.98
11.97
11.96
11.95
11.94
11.93
11.92
11.91
11.9
230
240
250
260
270
280
Temperature (K)

CH_4 (0.101MPa) Ning 2015
CH_4 (20MPa) Ning 2015
CH_4 (50MPa) Ning 2015
CH_4 (20MPa) Klapproth 2003
CH_4 (50MPa) Klapproth 2003
CH_4 (0.101MPa) Shpakov 1998
CH_4 (0.101MPa) Ogienko 2006
CH_4 (20MPa) This work
CH_4 (50MPa) This work
CO_2 (0.101MPa) Ning 2015
CO_2 (20MPa) Ning 2015
CO_2 (50MPa) Ning 2015
CO_2 (50MPa) This work
85% CH_4+15% CO_2 (20MPa) Ning 2015
85% CH_4+15% CO_2 (50MPa) Ning 2015
50% CH_4+50% CO_2 (50MPa) Ning 2015
75% CH_4+25% CO_2 (50MPa) This work
85% CH_4+15% CO_2 (50MPa) This work
37% CH_4+63% CO_2 (50MPa) This work

Fig. 10 Variation of lattice parameter with pressure and temperature for the CH_4/CO_2 hydrate system (Klapproth et al., 2003; Kondori et al., 2020; Ning et al., 2015; Ogienko et al., 2006; Shpakov et al., 1998).

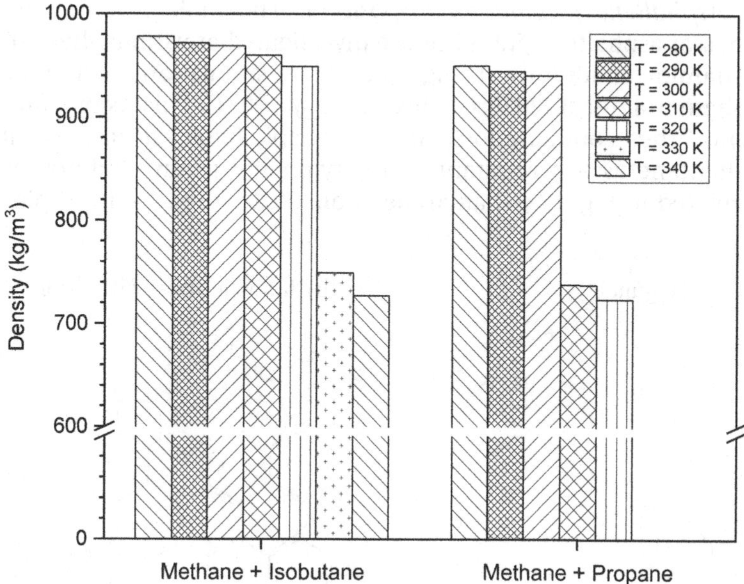

Fig. 11 Impact of temperature on the mixture density for gas hydrate structure II on the basis of MD outputs (Kondori et al., 2019b).

The effect of temperature on the magnitude of density for two hydrate systems including methane+isobutane and methane +propane is depicted in Fig. 11, where the operating pressure is 20MPa. According to Fig. 11, increasing temperature leads to a reduction in the density of the mixture. This density drop is more noticeable after hydrate dissociation occurrence; it should be noted that the density variation is minor prior to hydrate decomposition. It was also found that there is a good match between MD density prediction and real data (Kondori et al., 2019b).

Summarizing this case study, the MD simulation technique is able to simulate the stability and decomposition of gas hydrates in the molecular scale at different composition and thermodynamic conditions. The bubble formation and growth of methane and carbon dioxide molecules are observed and analyzed after decomposition using the MD technique. The results can demonstrate the instability and decomposition of the gas hydrate structure by changing temperature, pressure, cage occupancies, composition, and adding inhibitors. MD simulation proves that dissociation of gas hydrates occurs in two stages: first, water molecules in cages are continuously vibrating, and guest molecules rotate in the center of the cages. Then, the gas hydrates' structure is decomposed, and the guest gases move from the center of decomposed cages to create a gas cluster (Kondori et al., 2017; Kondori et al., 2019a; Kondori et al., 2020).

Asphaltene precipitation/deposition. The molecular interactions of asphaltene/inhibitor are investigated at various thermodynamic conditions using the MD simulation technique (Ghamartale et al., 2020; Ghamartale et al., 2021). This section presents the configurations of two asphaltene structures in *n*-heptane as a precipitant and octylphenol as an inhibitor, as depicted in Fig. 12 (Ghamartale et al., 2020). Figs. 13–15 display

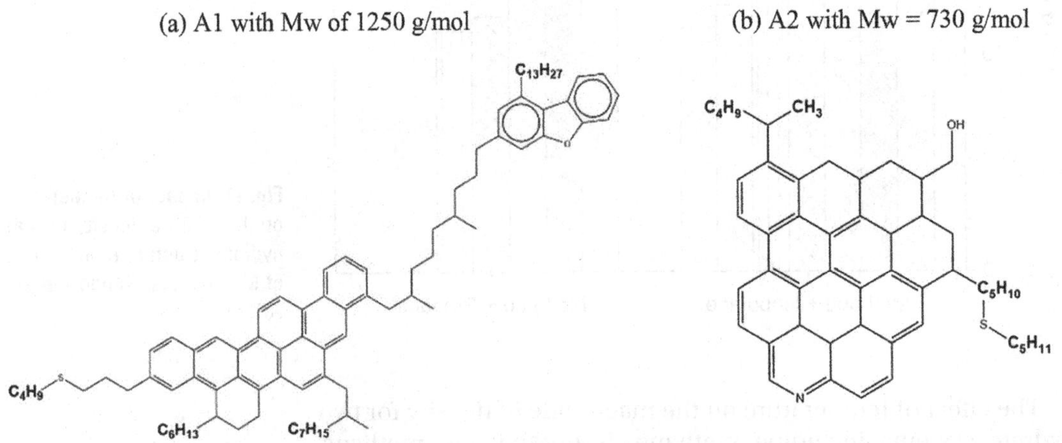

(a) A1 with Mw of 1250 g/mol

(b) A2 with Mw = 730 g/mol

Fig. 12 (A) archipelago-type asphaltene structure and (B) continental-type asphaltene structure. Modified after Ghamartale, A., Zendehboudi, S., & Rezaei, N. (2020). New molecular insights into aggregation of pure and mixed asphaltenes in the presence of n-Octylphenol inhibitor. *Energy & Fuels.*

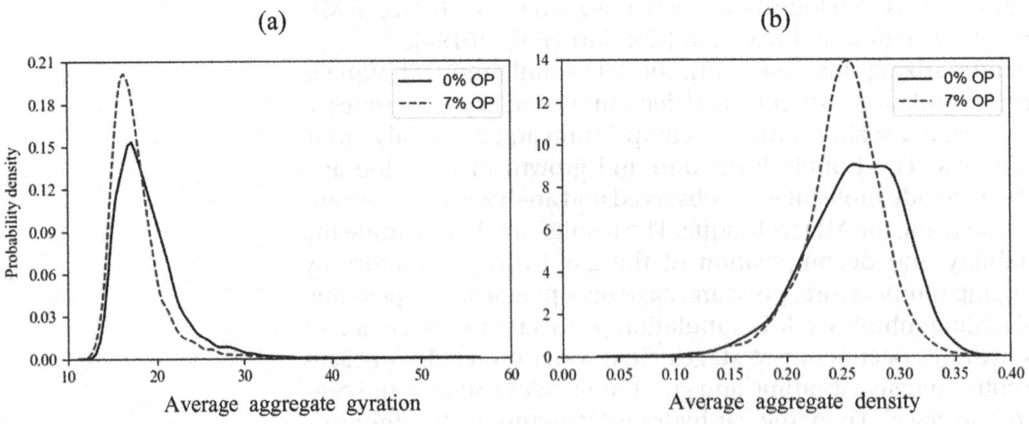

Fig. 13 (A) average aggregate gyration radius and (B) average aggregate density for the simulation case of A1/OP/*n*C7. Modified after Ghamartale, A., Zendehboudi, S., & Rezaei, N. (2020). New molecular insights into aggregation of pure and mixed asphaltenes in the presence of n-Octylphenol inhibitor. *Energy & Fuels.*

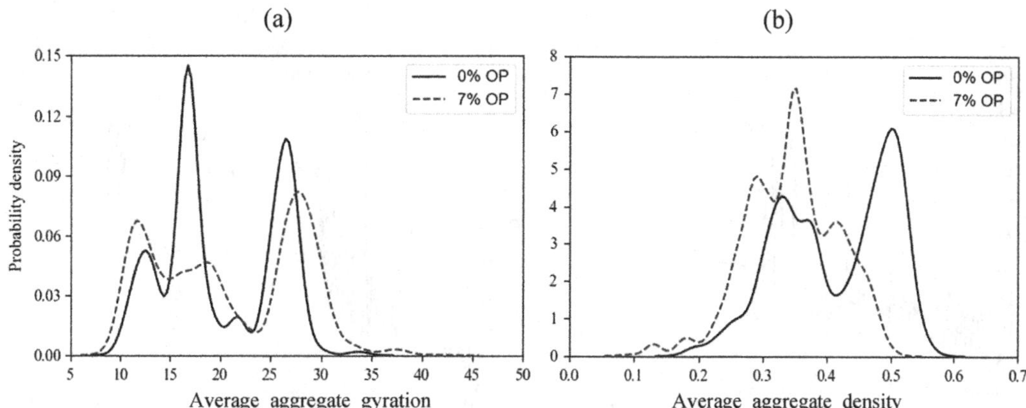

Fig. 14 (A) average aggregate gyration radius and (B) average aggregate density based on the MD simulation of the A2/OP/*n*C7 system. Modified after Ghamartale, A., Zendehboudi, S., & Rezaei, N. (2020). New molecular insights into aggregation of pure and mixed asphaltenes in the presence of n-Octylphenol inhibitor. *Energy & Fuels*.

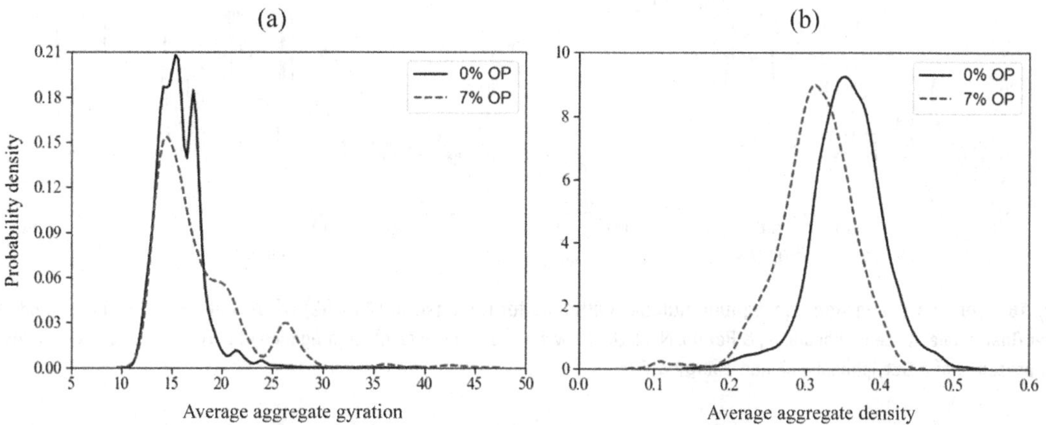

Fig. 15 (A) average aggregate gyration radius and (B) average aggregate density on the basis of the MD simulation of the (A1+A2)/OP/*n*C7 system. Modified after Ghamartale, A., Zendehboudi, S., & Rezaei, N. (2020). New molecular insights into aggregation of pure and mixed asphaltenes in the presence of n-Octylphenol inhibitor. *Energy & Fuels*.

the probability of aggregate gyration radius and density for three systems, that is, A1/OP/*n*C7, A2/OP/*n*C7, and (A1+A2)/OP/*n*C7, at atmospheric pressure and temperature. Fig. 16 indicates the influence of inhibitor concentration on asphaltene aggregation for the system of (A1+A2)/OP/*n*C7 at atmospheric pressure and temperature. The impacts of pressure and temperature on the

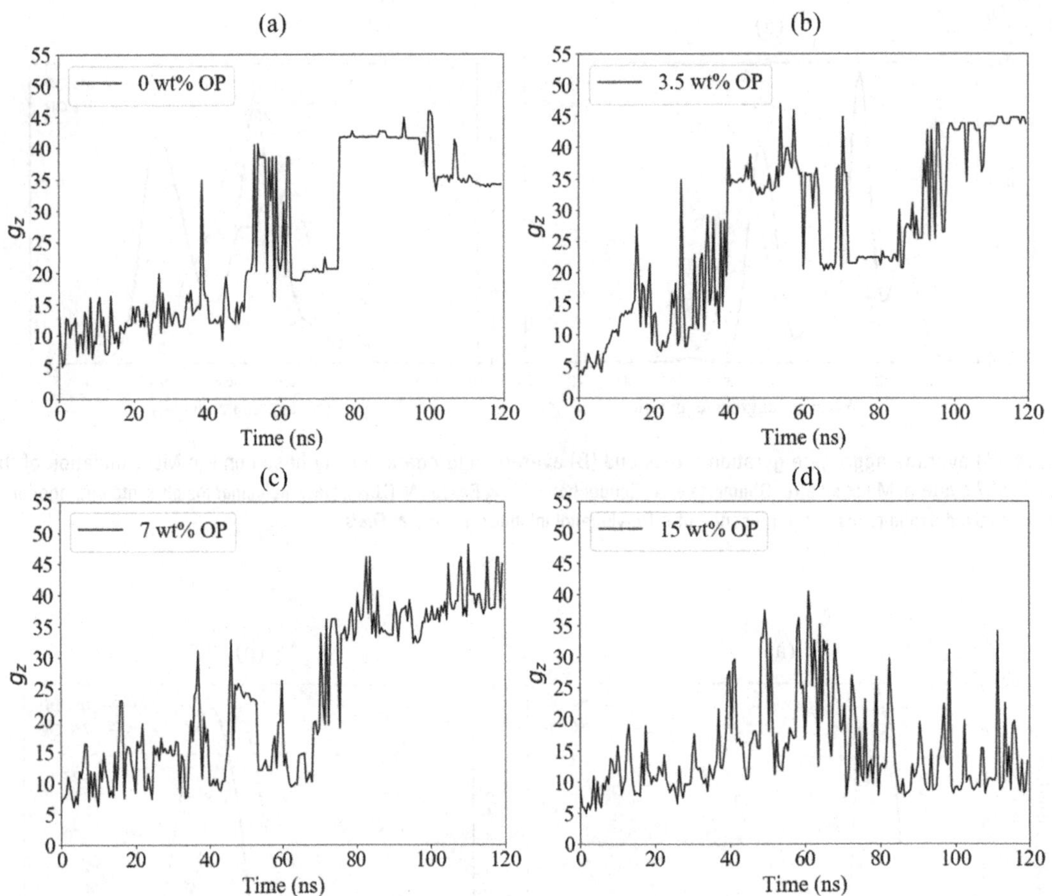

Fig. 16 Trends of *z*-average aggregation number with time for the case of (A1 + A2)/OP/nC_7 at 300 K and 1 bar. Modified after Ghamartale, A., Zendehboudi, S., & Rezaei, N. (2020). New molecular insights into aggregation of pure and mixed asphaltenes in the presence of n-Octylphenol inhibitor. *Energy & Fuels.*

efficiency of the inhibitor to restrict asphaltene aggregation are illustrated in Figs. 17 and 18.

Before adding the inhibitor, the average gyration radius of A1 aggregates is unimodal, with a peak at 17 Å (Fig. 12A). After OP addition, the peak shifts to the smaller gyration size and higher probability. There are two possibilities: either the inhibitor breaks the aggregates or squeezes them. The aggregate density evaluation confirms the aggregate breakdown hypothesis since Fig. 13B shows a higher probability for the aggregates with low density after inhibitor addition, while if the aggregates were squeezed, the density value should be higher after inhibitor addition (Ghamartale et al., 2020; Ghamartale et al., 2021).

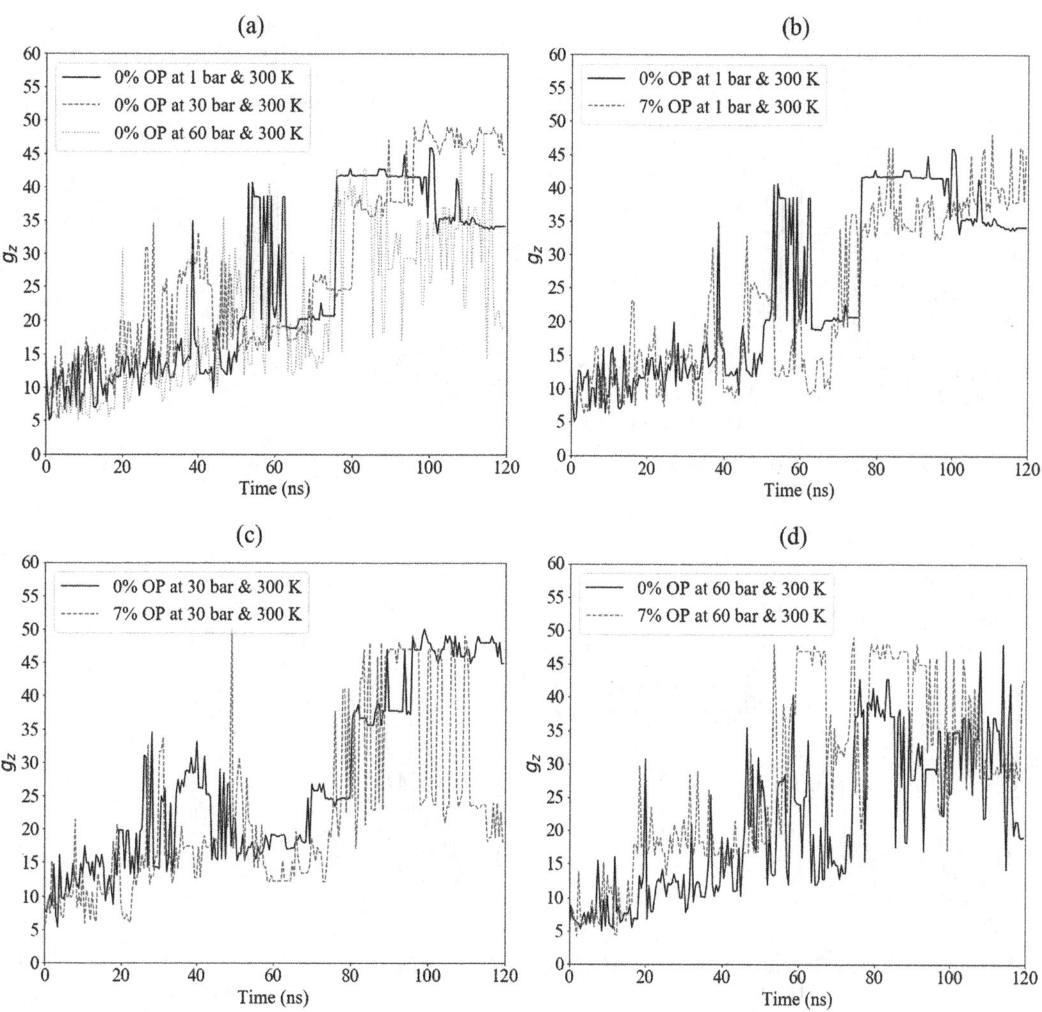

Fig. 17 *z*-average aggregation number versus time for (A1 + A2)/OP/nC_7 case at 1 bar, 30 bar, and 60 bar without and with OP when the temperature is 300 K. Modified after Ghamartale, A., Zendehboudi, S., & Rezaei, N. (2020). New molecular insights into aggregation of pure and mixed asphaltenes in the presence of n-Octylphenol inhibitor. *Energy & Fuels.*

The average gyration radius of the A2 aggregates is multimodal with peaks at 13, 17, and 26 Å in the absence of the inhibitor. In comparison, it becomes bimodal in the presence of the inhibitor (peaks at 11 and 28 Å), as shown in Fig. 14A. Hence, the inhibitor existence removes the medium-size aggregates. Since the plot is an average over the simulation time, it can be interpreted that the inhibitor keeps the aggregates small in the early time, while the inhibitor makes aggregation much worse in the later time.

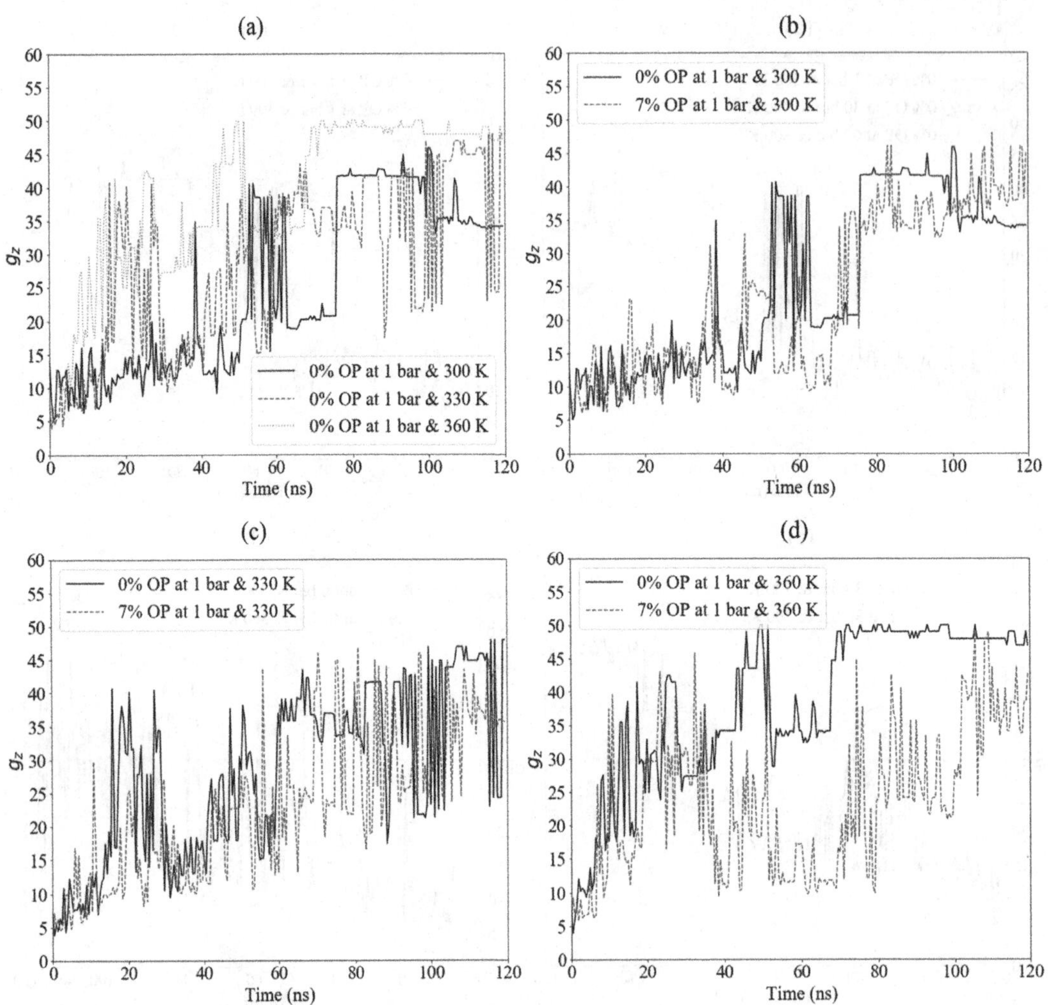

Fig. 18 Variation of z average aggregation number with time for (A1 + A2)/OP/nC_7 case at a pressure of 1 bar and the temperature range of 300–360 K in the presence and absence of the inhibitor. Modified after Ghamartale, A., Zendehboudi, S., & Rezaei, N. (2020). New molecular insights into aggregation of pure and mixed asphaltenes in the presence of n-Octylphenol inhibitor. *Energy & Fuels*.

This behavior can be explained as the aggregate density decreases whether the aggregate gyration radius increases or decreases (Fig. 14B). In the early time, inhibitors attach the asphaltene either by van der Waals or hydrogen bond. However, the inhibitors do not fully inactivate asphaltenes, and asphaltenes can approach and attach to each other with different angles. The presence of the inhibitor in the asphaltene aggregates confirms reduction in the aggregate density even with larger aggregates. The inhibitor alkyl chain provides a steric boundary in the aggregates and decreases the stability of aggregates (Ghamartale et al., 2020).

Fig. 15 depicts the gyration radius results for a system that includes both A1 and A2 structures with/without inhibitor. The average gyration radius of the asphaltene aggregates is between 12 and 20Å in the absence of the inhibitor, while a new peak between 25 and 30Å appears in the gyration radius curve when the inhibitor is present in the system (Ghamartale et al., 2020). Nevertheless, the aggregate density reduces when the inhibitor exists in the system compared to the case without the inhibitor (Fig. 15B). The aggregate size does not increase significantly, and the density is reduced, which confirms the inhibitor gets involved in the aggregate structure.

Since the asphaltene is composed of various types and structures, the system tends toward a more realistic situation if polydispersity is considered in the asphaltene structure. Therefore, we conduct the rest of the simulation for (A1 + A2)/OP/*n*C7. In the following section, we report the results of evaluating the effect of inhibitor concentration, pressure, and temperature on the inhibitor efficiency (Ghamartale et al., 2020).

Fig. 16 displays the *z*-average aggregation number for the system with 0–15 wt% inhibitor concentration. Although the low concentration of the inhibitor does not change the aggregation trend significantly (panels a and b of Fig. 16), the medium-to-high inhibitor concentration reduces the asphaltene aggregation substantially, which is more noticeable at 15 wt% inhibitor concentration (Ghamartale et al., 2020).

Fig. 17A depicts the *z*-average aggregation number for (A1 +A2)/*n*C7 when pressure increases from 1 bar to 60 bar at 300 K. In this case, the fluctuation of *z*-average aggregation number increases with increasing pressure. This demonstrates the instability of aggregates in higher pressure conditions, which can be confirmed from a thermodynamic viewpoint (Ghamartale et al., 2020). The inhibitor is not appreciably effective at 1 bar and 60 bar, while it reduces the aggregation stability at 30 bar. The analysis of MD results shows that the asphaltene-asphaltene attraction energy is notably higher than asphaltene-inhibitor attraction energy at 1 bar, making the inhibitor ineffective. At 60 bar, the asphaltene-asphaltene attraction energy is declined significantly due to incremental pressure, making the thermodynamic situation less favorable for aggregation. Hence, the inhibitor efficiency is meaningless when there is no aggregation. Therefore, the inhibitor has the highest efficiency at 30 bar (Ghamartale et al., 2020; Ghamartale et al., 2021).

Based on the thermodynamic prospect, the asphaltene-*n*C7 mixture tends toward the bubble point when the temperature is increased from 300 to 360 K at 1 bar. At the bubble point, the amount of precipitable asphaltene reaches its maximum. Hence, the MD results in the absence of the inhibitor (shown in Fig. 18A) support the thermodynamics behavior. Molecular movement increases at a higher temperature; considering the repulsion and attraction interaction of *n*C7 and inhibitor, respectively, inhibitor molecules have a better chance of finding a suitable angle and forming bonds with asphaltene molecules. Fig. 18B–D confirms the hypothesis, stating the higher efficiency of the inhibitor when the temperature is increased (Ghamartale et al., 2020; Ghamartale et al., 2021).

Summarizing this case study, understanding the asphaltene-inhibitor interaction mechanisms is essential to design a suitable inhibitor for target crude oil. The MD simulation provides a better understanding of active forces and mechanisms in the asphaltene precipitation in the presence of inhibitors. The results exhibit that the presence of the inhibitor molecules keeps the aggregates of archipelago asphaltene in a smaller size, while it behaves differently for the continental structure. The inhibitor restricts asphaltene aggregation in the early time for A2. However, the small aggregates agglomerate later in the simulation. Since the inhibitors are included in the aggregates, the aggregate gyration radius grows, while their density is lower compared to the case with no inhibitor. Increasing the inhibitor concentration and temperature improves the inhibitor's efficiency and reduces the asphaltene aggregation. The inhibitor shows the highest efficiency at the medium pressure (e.g., 30 bar) (Ghamartale et al., 2020; Ghamartale et al., 2021).

CO_2 capture using ionic liquids. The MD simulation tool is used to determine the transport properties of CO_2 in the bulk of [Bmim][BF4] and [Bmim][Ac] ionic liquids (ILs) and at the interface of gas-IL in the presence and absence of solvents (Aghaie et al., 2020a; Aghaie et al., 2020b). The important absorption mechanisms during carbon capture with ILs are explored using this strong modeling strategy. Fig. 19 demonstrates the influence of water content on the excess molar volume as well as molar enthalpy of the IL/W mixture at 300 K and 1 bar. According to Fig. 19, the minimum excess molar enthalpy and molar volume are seen around $x_w = 0.7$ and 0.5, respectively, highlighting strong hydrogen bonding. i.e., the high number of hydrogen bonds as a function of time exists between $[Ac]^-$ and H_2O at the range of 05–0.7 mol/mol (Aghaie et al., 2020a).

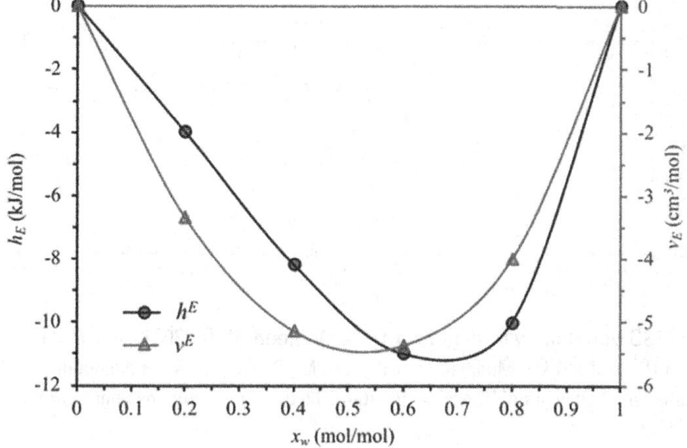

Fig. 19 Impact of water on excess molar volume (v^E) and excess molar enthalpy (h^E) at $T=300$ K and $P=1$ bar for the IL/W system. Modified after Aghaie, M., Rezaei, N., & Zendehboudi, S. (2020a). Effect of water on molecular and transport phenomena behaviors of [Bmim][Ac]/water/CO2, using molecular dynamics strategy. *The Journal of Physical Chemistry B 124(34)*, 7368–7378.

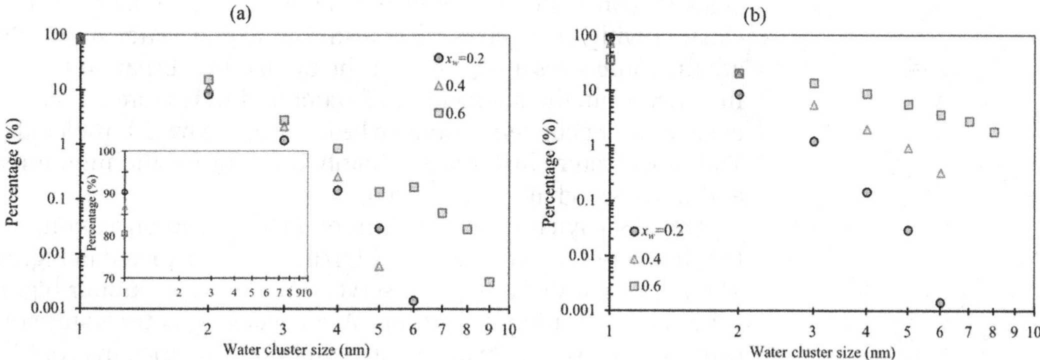

Fig. 20 Influence of water content on the water cluster size profile for two different systems (A) IL/W and (B) CO_2/IL/W mixtures at $T=300$ K and $P=1$ bar. Modified after Aghaie, M., Rezaei, N., & Zendehboudi, S. (2020a). Effect of water on molecular and transport phenomena behaviors of [Bmim][Ac]/water/CO2, using molecular dynamics strategy. *The Journal of Physical Chemistry B 124(34)*, 7368–7378.

Fig. 20 represents the size of the water cluster based on the water content variations. As it is clear, the water clusters have 2–5 water molecules at $x_w=0.2$. The water cluster contains more than eight water molecules at greater mole fractions (e.g., 0.4 and 0.6). Note that lower water concentration implies lower probability of cluster formation. Fig. 20B illustrates the cluster size of the ternary system, made of CO_2, IL, and water components. It is evident that higher clusters are formed in comparison with the binary mixture of IL/W. Furthermore, it is proven that larger clusters are formed at higher water concentrations for the mentioned ternary system; it implies that CO_2 molecules may attach to the IL

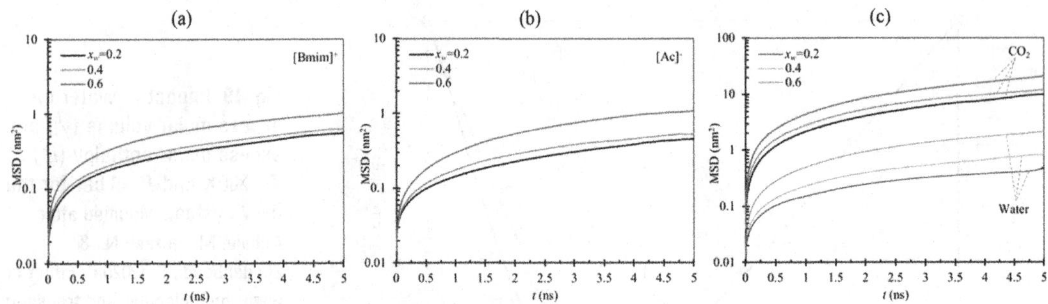

Fig. 21 Impact of water mole fraction on MSD variation with time for CO_2/IL/W blend at $T=300$ K and $P=1$ bar: (A) $[Bmim]^+$ (cation), (B) $[Ac]^-$ (anion), and (C) water/CO_2. Modified after Aghaie, M., Rezaei, N., & Zendehboudi, S. (2020a). Effect of water on molecular and transport phenomena behaviors of [Bmim][Ac]/water/CO2, using molecular dynamics strategy. *The Journal of Physical Chemistry B 124(34)*, 7368–7378.

ions and avoid water molecules in the closeness of the IL molecules. Accordingly, the water molecules are more likely to form clusters with themselves. It is also found that water molecules are the same sizes at $x_W=0.2$ for the binary and ternary mixtures. In conclusion, the association of water and IL is more noticeable at lower water concentrations to be interrupted by CO_2 molecules. The most water cluster sizes mainly account for the monomers and dimers (Aghaie et al., 2020a).

The MSD dynamic interactions of the water, anion, cation, and CO_2 in the ternary system (e.g., CO_2/IL/W) are depicted in Fig. 21. The molecular diffusion is observed for all species under higher water concentration conditions. As expected, greater water concentrations would lead to a reduction in the hydrophilic [Bmim][Ac] IL viscosity; thus, the movement of the molecules becomes quicker.

Fig. 22 depicts the number density along z direction at various pressures (10 bar and 40 bar) for CO_2, anion $[BF4]^-$, and cation $[Bmim]^+$, where the temperature is 300 K, based on the MD results up to 8 ns. At the interface of the cation and anion, the number density experiences a significant increase and fluctuations, as seen in Fig. 22. Increasing pressure causes more variations in the number density of CO_2, compared to the ions (Aghaie et al., 2020b).

Fig. 23 displays the variation of mean square displacement (MSD) with time for $[BF_4]^-$, $[Bmim]^+$, and CO_2 in $[Bmim][BF_4]$/CO_2 system at two different temperatures of 300 and 313 K. There is almost a linear relationship between the MSD and time, implying the importance of the normal molecular diffusion over the absorption process of CO_2 in IL. As the cations have higher diffusivity than anions, greater MSD values are attained for

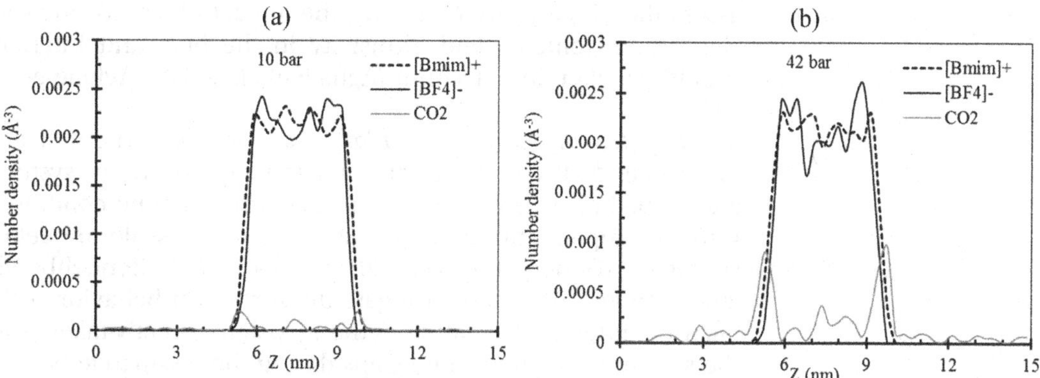

Fig. 22 Influence of pressure on the number density distribution for $[BF_4]^-$, $[Bmim]^+$, and CO_2, in the mixture of CO_2/IL: (A) $P=10$ bar and (B) $P=42$ bar (Aghaie et al., 2020b).

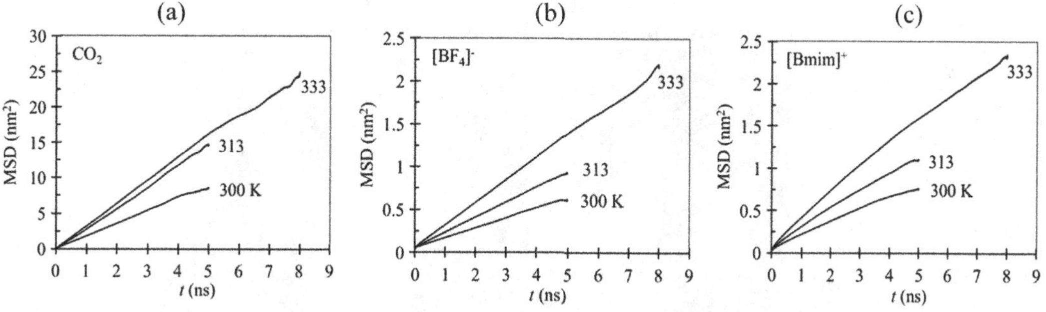

Fig. 23 Effect of time and temperature on the MSD of (A) CO_2, (B) anion, and (C) cation in the bulk mixture of [Bmim] $[BF_4]/CO_2$. Modified after Modified after Aghaie, M., Rezaei, N., & Zendehboudi, S. (2020b). New insights into bulk and interface properties of [Bmim][Ac]/[Bmim][BF4] ionic liquid/CO2 systems—Molecular dynamics simulation approach. *Journal of Molecular Liquids, 317*, 113497.

the cation of $[Bmim]^+$, compared to the anion of $[BF_4]^-$, as illustrated in Fig. 23. It is also found that CO_2 has higher magnitudes of diffusivity, compared to the ions. Thus, it can be concluded that CO_2 will travel faster in the system (Aghaie et al., 2020b).

Summarizing this case study, the MD simulation can be a proper modeling approach for costly and time-consuming laboratory works to obtain the properties of ILs and mixtures. MD simulation can precisely determine the IL's thermophysical characteristics, such as viscosity, enthalpy, entropy, thermal conductivity, and heat capacity. Also, it can forecast the transport parameters (e.g., interaction energies and diffusion coefficient). The force-field plays a vital role in finding the IL properties and enables the MD simulation tool to design the desired ILs. To assess the CO_2 absorption ability, the impact of structuring species, water, toluene, and diffusivity in the bulk and interface should be taken into account (Aghaie et al., 2020a; Aghaie et al., 2020b).

Oil/water/surfactant molecular behavior. The structural and interfacial features of water/surfactant/hydrocarbon systems are studied at different process and thermodynamic conditions through implementing meso- and molecular-scale modeling strategies (Goodarzi et al., 2019; Goodarzi & Zendehboudi, 2019). An approach to investigate the surfactant behavior at the oil/water interface is analysis of the snapshots directly taken from the system at different time steps during the dissipative particle dynamics (DPD) simulation. The snapshots of the oil/surfactant/water molecules for various water/oil ratios, salinty percentages, and surfactants concentrations are given in Fig. 24. The

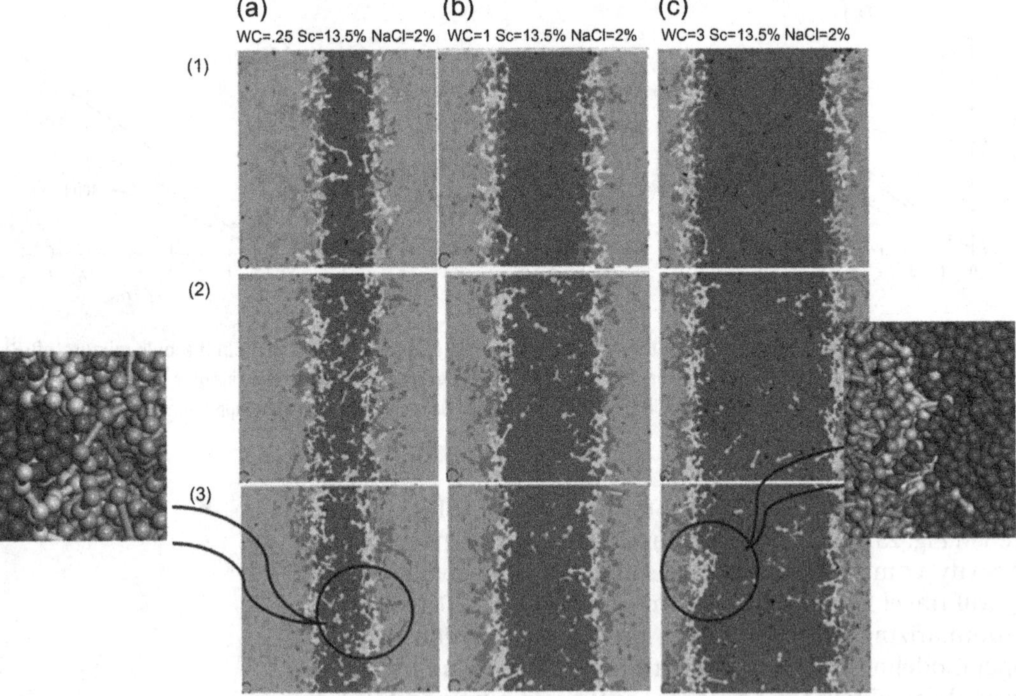

Fig. 24 Pictures of the oil/surfactant/water case simulation at various WC ratios, surfactant concentrations (SC), and salinities: (A) NaCl = 2, Sc = 13.5, WC = 0.25; (B) NaCl = 2, Sc = 13.5, WC = 1; and (C) NaCl = 2, Sc = 13.5, WC = 3 at three various stages including (1) primary configuration, (2) half of simulation time, and (3) final structure. Water beads are indicated in blue, salt beads in yellow, surfactant tail in green, surfactant head in pink, and oil beads in red. Modified after Goodarzi, F., & Zendehboudi, S. (2019). Effects of salt and surfactant on interfacial characteristics of water/oil systems: Molecular dynamic simulations and dissipative particle dynamics. *Industrial & Engineering Chemistry Research 58 (20)*, 8817–8834.

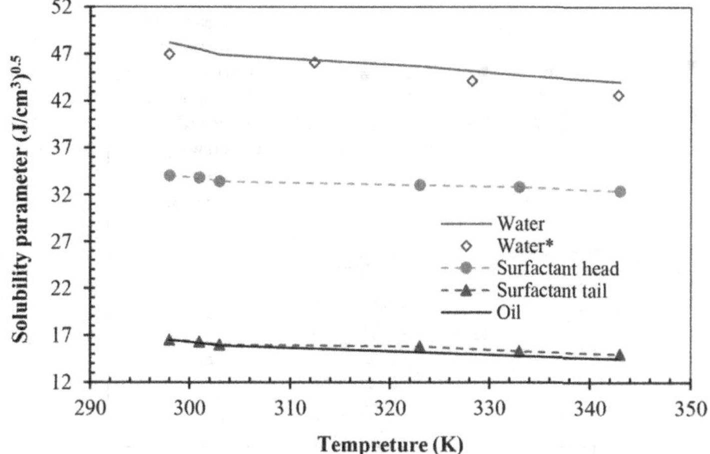

Fig. 25 Solubility parameter versus temperature for molecules existing in the system. Modified after Goodarzi, F., & Zendehboudi, S. (2019). Effects of salt and surfactant on interfacial characteristics of water/oil systems: Molecular dynamic simulations and dissipative particle dynamics. *Industrial & Engineering Chemistry Research 58 (20)*, 8817–8834.

surfactant heads and oil molecules are nearly miscible. Consequently, the surfactant head beads will bend and move toward the oil phase. The same phenomenon appears with the water molecules and surfactant head group. As a polar molecule, water has the tendency to attract the surfactant head group through dipole-dipole bonding. The NaCl molecules are also dispersed into the water phase where the electrostatic forces play an important role (Goodarzi & Zendehboudi, 2019).

The thermodynamic behaviors of the molecules can be determined directly from the bead chemical characteristics. Hence, the energy of mixing and corresponding chi parameter will be calculated by outlining the molecular structure and the type of forcefield. Fig. 25 reports the solubility parameter values of different components involved in the system as a function of temperature (Goodarzi & Zendehboudi, 2019).

One of the crucial roles of surfactants is to lower the interfacial tension (IFT) in water/oil systems. The interfacial characteristics of water and oil mixtures are the key parameters to be considered in different industrial operations such as chemical enhanced oil recovery (EOR) and pharmaceutical products (Case et al., 2017). To assess the overall effectiveness of surfactants, their capability to decrease IFT and the adsorption tendency of the head and tail groups is investigated. Fig. 26 illustrates the IFT variation as a function of time for different oil and water systems. The most nonpolar molecule in the system is dodecane, which has the highest solubility parameter difference with water. Therefore, a mixture of the two phases exhibits high interfacial energies at the interface. At the same time, benzene as a cyclic hydrocarbon with

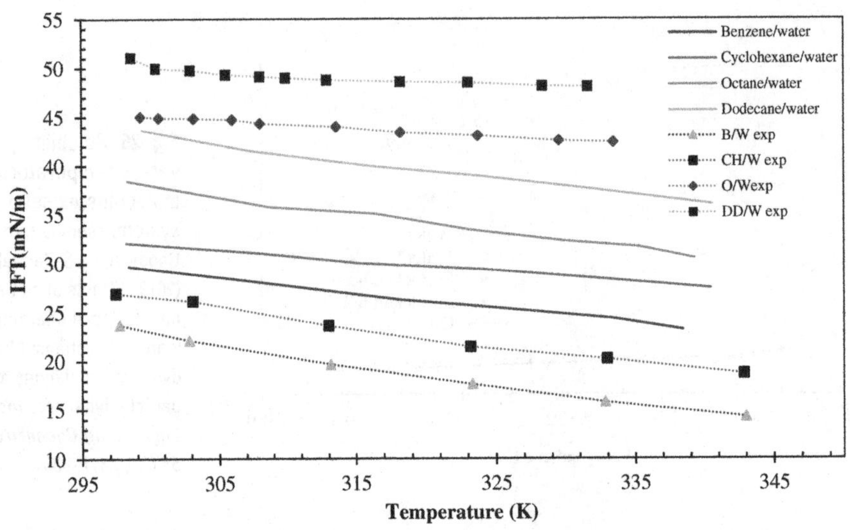

Fig. 26 Variations of interfacial tension with temperature for various oil/water systems based on the experimental data and DPD results. Modified after Goodarzi, F., & Zendehboudi, S. (2019). Effects of salt and surfactant on interfacial characteristics of water/oil systems: Molecular dynamic simulations and dissipative particle dynamics. *Industrial & Engineering Chemistry Research 58 (20)*, 8817–8834.

double bonds shows more polarity, highlighting its similarity to water in terms of solubility compared to other compounds. Based on Fig. 26, the temperature is a main factor affecting the IFT of oil/water molecules, regardless of the structure of the oil (Goodarzi & Zendehboudi, 2019).

The impact of surfactant concentration on the IFT values is investigated, as seen in Fig. 27. As it is clear from Fig. 27, the IFT experiences a noticeable reduction for both benzene and octane hydrocarbon/surfactant/water systems when the number of surfactant molecules at the interface increases. Introducing more surfactants to the system causes a decrease in the surface energy between water and oil beads at the interface due to the created dipole-dipole bonding between the linear part of the surfactant and the nonpolar oil molecules as well as the establishment of hydrogen and polar bonds between the surfactant and water molecules. The energy proceeds to decrease until the surfactant molecules saturate the interface. Addition of more surfactants does not affect the magnitude of IFT anymore, after the saturation condition is maintained (Goodarzi & Zendehboudi, 2019).

To examine the effect of salt in the oil/surfacant/water systems, the IFT values are calculated at different surfactant concentrations and water-cuts for different salinity percentages, as

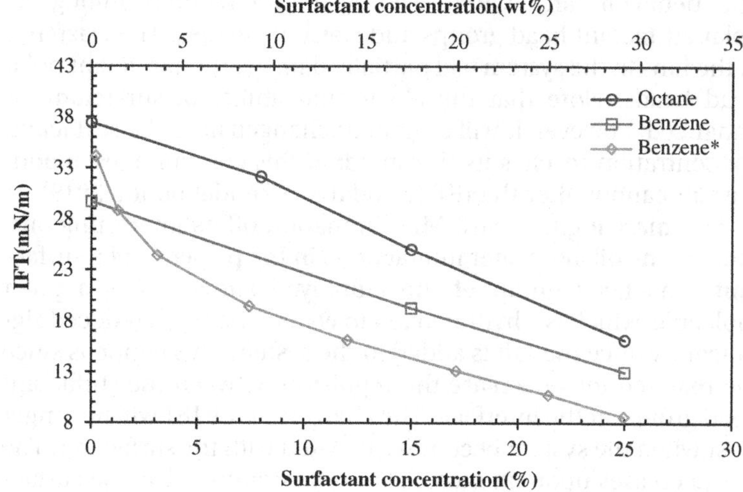

Fig. 27 Interfacial tension against surfactant concentration for different hydrocarbons including octane and benzene. Modified after Goodarzi, F., & Zendehboudi, S. (2019). Effects of salt and surfactant on interfacial characteristics of water/oil systems: Molecular dynamic simulations and dissipative particle dynamics. *Industrial & Engineering Chemistry Research 58 (20)*, 8817–8834.

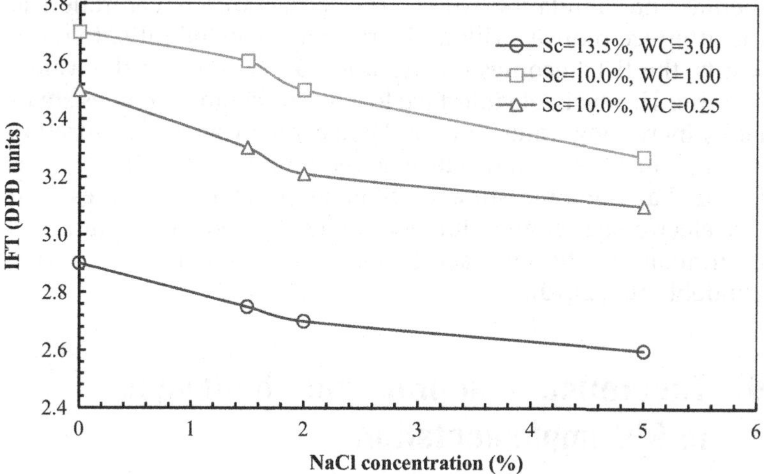

Fig. 28 Influence of salinity on the IFT extent at various water/oil ratios and surfactant concentrations in the system. Modified after Goodarzi, F., & Zendehboudi, S. (2019). Effects of salt and surfactant on interfacial characteristics of water/oil systems: Molecular dynamic simulations and dissipative particle dynamics. *Industrial & Engineering Chemistry Research 58 (20)*, 8817–8834.

illustrated in Fig. 28. As it is evident from the data, the IFT falls substantially upon addition of even a small amount of salt to the system. The IFT reduction is a function of the number of surfactant molecules absorbed at the water/oil interface. Adding salt to the mixture reduces the electrostatic repulsion among surfactant molecules. Also, the presence of the electrolyte would lead to reducing the electric double layer around the surfactant head; consequently, the surfactant molecules tend to migrate at the interface, and the IFT reduces as well. It is worth noting that IFT remains almost constant with adding more salt to the system.

This behavior can be addressed by the interactions among the polar surfactant head groups and NaCl molecules. The existence of the ions in the system will partially damage the surfactant polar head beads. Note that the absorption ability of surfactants is enhanced; however, it will remain unchanged after the surfactant concentration reaches its threshold; in this condition, even adding salt cannot alter the IFT (Goodarzi & Zendehboudi, 2019).

Summarizing this part, MD simulation offers interesting outputs on the oil and water interactions in the presence of a surfactant. The head group of the employed surfactant is a polar molecule, which is a hydrated ion in water. The IFT is reduced significantly once the salt is added to the system; this happens since the released ions decrease the repulsion between the surfactant head groups at the interface. The changes of the IFT will no longer exist when the system becomes saturated with the surfactant. The IFT decreases upon an increase in temperature. Also, the surfactant is stretched at higher temperatures based on the radius of gyration. This trend has been previously confirmed with experimental and simulation works. The surfactant concentration at the interface reduces with an increasing water/oil ratio. In other words, the IFT becomes greater upon an increase in the water/oil ratio. The available interface for the surfactant becomes smaller by increasing water content. Hence, high water concentration at the interface would cause a reduction in the IFT. It was found that nonionic surfactants generate stable mixtures, and the electrolytes cannot impact them; consequently, they will accumulate at the interface (Goodarzi et al., 2019; Goodarzi & Zendehboudi, 2019).

9 Theoretical and practical challenges in MD implementation

Researchers have been studying physical sciences using experimental methods and modeling/theoretical approaches for many years. Despite the acceptable performance of these methods to investigate and evaluate the phenomena, there are certain limitations associated with using them. Generally, modeling/theoretical methods are based on nonlinear and complex mathematical relations. Hence, it is necessary to simplify assumptions to reduce the involved complexities while solving them. Therefore, these methods are valid for cases sufficiently simplified to a level at which the governing equations of the mathematical models can be solved and cannot provide reliable results for complex systems. Experimental research investigations facilitate observation and measurements to understand the system's behaviors in qualitative and/or quantitative manners. However, there are numerous justifications and phenomena to explain the results and processes. Hence, modeling/simulation methods can cover a part of limitations of experimental methods and assist in exploring the unknown aspects.

From the mid-20th century, the advent of high-speed computers and their application in science and engineering disciplines introduced "computer experiments" to predict experimental results using modeling/theoretical approaches. The high-speed algorithmic computations in these computers enable researchers to simulate experimental runs in actual scale and various conditions. In addition, it is possible to consider more complexities in the system and decrease simplifying assumptions (Seyyedattar et al., 2019; Springer, 2006; Zendehboudi et al., 2018). For instance, we can investigate weather forecasts and chemical reaction kinetics inside the stars using experimental methods. Conducting experiments such as an explosion or flight-related tests is feasible but can be too dangerous. Moreover, considering all conditions (e.g., high pressure and temperature) during an experimental test is technically challenging, and we need to provide expensive materials/equipment, resulting in high research expenditures. In addition, the results of experimental techniques are generally obtained at the macroscopic scale and cannot present any information about the interaction forces in molecules. Hence, MD simulation methods can remove the limitations and be used to replace and/or explain experiments (Seyyedattar et al., 2019; Springer, 2006; Zendehboudi et al., 2018).

In fact, it is important to choose the suitable computing hardware for running the MD simulation. GPUs appear to be an attractive choice since the simulation process is fast with relatively low costs. On the other hand, the supercomputers and/or central processing units (CPUs) can handle fast simulation runs. Another important aspect is to use a proper force-field. OPLS, CHARMM, AMBER, and Universal are the most common force-fields (Hess et al., 1997; Hockney et al., 1974; Ning et al., 2015; Seyyedattar et al., 2019; Springer, 2006). They have similar functional forms, but there are some drawbacks and advantages with each one. For example, CHARMM force-field is appropriate for drug-like ligands, lipids, and proteins where its parameters are extensively optimized and validated for such systems (Hockney et al., 1974; Klapproth et al., 2003; Shpakov et al., 1998). There are a variety of software packages that can be used for MD implementation, including Materials Studio, CHARMM, AMBER, NAMD, and GROMACS (Case et al., 2017; Eastman et al., 2017; Ghamartale et al.,

2020; Ogienko et al., 2006; Phillips et al., 2005). Multiple force-fields can be incorporated in most of advanced MD simulators. The simulation packages generally follow the similar computational approach; but they show different performances in mapping to various hardwares and have different features such as support for coarse-grained simulations, temperature and pressure control schemes, and enhanced sampling methods (Seyyedattar et al., 2019; Springer, 2006).

Another important stage in MD simulation is preparation of the molecular system by adding target solvents, creating missing atoms, and finding suitable values for force-field parameters (Seyyedattar et al., 2019; Zendehboudi et al., 2018).

The main challenges in the MD simulation process are selection of type of simulation and analysis of MD simulation results. A significant amount of data including positions and velocities of many atoms (e.g., 150,000) over several time steps might be produced. Thus, it would be very challenging to identify the most relevant and important data (e.g., the interaction energy) among huge generated data/information (Seyyedattar et al., 2019; Zendehboudi et al., 2018).

In general, the analysis process of MD outputs needs a careful integration of quantitative analysis and visual analysis using appropriate simulation packages. In some cases, customized analysis programs or scripts should be coded to perform specific tasks in the analysis process (Seyyedattar et al., 2019; Zendehboudi et al., 2018).

Even minor changes in initial simulation conditions may result in extensively different MD simulation trajectories. Therefore, several MD simulation runs should be performed at each condition.

There are some limitations with MD simulations. First, the force-fields are inherently approximate (Lindorff-Larsen et al., 2012; Seyyedattar et al., 2019). Second, during typical MD simulation runs, formation and/or breakage of covalent bonds do not occur (Goh et al., 2014). Third, accurate experimental data and good molecular models are required to attain accurate simulation findings. Thus, the design of an MD simulation work is strongly affected by the availability of experimental data and structures.

It is not an easy task to utilize several GPUs to speed up a single simulation run. It should be noted that the system design has a key impact on simulation timescale requirements, irrespective of the type of simulation method (Hertig et al., 2016; Seyyedattar et al., 2019).

Furthermore, there are various enhanced sampling strategies that enable MD simulations to capture longer-timescale events;

however, no single technique is available to solve timescale limitations. In fact, different sampling approaches are suitable in different condition/situations (Bernardi et al., 2015; Seyyedattar et al., 2019) .The available techniques can be tuned to handle long timescales, but their accuracy might be reduced (de Oliveira et al., 2006; Seyyedattar et al., 2019; Zendehboudi et al., 2018).

To conduct accurate and reliable MD results, one should design appropriate simulation structures, including careful preparation of the simulation system, collection of relevant literature information and experimental data, identifying equilibrium conditions, validation phase, and adequate understanding of theoretical aspects, and molecular structures, and interactions.

10 Current status and future prospects of MD technique

The MD simulation technique is a powerful tool in modern computational physics for investigating and analyzing the processes at the atomic level that cannot be studied using experimental tests. Molecular structure, molecular/atoms movement, interaction force, physical properties, and process performance are the main results of MD simulation runs that provide information to calculate/estimate the properties and features of target processes at the actual scale. In recent years, MD simulation methods have been used as a valuable tool to simulate the chemical/petroleum processes. It is worth mentioning that considering all elements/conditions in the trajectory box of MD simulation is challenging; hence, the MD simulation is limited to simpler processes. Improvements in computational power and innovative modeling approaches assist in overcoming these problems and pave the way for future simulation studies. For instance, it will be possible to consider all fluid components and surfaces in separation membranes while simulating transport phenomena in porous media to achieve reliable results. In addition, increasing the simulation time/scale is another aspect that has been considered in recent years to increase the precision and strength of MD simulation approach in the future.

A review of the literature reveals that MD simulation is a powerful research tool to investigate the processes in the chemical, biochemical, and petroleum industries from upstream to downstream. This technique enables us to assess reaction conversion, separation efficiency, and thermodynamic equilibrium; to investigate the reaction and interfacial phenomena mechanisms; and to design new strategies, chemicals, and processes for performance improvement, considering process and thermodynamic conditions, fluid characteristics, and surface properties. A variety of the research topics addressed by the MD simulation shows that this modeling approach has an excellent potential to simulate the various processes in detail and improve theoretical and practical knowledge toward performance optimization.

Despite promising MD simulation results in chemical/petroleum engineering, it is necessary to design new methodologies and workflows to define real fluids, surfaces, conditions, and multiphase systems considering the low cost and time-efficient manners. Although the structure of actual molecules can be defined in most cases, considering all atoms during MD simulation for a complex system is commonly challenging. In addition, limitations in size and time prevent us from obtaining accurate outputs despite considerable advances in computational resources and efficient algorithms. The importance of mixture definition based on EOS results during MD simulation is an ideal example of the sensitivity of simulation results to each component in the system. At the same time, most research works attempt to simplify the fluids and components to reduce the time and cost of simulation; they try to introduce new and effective procedures for this important issue. However, further research is required to develop useful guidelines to simplify the processes.

The application of MD simulation in different sciences has attracted considerable attention, especially for the simulation of fluids, solids, and interfacial properties. This simulation tool enables researchers/scholars to calculate or estimate valuable information related to the system, such as influential parameters, complex phenomena at atomic/molecular scales, and physical/chemical interaction forces. In addition, this method can be employed under a wide range of temperatures and pressures for simulating a variety of processes such as catalytic reactions and nanomaterial synthesis. It is possible to design new and expensive materials (e.g., catalysts) and investigate their performance in different reactions using MD simulation techniques; this tool also helps us obtain some preliminary information before conducting experimental tests. In addition, MD simulation is very beneficial to study the formation of solids such as coke, asphaltene, wax, and hydrate, which are common issues in the chemical and petroleum industries. It is worth noting that the combination of MD simulation with advanced numerical and analytical modeling approaches provides a new pathway to study the processes without the need to oversimplify assumptions and identify their behaviors at the atomic scale.

References

Abraham, M. J., et al. (2015). GROMACS: High performance molecular simulations through multi-level parallelism from laptops to supercomputers. *SoftwareX, 1*, 19–25.

Aghaie, M., Rezaei, N., & Zendehboudi, S. (2020a). Effect of water on molecular and transport phenomena behaviors of [Bmim][Ac]/water/CO2, using molecular dynamics strategy. *The Journal of Physical Chemistry B, 124*(34), 7368–7378.

Aghaie, M., Rezaei, N., & Zendehboudi, S. (2020b). New insights into bulk and interface properties of [Bmim][Ac]/[Bmim][BF4] ionic liquid/CO2 systems—Molecular dynamics simulation approach. *Journal of Molecular Liquids, 317*, 113497.

Allen, M. P. (2004). Introduction of molecular dynamics simulation. In N. Attig, K. Binder, H. Grubmuller, & K. Kremer (Eds.), *NIC Series: Vol. 23. Computational soft matter: From synthetic polymers to proteins, Lecture notes* (pp. 1–28). Julich: John von Neumann Institute for Computing.

Allen, M. P., & Tildesley, D. J. (2017). *Computer simulation of liquids* (2nd ed.). Oxford: Oxford University Press.

Amjad-Iranagh, S., et al. (2015). Asphaltene solubility in common solvents: A molecular dynamics simulation study. *The Canadian Journal of Chemical Engineering, 93*(12), 2222–2232.

Berendsen, H. J., et al. (1984). Molecular dynamics with coupling to an external bath. *The Journal of Chemical Physics, 81*(8), 3684–3690.

Bernardi, R. C., Melo, M. C. R., & Schulten, K. (2015). Enhanced sampling techniques in molecular dynamics simulations of biological systems. *Biochimica et Biophysica Acta, 1850*, 872–877.

Bondor, P. L., Hite, J. R., & Avasthi, S. (2005). Planning EOR projects in offshore oil fields. In *SPE Latin American and Caribbean petroleum engineering conference, Rio de Janeiro, Brazil.*

Brooks, B. R., Bruccoleri, R. E., Olafson, B. D., States, D. J., Swaminathan, S., & Karplus, M. (1983). CHARMM: A program for macromolecular energy, minimization, and dynamics calculations. *Journal of Computational Chemistry, 4*(2), 187–217.

Bussi, G., Donadio, D., & Parrinello, M. (2007). Canonical sampling through velocity rescaling. *The Journal of Chemical Physics, 126*(1), 014101.

Case, D. A., Cerutti, D. S., Cheatham, T. E., III, Darden, T. A., Duke, R. E., Giese, T. J., Gohlke, H., Goetz, A. W., Greene, D., Homeyer, N., et al. (2017). *AMBER 2017*. San Francisco: University of California.

Chapman, J. B. J. (2018). *Improving the functional control of ferroelectrics using insights from atomistic modelling*. Doctoral thesis (Eng.D), UCL (University College London).

Chellehbari, Y. M., Amin, J. S., & Zendehboudi, S. (2021). How does a microfluidic platform tune the morphological properties of polybenzimidazole nanoparticles? *Journal of Physical Chemistry B* (in press).

Cuendet, M. (2008). *Molecular dynamics simulation: A short introduction.*

Darden, T., York, D., & Pedersen, L. (1993). Particle mesh Ewald: An N·log (N) method for Ewald sums in large systems. *The Journal of Chemical Physics, 98*(12), 10089–10092.

de Oliveira, C. A. F., Hamelberg, D., & McCammon, J. A. (2006). On the application of accelerated molecular dynamics to liquid water simulations. *The Journal of Physical Chemistry. B, 110*, 22695–22701.

Eastman, P., Swails, J., Chodera, J. D., McGibbon, R. T., Zhao, Y., Beauchamp, K. A., Wang, L. P., Simmonett, A. C., Harrigan, M. P., Stern, C. D., et al. (2017). OpenMM 7: Rapid development of high performance algorithms for molecular dynamics. *PLoS Computational Biology, 13*, e1005659.

Ebrahimi, D. (2014). *Multiscale modeling of clay-water systems.* Massachusetts Institute of Technology.

Ebro, H., Kim, Y. M., & Kim, J. H. (2013). Molecular dynamics simulations in membrane-based water treatment processes: A systematic overview. *Journal of Membrane Science, 438*, 112–125.

Essmann, U., et al. (1995). A smooth particle mesh Ewald method. *The Journal of Chemical Physics, 103*(19), 8577–8593.

Evans, D. J., & Holian, B. L. (1985). The nose–hoover thermostat. *The Journal of Chemical Physics, 83*(8), 4069–4074.

Fang, T., et al. (2018). Study on the asphaltene precipitation in CO2 flooding: A perspective from molecular dynamics simulation. *Industrial & Engineering Chemistry Research, 57*(3), 1071–1077.

Fu, C.-F., & Tian, S. X. (2011). A comparative study for molecular dynamics simulations of liquid benzene. *Journal of Chemical Theory and Computation, 7*(7), 2240–2252.

Gao, F., et al. (2014). Molecular dynamics simulation: The behavior of asphaltene in crude oil and at the oil/water interface. *Energy & Fuels, 28*(12), 7368–7376.

Gear, C. W. (1971). *Numerical initial value problems in ordinary differential equations.* Prentice Hall PTR.

Geng, C.-Y., et al. (2010). Molecular dynamics simulation on the decomposition of type SII hydrogen hydrate and the performance of tetrahydrofuran as a stabiliser. *Molecular Simulation, 36*(6), 474–483.

Ghamartale, A., Zendehboudi, S., & Rezaei, N. (2020). New molecular insights into aggregation of pure and mixed asphaltenes in the presence of n-Octylphenol inhibitor. *Energy & Fuels.*

Ghamartale, A., Zendehboudi, S., Rezaei, N., & Chatzis, I. (2021). Effects of inhibitor concentration and thermodynamic conditions on n-octylphenol-asphaltene molecular behaviours. *Journal of Molecular Liquids, 340*, 116897.

Gibbs, J. W. (1902). *Elementary principles in statistical mechanics.* Charles Scribner's Sons.

Goh, G. B., Hulbert, B. S., Zhou, H., & Brooks, C. L., 3rd. (2014). Constant pH molecular dynamics of proteins in explicit solvent with proton tautomerism. *Proteins, 82*, 1319–1331.

Goodarzi, F., Kondori, J., Rezaei, N., & Zendehboudi, S. (2019). Meso-and molecular-scale modeling to provide new insights into interfacial and structural properties of hydrocarbon/water/surfactant systems. *Journal of Molecular Liquids, 295*, 111357.

Goodarzi, F., & Zendehboudi, S. (2019). Effects of salt and surfactant on interfacial characteristics of water/oil systems: Molecular dynamic simulations and dissipative particle dynamics. *Industrial & Engineering Chemistry Research, 58*(20), 8817–8834.

Goual, L., & Sedghi, M. (2015). Role of ion-pair interactions on asphaltene stabilization by alkylbenzenesulfonic acids. *Journal of Colloid and Interface Science, 440*, 23–31.

Goual, L., et al. (2014). Asphaltene aggregation and impact of alkylphenols. *Langmuir, 30*(19), 5394–5403.

GROMACS. *Steps to perform a simulation.* [Online]. Available http://www.gromacs.org/Documentation/How-tos/Steps_to_Perform_a_Simulation. Accessed 13 October 2021.

Haile, J. M. (1992). *Molecular dynamics simulation—Elementary methods.* New York: John Wiley and Sons, Inc.

Headen, T. F., & Boek, E. S. (2011a). Potential of mean force calculation from molecular dynamics simulation of asphaltene molecules on a calcite surface. *Energy & Fuels, 25*(2), 499–502.

Headen, T. F., & Boek, E. S. (2011b). Molecular dynamics simulations of asphaltene aggregation in supercritical carbon dioxide with and without limonene. *Energy & Fuels, 25*(2), 503–508.

Headen, T. F., Boek, E. S., & Skipper, N. T. (2009). Evidence for asphaltene nanoaggregation in toluene and heptane from molecular dynamics simulations. *Energy & Fuels, 23*(3), 1220–1229.

Headen, T. F., & Hoepfner, M. P. (2019). Predicting asphaltene aggregate structure from molecular dynamics simulation: Comparison to neutron total scattering data. *Energy & Fuels, 33*(5), 3787–3795.

Headen, T., et al. (2017). Simulation of asphaltene aggregation through molecular dynamics: Insights and limitations. *Energy & Fuels, 31*(2), 1108–1125.

Hertig, S., Latorraca, N. R., & Dror, R. O. (2016). Revealing atomic-level mechanisms of protein allostery with molecular dynamics simulations. *PLoS Computational Biology, 12*, e1004746.

Hess, B., et al. (1997). LINCS: A linear constraint solver for molecular simulations. *Journal of Computational Chemistry, 18*(12), 1463–1472.

Hockney, R. W., Goel, S., & Eastwood, J. (1974). Quiet high-resolution computer models of a plasma. *Journal of Computational Physics, 14*(2), 148–158.

Hoover, W. G. (1985). Canonical dynamics: Equilibrium phase-space distributions. *Physical Review A, 31*(3), 1695.

Ippoliti, E. (2011). *What is molecular dynamics?*. German Research School for Simulation Sciences.

Jian, C., Tang, T., & Bhattacharjee, S. (2013). Probing the effect of side-chain length on the aggregation of a model asphaltene using molecular dynamics simulations. *Energy & Fuels, 27*(4), 2057–2067.

Jian, C., et al. (2018). A molecular dynamics study of the effect of asphaltenes on toluene/water interfacial tension: Surfactant or solute? *Energy & Fuels, 32*(3), 3225–3231.

Karplus, M., & McCammon, J. A. (2002). Molecular dynamics simulations of biomolecules. *Nature Structural Biology, 9*(9), 646–652.

Khalaf, M. H., & Mansoori, G. A. (2018). A new insight into asphaltenes aggregation onset at molecular level in crude oil (an MD simulation study). *Journal of Petroleum Science and Engineering, 162*, 244–250.

Khalaf, M. H., & Mansoori, G. A. (2019). Asphaltenes aggregation during petroleum reservoir air and nitrogen flooding. *Journal of Petroleum Science and Engineering, 173*, 1121–1129.

Klapproth, A., et al. (2003). Structural studies of gas hydrates. *Canadian Journal of Physics, 81*(1–2), 503–518.

Kondori, J., James, L., & Zendehboudi, S. (2020). Molecular scale modeling approach to evaluate stability and dissociation of methane and carbon dioxide hydrates. *Journal of Molecular Liquids, 297*, 111503.

Kondori, J., Zendehboudi, S., & Hossain, M. E. (2017). A review on simulation of methane production from gas hydrate reservoirs: Molecular dynamics prospective. *Journal of Petroleum Science and Engineering, 159*, 754–772.

Kondori, J., Zendehboudi, S., & James, L. (2019a). New insights into methane hydrate dissociation: Utilization of molecular dynamics strategy. *Fuel, 249*, 264–276.

Kondori, J., Zendehboudi, S., & James, L. (2019b). Molecular dynamic simulations to evaluate dissociation of hydrate structure II in the presence of inhibitors: A mechanistic study. *Chemical Engineering Research and Design, 149*, 81–94.

Kuznicki, T., Masliyah, J. H., & Bhattacharjee, S. (2008). Molecular dynamics study of model molecules resembling asphaltene-like structures in aqueous organic solvent systems. *Energy & Fuels, 22*(4), 2379–2389.

Lennard-Jones, J. E. (1924). On the determination of molecular fields. II. From the equation of state of a gas. *Proceedings of the Royal Society of London, 106*(738), 463–477.

Li, K., et al. (2018). Dissociation mechanism of propane hydrate with methanol additive: A molecular dynamics simulation. *Computational and Theoretical Chemistry, 1123*, 79–86.

Lindahl, A., Hess, S. V. D., & Spoel, V. D. (2019). *GROMACS 2019.3 Source code.*

Lindorff-Larsen, K., Maragakis, P., Piana, S., Eastwood, M. P., Dror, R. O., & Shaw, D. E. (2012). Systematic validation of protein force fields against experimental data. *PLoS One, 7*, e32131.

Liou, M.-S. (1995). An extended Lagrangian method. *Journal of Computational Physics, 118*(2), 294–309.

Lowry, E., Sedghi, M., & Goual, L. (2016). Novel dispersant for formation damage prevention in CO2: A molecular dynamics study. *Energy & Fuels, 30*(9), 7187–7195.

Lowry, E., Sedghi, M., & Goual, L. (2017). Polymers for asphaltene dispersion: Interaction mechanisms and molecular design considerations. *Journal of Molecular Liquids, 230*, 589–599.

Lv, G., et al. (2017). The properties of asphaltene at the oil-water interface: A molecular dynamics simulation. *Colloids and Surfaces A: Physicochemical and Engineering Aspects, 515*, 34–40.

Mcquarrie, D. (1965). *Statistical mechanics.* 1305 Walt Whitman Rd, STE 300, Melville, NY 11747–4501 USA: Amer Inst Physics.

Mehana, M., Fahes, M., & Huang, L. (2019). Asphaltene aggregation in oil and gas mixtures: Insights from molecular simulation. *Energy & Fuels, 33*(6), 4721–4730.

Meller, J. (2001). Molecular dynamics. In *Encyclopedia of life sciences* Max-Born-Institut (MBI) im Forschungsverbund Berlin e.V.

Mikami, Y., et al. (2013). Molecular dynamics simulations of asphaltenes at the oil–water interface: From nanoaggregation to thin-film formation. *Energy & Fuels, 27*(4), 1838–1845.

Mohammed, S., & Gadikota, G. (2019a). The role of calcite and silica interfaces on the aggregation and transport of asphaltenes in confinement. *Journal of Molecular Liquids, 274*, 792–800.

Mohammed, S., & Gadikota, G. (2019b). The influence of CO2 on the structure of confined asphaltenes in calcite nanopores. *Fuel, 236*, 769–777.

Mohammed, S., & Mansoori, G. A. (2018a). Effect of CO2 on the interfacial and transport properties of water/binary and asphaltenic oils: Insights from molecular dynamics. *Energy & Fuels, 32*(4), 5409–5417.

Mohammed, S., & Mansoori, G. A. (2018b). Molecular insights on the interfacial and transport properties of supercritical CO2/brine/crude oil ternary system. *Journal of Molecular Liquids, 263*, 268–273.

"Molecular dynamics—Statistical ensembles," [Online]. Available: http://www.quimica.urv.es/~bo/MOLMOD/General/Dynamics/Ensembles.html. [Accessed 10 December 2021].

Naeem, R. (2017). *Lennard-Jones potential.* The LibreTexts. [Online]. Available https://chem.libretexts.org/. Accessed 05 December 2021.

Ning, F., et al. (2015). Compressibility, thermal expansion coefficient and heat capacity of CH 4 and CO2 hydrate mixtures using molecular dynamics simulations. *Physical Chemistry Chemical Physics, 17*(4), 2869–2883.

Nosé, S. (1984). A molecular dynamics method for simulations in the canonical ensemble. *Molecular Physics, 52*(2), 255–268.

Ogienko, A. G., et al. (2006). Gas hydrates of argon and methane synthesized at high pressures: Composition, thermal expansion, and self-preservation. *The Journal of Physical Chemistry B, 110*(6), 2840–2846.

Pacheco-Sánchez, J., Zaragoza, I., & Martínez-Magadán, J. (2003). Asphaltene aggregation under vacuum at different temperatures by molecular dynamics. *Energy & Fuels, 17*(5), 1346–1355.

Pacheco-Sánchez, J., Zaragoza, I., & Martínez-Magadán, J. (2004). Preliminary study of the effect of pressure on asphaltene disassociation by molecular dynamics. *Petroleum Science and Technology, 22*(7–8), 927–942.

Panizon, E., Bochicchio, D., Monticelli, L., & Rossi, G. (2015). MARTINI coarse-grained models of polyethylene and polypropylene. *The Journal of Physical Chemistry. B, 119*, 8209–8216.

Parrinello, M., & Rahman, A. (1982). Strain fluctuations and elastic constants. *The Journal of Chemical Physics, 76*(5), 2662–2666.

Phillips, J. C., Braun, R., Wang, W., Gumbart, J., Tajkhorshid, E., Villa, E., Chipot, C., Skeel, R. D., Kalé, L., & Schulten, K. (2005). Scalable molecular dynamics with NAMD. *Journal of Computational Chemistry, 26*, 1781–1802.

S.L. Roux and V. Petkov, "Model Box Periodic Boundary Conditions—P.B.C.," Central Michigan University, [Online]. Available: http://isaacs.sourceforge.net/index.html. [Accessed 10 November 2021].

Sakemoto, R., et al. (2010). Clathrate hydrate crystal growth at the seawater/hydrophobic–guest–liquid Interface. *Crystal Growth & Design, 10*(3), 1296–1300.

Schäfer, H., Mark, A. E., & van Gunsteren, W. F. (2000). Absolute entropies from molecular dynamics simulation trajectories. *The Journal of Chemical Physics, 113*, 7809–7817.

Sedghi, M., & Goual, L. (2016). Molecular dynamics simulations of asphaltene dispersion by limonene and PVAc polymer during CO 2 flooding. In *SPE international conference and exhibition on formation damage control*Society of Petroleum Engineers.

Sedghi, M., et al. (2013). Effect of asphaltene structure on association and aggregation using molecular dynamics. *The Journal of Physical Chemistry B, 117*(18), 5765–5776.

Seyyedattar, M., Zendehboudi, S., & Butt, S. (2019). Molecular dynamics simulations in reservoir analysis of offshore petroleum reserves: A systematic review of theory and applications. *Earth-Science Reviews, 192*, 194–213.

Shimada, J., Kaneko, H., & Takada, T. (1994). Performance of fast multipole methods for calculating electrostatic interactions in biomacromolecular simulations. *Journal of Computational Chemistry, 15*(1), 28–43.

Shpakov, V., et al. (1998). Elastic module calculation and instability in structure I methane clathrate hydrate. *Chemical Physics Letters, 282*, 107–114.

Silva, H. S., et al. (2016). Molecular dynamics study of nanoaggregation in asphaltene mixtures: Effects of the N, O, and S heteroatoms. *Energy & Fuels, 30*(7), 5656–5664.

Smirnov, G. S., & Stegailov, V. V. (2012). Melting and superheating of sI methane hydrate: Molecular dynamics study. *Journal of Chemical Physics, 136*(4).

Sodero, A. C., et al. (2016). Investigation of the effect of sulfur heteroatom on asphaltene aggregation. *Energy & Fuels, 30*(6), 4758–4766.

Song, S., et al. (2018). Molecular dynamics study on aggregating behavior of asphaltene and resin in emulsified heavy oil droplets with sodium dodecyl sulfate. *Energy & Fuels, 32*(12), 12383–12393.

Springer, Czichos, H., Saito, T., & Smith, L. (Eds.). (2006). Chapter 17: Molecular dynamics. In *Springer handbook of materials measurement methods* (pp. 915–952). Springer.

Stote, R., Dejaegere, A., Kuznetsov, D., & Falquet, L. (1999). *CHARMM molecular dynamics simulation*. Swiss Institute of Bioinformatics. [Online]. Available http://www.ch.embnet.org/. Accessed 05 December 2021.

Stote, R. H., States, D. J., & Karplus, M. (1991). On the treatment of electrostatic interactions in biomolecular simulation. *Journal de Chimie Physique et de Physico-Chimie Biologique, 88*, 2419–2433.

Sun, D., & Zhou, J. (2013). Ionic liquid confined in Nafion: Toward molecular-level understanding. *AICHE Journal, 59*, 2630–2639.

Swope, W. C., et al. (1982). A computer simulation method for the calculation of equilibrium constants for the formation of physical clusters of molecules: Application to small water clusters. *The Journal of Chemical Physics, 76*.

Tazikeh, S., Kondori, J., Zendehboudi, S., Amin, J. S., & Khan, F. (2021). Molecular dynamics simulation to investigate the effect of polythiophene-coated Fe3O4 nanoparticles on asphaltene precipitation. *Chemical Engineering Science, 237*, 116417.

Tirjoo, A., et al. (2019). Molecular dynamics simulations of asphaltene aggregation under different conditions. *Journal of Petroleum Science and Engineering, 177*, 392–402.

Ungerer, P., Nieto-Draghi, C., Rousseau, B., Ahunbay, G., & Lachet, V. (2007). Molecular simulation of the thermophysical properties of fluids: From understanding toward quantitative predictions. *Journal of Molecular Liquids, 134*(1–3), 71–89.

Verlet, L. (1967). Computer "experiments" on classical fluids. I. Thermodynamical properties of Lennard-Jones molecules. *Physical Review, 159*(1), 98.

Wan, L. H., et al. (2009). Molecular dynamics simulation of methane hydrate dissociation process in the presence of thermodynamic inhibitor. *Wuli Huaxue Xuebao/Acta Physico-Chimica Sinica, 25*(3), 486–494.

Yaseen, S., & Mansoori, G. A. (2017). Molecular dynamics studies of interaction between asphaltenes and solvents. *Journal of Petroleum Science and Engineering, 156*, 118–124.

Yaseen, S., & Mansoori, G. A. (2018). Asphaltene aggregation due to waterflooding (a molecular dynamics study). *Journal of Petroleum Science and Engineering, 170*, 177–183.

York, D. M., Darden, T. A., & Pedersen, L. G. (1993). The effect of long-range electrostatic interactions in simulations of macromolecular crystals: A comparison of the Ewald and truncated list methods. *The Journal of Chemical Physics, 99*(10).

Zendehboudi, S., Rezaei, N., & Lohi, A. (2018). Applications of hybrid models in chemical, petroleum, and energy systems: A systematic review. *Applied Energy, 228*, 2539–2566.

Zhang, J., & Pan, Z. (2011). Effect of potential energy on the formation of methane hydrate. *Journal of Petroleum Science and Engineering, 76*(3), 148–154.

Zhang, H., et al. (2002). Molecular dynamics simulations on the adsorption and surface phenomena of simple fluid in porous media. *Chemical Physics Letters, 366*(1–2), 24–27.

5

Single-event kinetic modeling of catalytic dewaxing on commercial Pt/ZSM-5

E. Turco Neto, Syed Ahmad Imtiaz, and S. Ahmed
Center for Risk, Integrity and Safety Engineering (C-RISE), Department of Process Engineering, Memorial University of Newfoundland, St. John's, NL, Canada

Nomenclature

a gas-liquid specific surface area (m_l^2/m_r^3)
A reactor cross-sectional area (m_r^2)
$\tilde{A}_l$ single-event frequency factor for l type elementary step (h^{-1})
$\tilde{A}_{pr/de}$ Van't Hoff frequency factor for protonation/deprotonation elementary step (kg/kmol)
$\tilde{A}_l^*$ composite frequency factor for l type elementary step (kmol/kg h)
C_i^G gas phase concentration of the lump i ($kmol/m_G^3$)
C_i^L liquid phase concentration of the lump i ($kmol/m_L^3$)
C_{Lm}^L bulk concentration of lump L_m in liquid phase ($kmol/m_r^3$)
$C_{H_2}^L$ bulk concentration of hydrogen in liquid phase ($kmol/m_r^3$)
C_{sat} surface concentration of total physisorption sites (kmol/kg)
C_t surface concentration of total active sites (kmol/kg)
E_l activation energy for l type elementary step (kJ/kmol)
F_i^G molar flow rate of lump i in gas phase (kmol/h)
F_i^L molar flow rate of lump i in liquid phase (kmol/h)
h Planck constant (J s)
k_B Boltzmann constant (J/K)
k rate constant ($1/h^{-1}$)
$\tilde{k}_{PCP}(m,w)$ single-event rate constant for PCP isomerization forming carbenium ion of w type from m type carbenium ion (h^{-1})
$\tilde{k}_{cr}(m,w,O)$ single-event rate constant for β-scission cracking forming carbenium ion of w type and olefin of O type from m type carbenium ion (h^{-1})
$\tilde{k}_l^*$ composite single-event rate constant (kmol/kgh)
$\tilde{K}_{DHij}$ single-event dehydrogenation equilibrium constant ($kmolkg/m_r^6$)
$\tilde{K}_{isom}^{(O_{ij}\leftrightarrow O_r)}$ single-event isomerization equilibrium constant between O_{ij} and reference olefin O_r (–)
$\tilde{K}_{pr/de}^{(O_r\leftrightarrow m_{ik})}$ single-event protonation equilibrium constant between reference olefin O_r and carbenium ion m_{ik} (kg/kmol)
k_G volumetric mass transfer film coefficient for gas phase (m_G^3/m_l^2 h)

Modeling of Chemical Process Systems. https://doi.org/10.1016/B978-0-12-823869-1.00006-5

k_L	volumetric mass transfer film coefficient for liquid phase (m_L^3/m_i^2 h)
$k_{o,i}$	overall mass transfer coefficient (m_r^3/m_i^2 h)
$K_i^{C,VLE}$	concentration-based equilibrium constant (m_L^3/m_G^3)
K_i^{VLE}	true equilibrium constant (–)
$K_{L,j}$	Langmuir adsorption constant for j-type paraffin (m_r^3/kmol)
n_e	number of single events (–)
N_i	interphase mass transfer flux of the *i*th lump (kmol/m_r^2 h_l)
N_{OBS}	number of observations
N_{RES}	number of responses
$R^{Cons}_{L_m,l_{m,w}}$	rate of consumption of m-type carbenium ions within lump L_m forming *w* type carbenium ions through *l* type elementary step (kmol/m_r^3 h)
$R^{Form}_{L_m,L_k,l_{m,w}}$	rate of formation of carbenium ions of *w* type in lump L_m from carbenium ions of m in lump L_k through *l* elementary steps (kmol/m_r^3 h)
$R^{Form}_{L_m,net}$	net rate of formation of lump L_m (kmol/m_r^3 h)
R	ideal gas constant (kJ/kmolK)
SSE	sum of squares error
T	temperature (K)
$y_{i,Lm}$	equilibrium composition of *i*th isomer within lump L_m (–)
y_{ij}	experimental value of the response
$\hat{y}_{ij}$	simulated value of the response
z	axial position (m_r)

Greek letters and symbols

σ_{gl}	global symmetry number (–)
ρ^G_{molar}	molar density of gas phase (kmol/m_G^3)
ρ^L_{molar}	molar density of liquid phase (kmol/m_L^3)
$\Delta H_{pr/de}$	protonation/deprotonation enthalpy change (kJ/kmol)
ΔH	enthalpy difference (kJ/kmol)
$\Delta \hat{S}$	intrinsic entropy difference (kJ/kmolK)

Superscripts

o	standard
#	activated complex

1 Introduction

High concentration of long-chain paraffins in diesel fuels have been a major concern for oil refineries located in countries with harsh winter seasons. Due to the high fusion temperature values of the heavy paraffinic species, the fuel can crystallize even when exposed to moderate temperatures, which can lead to wax formation in automotive filters. To overcome this problem, Exxon Mobil developed a new technological process similar to hydrocracking called catalytic dewaxing, which uses bifunctional shape-selective catalysts such as Pt/ZSM-22 or Pt/ZSM-5 to convert heavy paraffinic molecules through selective isomerization and cracking reactions carried out in trickle bed reactors (van Veen et al., 2008). In order to assess the performance of this process, the cloud point, which is the highest temperature at which the first wax crystal appears in the fuel, is the common metric used in industry to determine if the fuel cold flow specifications have been met (Rakoczy & Morse, 2013). There are strong evidences that by increasing the dewaxing conversion, the cloud point of the diesel product is reduced; however, high yield losses are observed due to the higher selectivity toward cracking products (van Veen et al., 2008).

In order to design and optimize a catalytic dewaxing reactor for industrial purposes, a kinetic model is necessary to predict the net rate of consumption of the heavy paraffinic components in a given diesel feedstock so that the continuity and energy equations can be solved. However, such feedstock is a complex mixture involving thousands of hydrocarbons, which would lead to a prohibitive number of kinetic parameters to be estimated (Ancheyta et al., 2005). To overcome this problem, several reactor models in refining engineering apply lumping techniques, which split the true boiling point (TBP) curve into pseudo-components representing common petroleum cuts so that the complexity of the mixture and the number of kinetic parameters are reduced (Ancheyta et al., 2005). Qader and Hill (1969) proposed the first lumped kinetic model reported for vacuum gas oil (VGO) hydrocracking using the most common boiling point cuts such as gasoline, middle distillate, and diesel as pseudo-components. Only one reaction was considered to represent the conversion of VGO, and a first-order kinetic model with respect to the feed concentration was proposed. Callejas and Martínez (1969) formulated a three-lump kinetic model for Maya residue hydrocracking considering first-order reactions converting the feedstock into lumps representing light oils and gases. Sánchez et al. (2005) introduced a 5-lump kinetic model for hydrocracking of heavy oils using first-order kinetics involving 5 pseudo-components: unconverted residue, vacuums gas oil, distillates, naphtha, and gases.

Even though the previous models are simple and can be easily implemented, their main disadvantage is related to their dependence on the feedstock composition. Therefore, a re-estimation of the kinetic parameters is necessary when the feedstock is changed (Kumar & Froment, 2007a). Also, the number of kinetic parameters quickly increases as more lumps is added to the kinetic model for better accuracy (Kumar & Froment, 2007a).

Another type of kinetic model available in the literature suitable to describe the hydrocracking of complex feedstocks is the continuum lumping technique, which considers the reaction mixture continuous with respect to the boiling point and molecular weight. This technique was developed by Laxminarasimhan et al. (1996), and the proposed kinetic model depends on the estimation of a characterization parameter from the feedstock TBP curve. It is also assumed that the material balances and kinetic model are functions of the reactivity of each component in the feedstock, which are represented by integro-differential equations. Even though this methodology involves less parameters to be estimated, the chemistry of the process is still not taken into consideration (Kumar & Froment, 2007a).

A mechanistic kinetic model was introduced by Froment and co-workers based on the single-event theory. According to this methodology, the rate constants are formulated based on the statistical thermodynamics and quantum chemistry allowing the effects of the molecular structure of reactant and activated complex to be factored out by introducing the number of single events in the rate constant formulation. The carbenium ion chemistry can be easily incorporated to the kinetic model by using a computer reaction network algorithm proposed by Clymans and Froment (1984) and Baltanas and Froment (1985) that generates all possible elementary steps commonly taking place on bifunctional catalysts, which are constituted by metal and acid sites responsible by hydrogenation/dehydrogenation and cracking/isomerization reactions, respectively. The main advantage in using this type of kinetic model relies on the independence of the kinetic parameters on the feedstock type, which allows the application of the hydrocracking model to cases involving feedstocks from different sources.

Baltanas et al. (1989) proposed a fundamental kinetic model for hydroisomerization of n-octanes into monomethylheptanes on Pt/US-Y zeolites. The reaction network was built through a computer algorithm, and an ideal hydrocracking process was also assumed by considering the rate-determining steps as the PCP (protonated cyclopropane) isomerization and β-scission cracking steps taking place at the acid sites. A total of 12 feedstock-independent kinetic parameters were estimated from experimental data for the gas-phase hydrocracking carried out in a Berty-type reactor. The results showed very good agreement between experimental and calculated rate of formation values for several reactant/product species.

Svoboda et al. (1995) determined the single-event kinetic parameters for hydroisomerization and hydrocracking of paraffins on Pt/US-Y zeolite in gas phase. The reaction network was generated through a computer algorithm proposed by Baltanas and Froment (1985). A total of 22 feedstock-independent Arrhenius parameters were estimated from experimental values of reaction rate for a feedstock containing pure n-octane and ideal hydrocracking conditions were confirmed based on the results. It was the first published work after Baltanas et al. (1989) that considered the formation of multibranched paraffins. Thermodynamic constraints involving a reference olefin were incorporated into the kinetic model formulation to further reduce the number of kinetic parameters. A total of 22 single-event rate constants were estimated from a wide range of feedstocks. The results showed that their influence on the kinetic parameters was statistically nonsignificant. The validation step was done by comparing experimental and calculated effluent flow rates.

Schweitzer et al. (1999) applied the single-event kinetics to model the hydrocracking of paraffins taking place in a three-phase reactor operating in trickle flow. All the elementary steps in metal and acid sites were included in the parameter estimation. The reaction network generation considered not only the PCP but also the protonated cyclobutane (PCB) isomerization steps. A total of 12 single-event parameters and 4 Langmuir adsorption coefficients were estimated from experimental data for cracking yields, selectivities, and mole distribution of the products using n-hexadecane as feedstock. The vaporization of the liquid phase along the reactor was also taken into account. The estimated kinetic parameters were used to simulate the hydrocracking of a Fisher-Tropsch effluent containing paraffins within C_{11}-C_{18} carbon range, and it was reported good predictions from the model. However, diffusional limitations in the experimental reactor were detected once the calculated outlet mole fractions for monoparaffins considerably deviated from the measured values.

Kumar and Froment (2007a) applied the single-event kinetics to model the nonideal hydrocracking of long-chain paraffins taking place in a trickle bed reactor. The feedstock considered for parameter estimation involved heavy paraffinic molecules up to C_{32}. An improved computer algorithm was used to generate the reaction network, and the rate of formation of the intermediate olefins at the metal sites was formulated based on the molecular (de)hydrogenation mechanisms. To further reduce the number of kinetic parameters related to the reactions taking place at the metal sites, the activation energy was considered to be only dependent on nature of the carbons forming the double bond, which led to five activation energies and one frequency factor to be estimated. Moreover, the catalyst surface was considered to be entirely surrounded by the liquid phase so that the concentration of sorbed species could be related to their actual concentration in that phase through Langmuir isotherms. To account for the effect of the nature of the carbenium ions on the activation energy, the Evan-Polanyi free-energy relationship was adopted. The posteriori-lumping technique was used to express the net rate of formation in terms of lumps selected based on carbon number and degree of branching. A total of 14 parameters were estimated from experimental data obtained from a tubular reactor using n-hexadecane. The authors studied the effects of the catalyst activity on the product distribution by carrying out reactor simulations and reported that as farther from the equilibrium the metal site elementary steps are, more the cracking steps are favored once the rate of hydrogenation of product olefins is lower.

The single-event kinetic modeling has been extensively applied to simulate and optimize other types of refining processes such as catalytic cracking in riser reactors (Dewachtere et al., 1999), alkylation (Martinis & Froment, 2006a, 2006b), and methanol-to-olefins process (Park & Froment, 2001a, 2001b); however, no studies using such mechanistic concept have been applied to model catalytic dewaxing reactors.

In this work, the single-event kinetic modeling approach is developed to describe the selective isomerization and cracking reactions carried out on a catalytic dewaxing bed loaded with commercial shape-selective Pt/ZSM-5 catalyst. The feedstock independent single-event rate constants were estimated using experimental data for n-hexadecane as feedstock. Finally, reactor simulations were carried out for different operating conditions and feedstock to study the effect of the main process parameters on the reactor performance.

2 Reactor modeling

2.1 Shape-selectivity effects

The main difference between the conventional hydrocracking and catalytic dewaxing processes lies on the type of catalyst used. Conventional hydrocracking reactors are usually loaded with wide-pore zeolite catalysts on which all molecules in the reaction mixture have the same probability to occupy a particular acid site (Gerasimov et al., 2015). It is also necessary that the isomerization reactions achieve a certain extension before the elementary cracking steps take place (Gerasimov et al., 2015).

On the other hand, shape-selective catalysts are used in catalytic dewaxing units and the ability of the reactant molecules to reach an acid site in the zeolite crystal depends on the shape and size of its molecular structure (Gerasimov et al., 2015). The most accepted theory used to explain the shape selectivity promoted by this type of catalyst is based on the pore-mouth and key-lock mechanisms (Laxmi Narasimhan et al., 2003).

Both mechanisms have been intensively adopted to formulate a kinetic model for hydroconversion of n-alkanes on Pt/H-ZSM-22 (Arroyo, 2000; Laxmi Narasimhan et al., 2003; Souverijns et al., 1998). In the pore-mouth mechanism, the physisorption mode of the intermediate olefins attempting to reach the acid sites determines whether acid-catalyzed reactions such as protonation, cracking, and isomerization are going to take place (Laxmi Narasimhan et al., 2003). As normal alkanes have straight carbon chains, they form olefins that are entirely physisorbed into the micropore of the zeolite crystal without any sterical hindrance. On the other hand, only one of straight ends of the branched alkanes is allowed to sorb into the pore mouth as the methyl branches block their full entrance into the pore (Laxmi Narasimhan et al., 2003). Therefore, branched carbenium ions are only formed at the pore mouth for a given physisorption mode that allows the entire double bond of the reacting olefinic specie to chemisorb at an acid site (Laxmi Narasimhan et al., 2003).

For US-Y zeolites, the protonation reactions are competitive once all molecules can access the acid sites; however, in shape-selective catalysts, these reactions are noncompetitive because the acid site located at the pore mouth can only be accessed by the particular olefin sorbed at this location, and thereby, this step becomes competitive (Laxmi Narasimhan et al., 2003). This mechanism explains the high yield of monobranched paraffins with a methyl group located at one of extremes in the carbon chain when shape-selective zeolites are used (Gerasimov et al., 2015).

The key-lock mechanism is used to explain the increase in the yield of dimethyl isomers as the size of reactant main chain increases. According to this theory, the carbon chain is stretched and two straight ends are simultaneously physisorbed at adjacent pore mouths (Laxmi Narasimhan et al., 2003). The experimental evidence confirming the existence of such mechanism lies on the formation of monobranched paraffinic molecules with a methyl group occupying a central position in the main chain and dimethyl isomers whose methyl branches are spaced by two or three carbon atoms (Laxmi Narasimhan et al., 2003). It is important to mention that geminal or vicinal dibranched paraffins are not formed due to steric hindrance effects.

In the single-event, microkinetic model developed by Laxmi Narasimhan et al. (2003) describe the hydroconversion of n-octane on Pt/H-ZSM-22, and the protonation/deprotonation equilibrium constant was the term in their rate expression that introduced the shape-selectivity effects to the kinetic model. Hence, the single-event kinetic parameters available in the literature for Pt/US-Y zeolites cannot be used to model catalytic dewaxing reactors, and they must be re-estimated when intended for such application.

Therefore, due to the pore size constraints found in Pt/ZSM-5 catalysts, less reactions are allowed to take place when compared to those in Pt/US-Y catalysts. Laxmi Narasimhan et al. (2003) suggested some assumptions to be made to incorporate the shape-selectivity effects when building the reaction network, which are listed as follows:

- Elementary steps involving tertiary carbenium ions are fully suppressed due to the transitional-state shape-selectivity effects (Laxmi Narasimhan et al., 2003).
- Methyl-shift reactions are not allowed to take place in shape selective catalysts due to steric hindrance effects (Laxmi Narasimhan et al., 2003).

2.2 Reaction mechanism

As the Pt/ZSM-5 is a bifunctional catalyst, the reaction mechanism based on the carbenium ion chemistry can be used to describe the chemical phenomena involved in the catalytic dewaxing process as it is a special case of hydrocracking. The

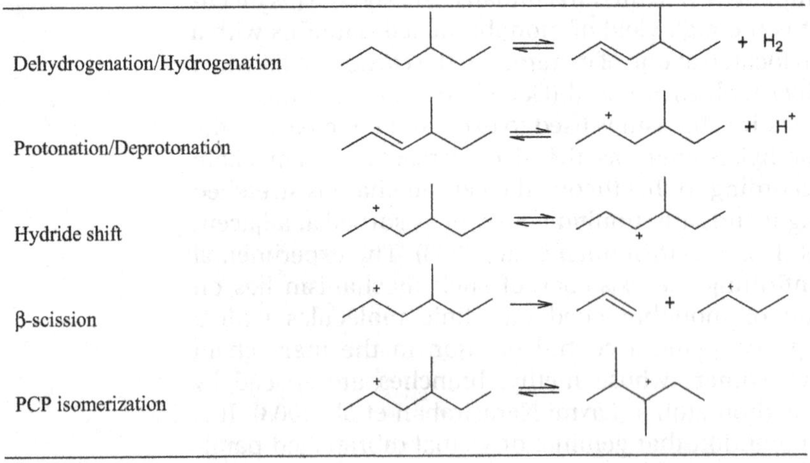

Scheme 1 Common elementary steps promoted by the metal and acid sites on Pt/ZSM-5 zeolites.

elementary steps taking place at both metal and acid sites on bifunctional Pt/ZSM-5 are shown in Scheme 1.

According to Kumar (2004), the following assumptions can be made based on experimental evidences:

- Only methyl branches are considered and the maximum number of branches attached to the main chain is three.
- The carbenium ions are usually classified based on its stability, and thereby, only secondary and tertiary carbenium ions can be formed from the elementary steps. Consequently, elementary reactions involving methyl and primary carbenium ions are excluded.
- The protonated cyclopropane (PCP) ring is the only intermediate structure considered to described the isomerization steps.

If the catalyst particle is properly loaded with platinum, the elementary steps taking place at the metal sites reach equilibrium, and the rate-determining steps are fully located at the acid sites. Laxmi Narasimhan et al. (2003) showed that these assumptions are also valid for shape-selective catalysts, and thereby, it is going to be considered in this work. Also, the hydride-shift elementary steps are assumed to be in pseudo-equilibrium once they are faster than those for PCP isomerization (Kumar, 2004).

2.3 Reaction network generation

The reaction network was generated in this work using the computer algorithm proposed by Baltanas and Froment (1985), which shape-selectivity effects were incorporated by imposing the constraints suggested by Laxmi Narasimhan et al. (2003). The number of elementary steps and species generated for a feedstock containing pure n-hexadecane are shown in Tables 1 and 2, respectively.

Based on this information, the single-event kinetic modeling can be applied to the catalytic dewaxing reactor model, which incorporates the mechanistic aspects of the chemical reactions to the reactor operation.

2.4 Single-event kinetic modeling

The single-event kinetic modeling was proposed by Froment and co-workers (Baltanas et al., 1989; Kumar, 2004; Kumar & Froment, 2007a; Svoboda et al., 1995) and has been extensively used to design, simulate, and optimize several industrial refining operations. Its formulation is based on the statistical thermodynamics and single-event theory, which proposes the rate constant expressed by Eq. (1) (Kumar & Froment, 2007a).

Table 1 Number of elementary steps per type generated by the reaction network for n-hexadecane.

Elementary step	Number of elementary steps
Dehydrogenation	5541
Deprotonation/protonation	7358
Hydride-shift	4738
PCP-isomerization	8085
β-Scission	1570
Hydrogenation	4983

Table 2 Number of species per type generated by the reaction network for n-hexadecane.

Molecular type	Number of species
Paraffin	891
Carbenium ion	3680
Olefin	4983

$$k = \left(\frac{\sigma_{gl}^{r}}{\sigma_{gl}^{\#}}\right)\left(\frac{k_B T}{h}\right)\exp\left(\frac{\Delta \widehat{S}^{0\#}}{R}\right)\exp\left(-\frac{\Delta H^{0\#}}{RT}\right) \tag{1}$$

The dependence of the rate constant on the feedstock was factored out through the ratio of global symmetry numbers represented by the first term in Eq. (1), which is denominated as number of single events (n_e) and accounts for the changes in the reactant molecular structure due to chemical reaction. The other intrinsic terms are lumped into a single-event rate constant ($\tilde{k}$), which is independent on the feedstock used (Kumar & Froment, 2007a). Therefore, the rate constant can be rewritten as

$$k = n_e \tilde{k} \tag{2}$$

This approach allows estimating a small set of kinetic parameters even if a large amount of elementary steps is taken into account by the kinetic model. They are identified based on the type of elementary step and nature of the reactants and/or product carbenium ions involved in either PCP isomerization or β-scission reactions. Also, thermodynamic constraints are used to further reduce the number of parameters to be estimated.

Under "ideal" hydrocracking conditions, the rate-determining steps are located at the acid sites, and thereby, dehydrogenation/hydrogenation reactions taking place at the metal sites are assumed to be in pseudo-equilibrium. In this way, the kinetic parameters are only the ones related to the isomerization and cracking steps.

For PCP isomerization elementary steps, the difference in energy levels between reactant and product carbenium ions is associated to the type of the carbon atom at which the positive charge is located, i.e., secondary or tertiary (Baltanas et al., 1989). On the other hand, the cracking steps are not only dependent on the nature of the ions but also on the type of olefin being produced, which can be classified as normal or branched (Govindhakannan, 2003). As the hydride-shift reactions are assumed to be in pseudo-equilibrium, their rate constants are not included in the parameter estimation. Based on the previous information, Kumar (2004) concluded that 13 single-event rate constants were necessary to model a conventional hydrocracking of a paraffinic feed. However, by adding the constraints imposed by shape-selectivity effects, less single-event rate constants are required, which are presented in Table 3.

Also, the Langmuir physisorption equilibrium constants are assumed to be only dependent on the nature of the paraffinic species (Govindhakannan, 2003).

2.5 Posteriori lumping

The amount of chemical species involved in the hydrocracking of petroleum fractions is extremely large and there is no analytical technique known so far that can measure their individual compositions. Therefore, the posteriori-lumping technique proposed by

Table 3 Single-event kinetic and Langmuir equilibrium constants.

Rate-determining step	Single-event kinetic parameters
PCP-isomerization	$\tilde{k}_{PCP}(s,s)$
β-Scission	$\tilde{k}_{cr}(s,s,no)$, $\tilde{k}_{cr}(s,s,io)$
Langmuir equilibrium constants	$K_{L,np}$, $K_{L,mp}$, $K_{L,dp}$, $K_{L,tp}$

Vynckier and Froment (1991) is applied to group the intermediate olefins, carbenium ions, and paraffins with same degree of branching and carbon number into a single lump with internal composition distribution governed by the thermodynamic equilibrium. Therefore, it is possible to describe the reaction mixture involved in the dewaxing of n-hexadecane using 45 lumps (including one for hydrogen), which makes the validation easier with experimental data obtained from modern analytical techniques such as chromatography.

2.6 Net rate of formation

The rate of consumption of a paraffinic lump L_m undergoing the elementary step of type l involving the reactant and product carbenium ions of type m and w (secondary or tertiary), respectively, can be obtained using the Eq. (3) (Kumar, 2004).

$$R^{Cons}_{L_m,l_{m,w}} = \mathrm{LCC}_{Lm,l_{m,w}} \frac{\tilde{k}^*_l(mw) K_{L,Lm} C^L_{Lm}}{C^L_{H_2}\left[1 + \sum_i K_{L,Lm} C^L_{Lm}\right]} \tag{3}$$

where the lumping coefficient of consumption (LCC) and composite single-event rate constant are defined by Eqs. (4) and (5), respectively (Kumar, 2004).

$$\mathrm{LCC}_{Lm,l_{m,w}} = \sum_{q=1}^{q_T} n_{e,q} y_{i,Lm} \frac{1}{n} \left(\frac{\sigma^{gl}_{Pi}}{\sigma^{gl}_{O_{ij}} \sigma^{gl}_{H_2}} \right) \sum_{j=1}^{n} \tilde{K}_{DHij} \tilde{K}^{(O_{ij} \leftrightarrow O_r)}_{isom} \tag{4}$$

$$\tilde{k}^*(mw) = \tilde{k}(mw) \tilde{K}^{(O_r \leftrightarrow m_{ik})}_{pr/de} C_{sat} C_t \tag{5}$$

It is important to note that the rate of consumption of a given lump represents the consumption of all the carbenium ions of type m within that lump being converted into ions of type w within other lumps through elementary step l. Similarly, the rate of formation of given lump L_m from k different lumps L_k and the lumping coefficient of formation (LCF) are expressed by Eqs. (6) and (7), respectively (Kumar, 2004).

$$R^{Form}_{L_m,L_k,l_{m,w}} = \sum_k \mathrm{LCF}_{L_m,L_k,l_{m,w}} \frac{\tilde{k}^*_l(mw) K_{L,Lk} C^L_{Lk}}{C^{liq}_{H_2}\left[1 + \sum_i K_{L,Lk} C^L_{Lk}\right]} \tag{6}$$

$$\mathrm{LCF}_{L_m,L_k,l_{m,w}} = \sum_{q=1}^{q_T} n_{e,q} y_{i,L_k} \frac{1}{n} \left(\frac{\sigma_{Pi}^{gl}}{\sigma_{O_{ij}}^{gl} \sigma_{H_2}^{gl}} \right) \sum_{j=1}^{n} \widetilde{K}_{DHij} \widetilde{K}_{isom}^{(O_{ij} \leftrightarrow O_r)} \tag{7}$$

The net rate of formation for the lump L_m is obtained by summing the difference between its rates of formation and consumption over all the elementary steps involved in the reaction network as shown by Eq. (8) (Kumar, 2004).

$$R_{L_m,net}^{Form} = \sum_{l=1}^{n_l} \left(R_{L_m}^{Form} - R_{L_m}^{Cons} \right) \tag{8}$$

To simplify the parameter estimation procedure, the composite single-event rate constant is expressed in Arrhenius form. This can be done by using the Arrhenius and Van't Hoff laws to express the single-event rate constant $\tilde{k}(m,w)$ and the protonation/deprotonation equilibrium coefficients $\tilde{K}_{pr/de}^{(O_r \leftrightarrow m_{ik})}$ as a function of temperature, which is presented by Eqs. (9) and (10), respectively (Kumar, 2004).

$$\tilde{k}_l(m,w) = \tilde{A}_l \exp\left(-\frac{E_l}{RT} \right) \tag{9}$$

$$\tilde{K}_{pr/de}^{(O_r \leftrightarrow m_{ik})} = \tilde{A}_{pr/de} \exp\left(-\frac{\Delta H_{pr/de}}{RT} \right) \tag{10}$$

Substituting Eqs. (9) and (10) into Eq. (5), the Arrhenius equation form of the composite single-event rate constant can be expressed by Eq. (11) (Kumar, 2004).

$$\tilde{k}_l^*(mw) = \tilde{A}_l \tilde{A}_{pr/de} C_{sat} C_t \exp\left(-\frac{(E_l + \Delta H_{pr/de})}{RT} \right) \tag{11}$$

Hence, the composite single-event frequency factors and activation energies that must be estimated are represented by Eqs. (12) and (13), respectively (Kumar, 2004).

$$\tilde{A}_l^* = \tilde{A}_l \tilde{A}_{pr/de} C_{sat} C_t \tag{12}$$

$$E_l^* = E_l + \Delta H_{pr/de} \tag{13}$$

Finally, the adsorption equilibrium coefficients can be expressed in terms of Van't Hoff law represented in generic form by Eq. (14).

$$K_{L,j} = K_{0,j} \exp\left(-\frac{\Delta H_{L,j}}{R} \frac{1}{T} \right) \tag{14}$$

Table 4 Composite rate and Langmuir adsorption parameters.

Rate-determining step	Parameters
PCP-isomerization	$\tilde{A}^{*}_{PCP(s,s)}$, $E^{*}_{PCP(s,s)}$
β-Scission	$\tilde{A}^{*}_{cr(s,s,no)}$, $\tilde{A}^{*}_{cr(s,s,io)}$, $E^{*}_{cr(s,s,no)}$, $E^{*}_{cr(s,s,io)}$
Langmuir equilibrium constants	$K_{0,np}$, $K_{0,mp}$, $K_{0,dp}$, $K_{0,tp}$, $\frac{\Delta H_{L,np}}{R}$, $\frac{\Delta H_{L,mp}}{R}$, $\frac{\Delta H_{L,dp}}{R}$, $\frac{\Delta H_{L,tp}}{R}$

Based on the rate constants shown in Table 3, a total of fourteen Arrhenius and Van't Hoff parameters must be estimated from experimental data, which are summarized in Table 4.

2.7 Reactor model

In this section, the material balance equations representing the isothermal reactor model are presented. For a trickle bed reactor, the continuity equations for a given lump in liquid and gas phases must account for the interphase mass transfer and the reaction kinetics. The assumptions used to simplify the reactor model are shown as follows:

- Isothermal operation.
- The gas and liquid phases flow downward in plug flow pattern.
- The gas phase is considered to be continuous while liquid is the dispersed phase.
- Trickle flow pattern is assumed for liquid phase.
- Due to the evaporation of light components and dissolution of hydrogen into the liquid phase, the gas-liquid interphase mass transfer of the species is modeled based on the two-film theory.
- As the hydrogenation/dehydrogenation elementary steps are considered in pseudo-equilibrium, the hydrogen concentration in liquid phase is limited by its solubility value.
- Negligible pressure drop across the bed.
- Negligible internal mass transfer resistance.
- Reactions take place in liquid phase only.

The interphase mass transfer flux was formulated based on the two-film theory, which is represented by Eq. (15) (Froment, Wild, & Bischoff, 2011; Kumar & Froment, 2007a).

$$N_i = k_{o,i}a\left(\frac{C_i^G}{K_i^{C,VLE}} - C_i^L\right) \tag{15}$$

The volumetric overall mass transfer coefficient can be calculated using the individual mass transfer film coefficients as given by Eq. (16) (Froment, Wild, & Bischoff, 2011; Kumar & Froment, 2007a).

$$\frac{1}{k_{o,i}a} = \frac{1}{K_i^{C,VLE}k_Ga} + \frac{1}{k_La} \tag{16}$$

The partitioning coefficient in Eqs. (15) and (16) can be expressed in terms of the true equilibrium coefficient and molar density for gas and liquid phases as shown in Eq. (17) (Kumar & Froment, 2007a).

$$K_i^{C,VLE} = K_i^{VLE}\frac{\rho_{molar}^G}{\rho_{molar}^L} \tag{17}$$

The true equilibrium coefficient was obtained using the Peng-Robinson equation of state (Peng & Robinson, 1976). The continuity equations for gas and liquid phases for a given lump can be expressed by Eqs. (18) and (19), respectively (Froment, Wild, & Bischoff, 2011).

$$\frac{1}{A}\frac{dF_i^G}{dz} = -k_{o,i}a\left(\frac{C_i^G}{K_i^{C,VLE}} - C_i^L\right) \text{at } z = 0, F_i^G = F_{i,0}^G \tag{18}$$

$$\frac{1}{A}\frac{dF_i^L}{dz} = k_{o,i}a\left(\frac{C_i^G}{K_i^{C,VLE}} - C_i^L\right) + R_{i,net}^{form} \quad \text{at } z = 0, F_i^L = F_{i,0}^L \tag{19}$$

The boundary conditions necessary to solve Eqs. (18) and (19) were obtained from VLE flash calculations as it was assumed that the initial mixture containing hydrogen and hydrocarbons flashes before reaching the catalytic bed. Also, as part of the liquid vaporizes along the reactor due to the high operating temperature, vapor-liquid equilibrium (VLE) calculations must be simultaneously performed while solving the set of differential equations. The calculation routine used to solve the continuity equations represented by Eqs. (18) and (19) is presented in Fig. 1.

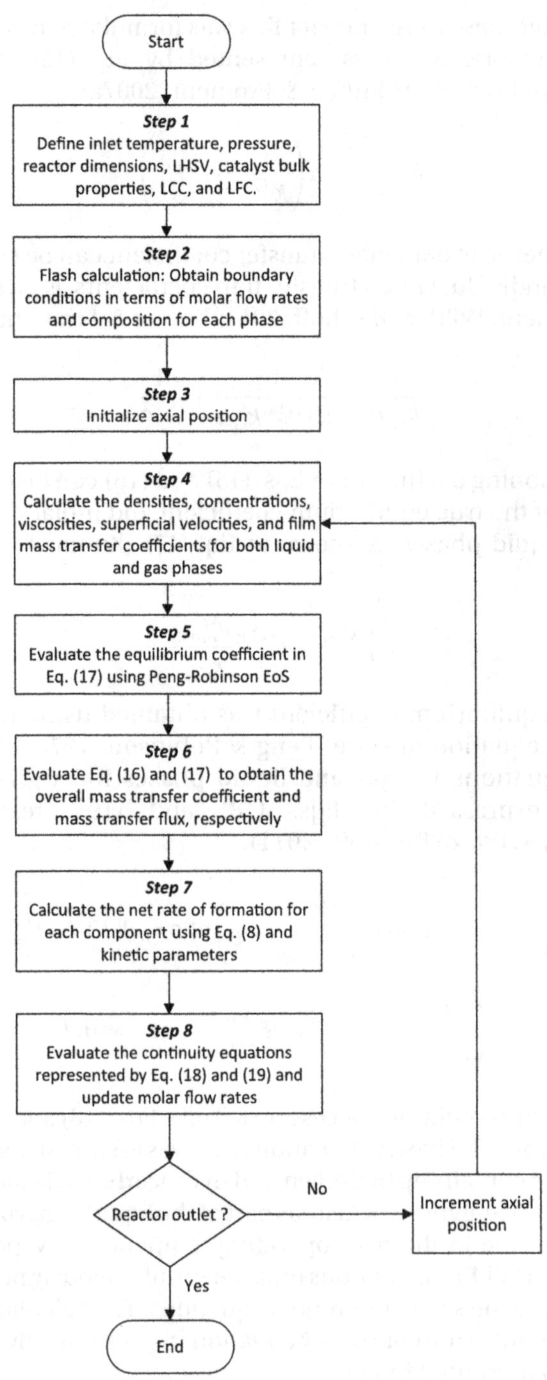

Fig. 1 Schematic representation of the calculation routine adopted to solve the reactor model.

2.8 Physical properties estimation

The physical and thermodynamic properties for the lumps/ paraffinic molecules were evaluated using different sources as indicated below:

- The gas and liquid densities were evaluated using the Peng-Robinson equation of state (Peng & Robinson, 1976) and suitable mixing rules.
- The critical properties for pure paraffinic molecules were calculated using the group contribution method proposed by Skander and Chitour (2007). Then, the mixing rules proposed in the API Technical Data Book (American Petroleum Institute, 1997) were applied to estimate the critical properties for the paraffinic lumps.
- The standard heat of formation and entropy for the pure paraffins were evaluated using the group contribution method developed by Domalski and Hearing (1988). To calculate these properties for the paraffinic lumps, the weighted average based on the paraffinic molar distribution within the lump was used. The equilibrium constants were evaluated by applying the Van't Hoff law.
- The viscosity for gas and liquid phases were evaluated using the correlation proposed by Twu (1985).
- The correlations proposed by Reiss (1967) and Sato et al. (1972) were used to evaluate the volumetric mass transfer film coefficients for the gas and liquid sides, respectively.
- The symmetry numbers for the paraffinic species were calculated using the group contribution method proposed by Song (2004).
- The number of single events was estimated using the rules proposed by Martinis and Froment (2006b) for the different types of elementary steps.

3 Results and discussion

3.1 Parameter estimation

In order to estimate the composite single-event rate and Langmuir equilibrium constants, the sum of squares error between calculated and experimental values of total conversion and isomerization yields was used as objective function for the minimization problem given by Eq. (20).

$$\min_{\theta} \mathrm{SSE}(\theta) = \sum_{i=1}^{N_{\mathrm{RES}}} \sum_{j=1}^{N_{\mathrm{OBS}}} \left(y_{ij} - \widehat{y}_{ij}\right)^2 \tag{20}$$

As the composite single-event kinetic parameters do not depend on the feedstock type, the experimental values for total conversion of n-hexadecane and isomerization yield were used to perform the parameter estimation. Those values were taken from Dougherty et al. (2001) who conducted catalytic activity studies feeding this component to a tubular reactor loaded with Pt/ZSM-5 catalyst. In total, 12 data points for six different temperatures were available. The authors carried out the shape-selective hydrocracking reactions in a ½″ tubular reactor loaded with 5.7g of catalyst and liquid hourly space velocity (LHSV) set to $0.4\,h^{-1}$. As hydrogen-to-hydrocarbon (H_2/HC) ratio was not provided, it was predicted through the parameter estimation in order to maintain the accuracy of model. Single regression imputation was used to obtain four additional data points, which regression curves were fit according to the available data and the total conversion and isomerization yield for two unavailable temperatures (285°C and 312°C) were estimated using the obtained regressed expressions.

The objective function represented by Eq. (18) is proven to be highly nonlinear (Govindhakannan, 2003; Kumar & Froment, 2007a), and consistent lower and upper bounds were imposed to the parameters based on thermodynamic constraints so that realistic values could be obtained. The initial guess for optimization was chosen based on the kinetic parameters obtained by Kumar (2004) for Pt/US-Y zeolites and the constrained interior-point algorithm available in the MATLAB optimization toolbox was used to minimize Eq. (18). This optimization framework adopted in this work is summarized in Fig. 2.

The estimated kinetic parameters are shown in Table 5.

The comparison between experimental and calculated values for total conversion and isomerization yield is shown in Figs. 3 and 4, respectively.

In order to test the model, small random numbers were added to each total conversion and isomerization yield data point and the parameters were re-estimated five times so that their standard deviation could be calculated, which are shown in Table 6.

The hydrogen-to-hydrocarbon ratio was found to be 2.016, which is a reasonable for a pilot reactor operation. It can also be observed from Fig. 3 that the total conversion profile is in well agreement with the literature data for conversion up to 306°C, and it is slightly underpredicted for higher temperatures. Furthermore, the simulated values for isomerization yield in Fig. 4 are overpredicted at low temperatures while slightly underpredicted at high temperatures. Finally, it can be noted from Tables 3–7 that

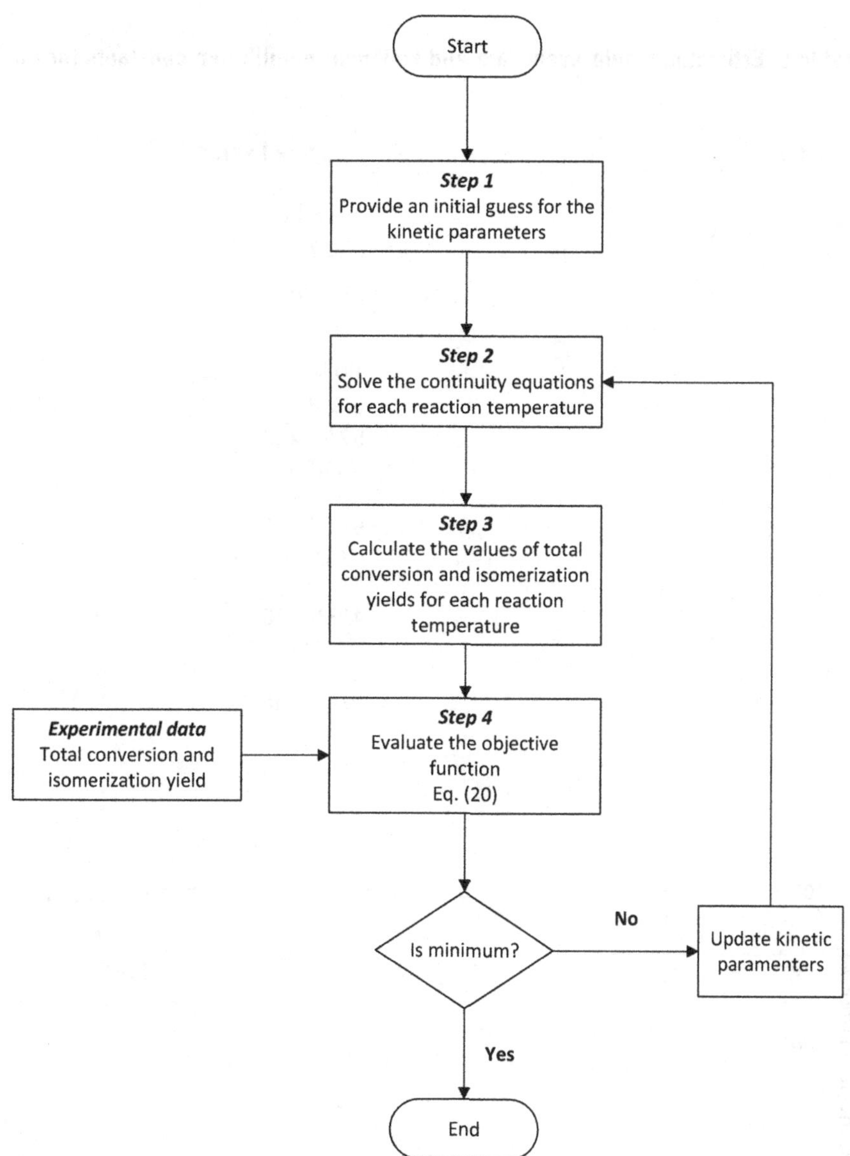

Fig. 2 Optimization framework for kinetic parameter estimation.

most of the parameters have not changed except those related to cracking steps when small random numbers were added to the data points and they were re-estimated. Several reasons can be attributed to explain those deviations. The first one is related to the fact that the assumptions made regarding to the

Table 5 Estimated single-event rate and Langmuir equilibrium constants for Pt/ZSM-5.

Parameter	Estimated value
$\tilde{A}^*_{PCP(s,s)}$	1.44×10^8
$E^*_{PCP(s,s)}$	9782.7
$\tilde{A}^*_{cr(s,s,no)}$	54485.91
$E^*_{cr(s,s,no)}$	1754.4
$\tilde{A}^*_{cr(s,s,io)}$	1989.6
$E^*_{cr(s,s,io)}$	8784.3
$K_{0,np}$	5.2703×10^{-5}
$\frac{\Delta H_{L,np}}{R}$	−4613.5
$K_{0,mp}$	6.8718×10^{-7}
$\frac{\Delta H_{L,mp}}{R}$	−6504.00
$K_{0,dp}$	4.9337×10^{-7}
$\frac{\Delta H_{L,dp}}{R}$	−0.3562
$K_{0,tp}$	1.4693×10^{-7}
$\frac{\Delta H_{L,tp}}{R}$	−296.32

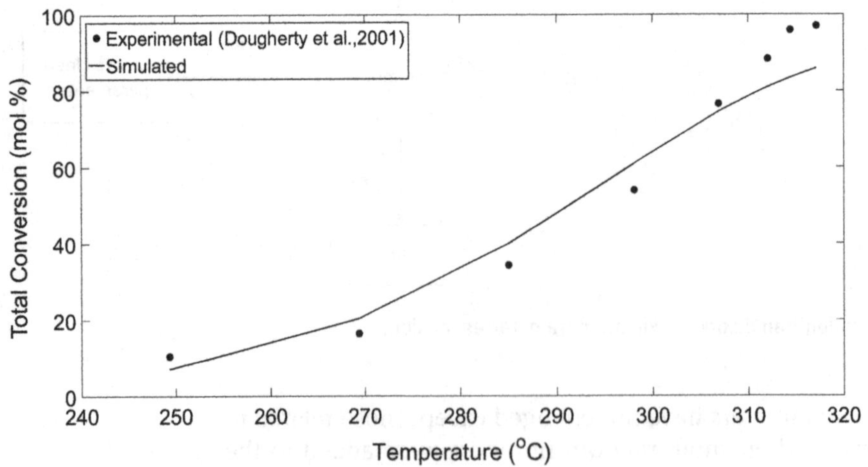

Fig. 3 Total conversion profile for n-hexadecane feedstock.

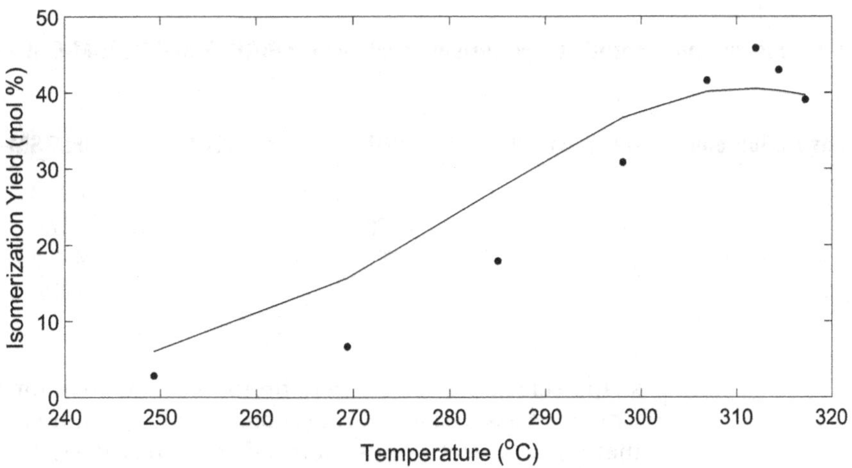

Fig. 4 Isomerization yield profile for n-hexadecane feedstock.

Table 6 Calculated standard deviation for each kinetic parameter.

Parameter	Standard deviation
$\tilde{A}^*_{PCP(s,s)}$	0
$E^*_{PCP(s,s)}$	0
$\tilde{A}^*_{cr(s,s,no)}$	8.73×10^{-4}
$E^*_{cr(s,s,no)}$	0.105
$\tilde{A}^*_{cr(s,s,io)}$	0.0171
$E^*_{cr(s,s,io)}$	3.49×10^{-3}
$K_{0,np}$	0
$\frac{\Delta H_{L,np}}{R}$	0
$K_{0,mp}$	0
$\frac{\Delta H_{L,mp}}{R}$	8.73×10^{-4}
$K_{0,dp}$	0
$\frac{\Delta H_{L,dp}}{R}$	6.98×10^{-3}
$K_{0,tp}$	1.39×10^{-2}
$\frac{\Delta H_{L,tp}}{R}$	0

Table 7 Langmuir physisorption equilibrium constants for Pt/US-Y and Pt/ZSM-5 at $T = 249.32°C$.

Physisorption equilibrium constant	Pt/US-Y (Kumar, 2004)	Pt/ZSM-5
$K_{L,np}$	0.0822	0.3604
$K_{L,mp}$	0.2664	0.1751
$K_{L,dp}$	3.2089	4.9371×10^{-7}
$K_{L,tp}$	68.5223	7.8805×10^{-7}

shape-selectivity effects may be too conservative for Pt/ZSM-5, which is classified as a 5-ring zeolite catalyst. It should be recalled that those assumptions were adopted based studies done on shape-selective catalysis involving Pt/ZSM-22 catalysts, which have smaller pores for being 10-ring zeolites. Therefore, some of the suppressed elementary steps involving tertiary carbenium ions either for PCP isomerization or β-scission might be important for the hydroconversion on Pt/ZSM-5. It was also realized through VLE calculations that most of the feedstock was flashed at high temperatures and the estimation of some physical properties such as viscosity became slightly inaccurate at such condition. As data-set used for parameter estimation was small, and the H_2/HC was also estimated, extra data points needed to be obtained by using simple regression imputation, which can be affecting the accuracy of the model. However, the profiles for total conversion and isomerization yield presented in this work are in agreement with those commonly reported in literature for conventional and shape-selective hydrocracking processes.

3.2 Effect of temperature

At low temperatures, the total conversion is low, and the PCP-isomerization reactions does not take place at enough extension to allow cracking reactions to emerge as they are secondary reactions. On the other hand, at high temperatures, the n-hexadecane is quickly converted to monobranched paraffins, which are then converted to multibranched isomers through secondary PCP-isomerization steps. Consequently, the cracking reactions rates are high enough to make the isomerization yield to decrease as temperature is further raised. Therefore, the isomerization yield should present a maximum value as observed in Fig. 4. A better view of this phenomena can be seen by analyzing the molar flow rate profiles across the reactor as shown in Fig. 5 for temperatures

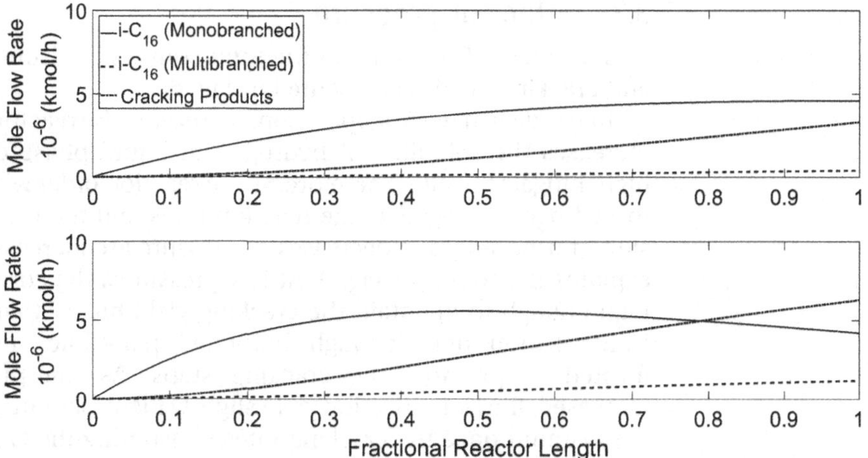

Fig. 5 Combined molar flow rate profiles for reactor temperatures at 298.03°C (top) and 317.12°C (bottom).

at 298.03°C and 317.12°C, respectively. These profiles were obtained by summing the gas phase mole flow rates for each class to those for liquid phase.

It can be observed that at low temperature the cracking products start forming at 20% of catalytic bed when enough monobranched C_{16} has been formed to allow β-scission reactions to take place. This is due to the fact that the cracking reactions can only form stable carbenium ions when the carbon located at the β position with respect to charged carbon atom is tertiary. As the formation of tertiary carbenium ions are suppressed in shape-selective catalyst, only few (s,s) PCP-isomerization steps are found in the reaction network forming dibranched and tribranched C_{16} isomers, which are more likely to crack and explains their small value of molar flow rate along the reactor. When the temperature is raised, the reaction rates are increased so that C_{16} isomers are quickly formed. However, the cracking reaction rates are also intensified so that their products become significant after 30% of the reactor length and they become faster than the PCP-isomerization one at around 80% of the reactor length, which explains the maximum value of molar flow rate for monobranched C_{16} isomers. Also, the quicker formation of multibranched C_{16} isomers contributes to the higher cracking reaction rates once they have higher probability to crack. The zero slope of the molar flow rate curves for cracking and multibranched C_{16} products indicates that they are secondary reaction products, which is related to the ideal hydrocracking assumption.

3.3 Effect of pressure

The effect of pressure on the total conversion, isomerization, and cracking yields is presented in Fig. 6.

In trickle bed reactor operation, increasing the reactor pressure increases the solubility of hydrogen in liquid phase due to its higher fugacity value. Therefore, as the reaction order with respect to hydrogen is negative, the reaction rates and the total conversion of n-hexadecane decreases as pressure are increased, which explains the trends in Fig. 6. At low pressures, the conversion is high enough to maintain the cracking yield higher than that for isomerization once enough branched paraffinic species are formed to promote the cracking steps. As the pressure is increased, the isomerization yield slightly increases until reaching a maximum once the cracking rates consuming the C_{16} isomers slow down. This trend continues until the cracking yield become significantly low and the isomerization steps dominates the total conversion behavior, which starts dropping due to the slower isomerization rates caused by the higher hydrogen solubility.

Therefore, the reactor simulations performed in this section indicate that reactor pressure must be carefully selected during industrial catalytic dewaxing operations so that isomerization yields are maximized and yield losses are avoided while keeping an acceptable level of total conversion.

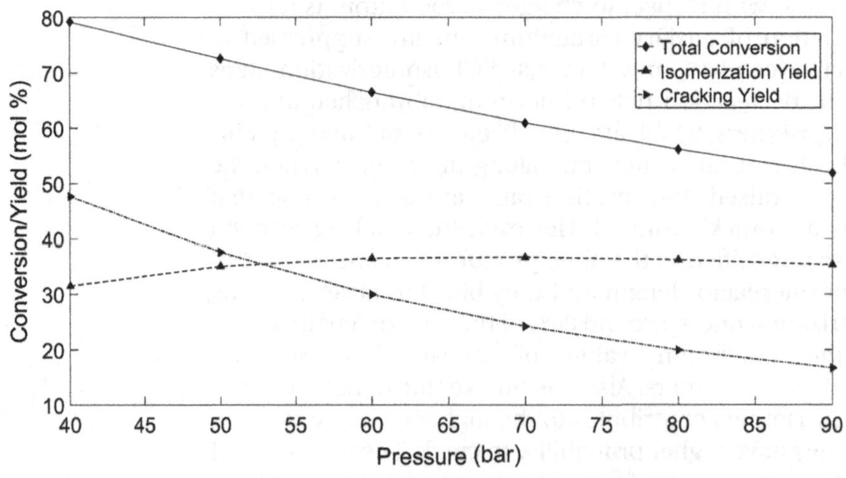

Fig. 6 Effect of pressure on total conversion, isomerization, and cracking yields for n-hexadecane feed at LHSV $=0.4\,h^{-1}$, $T=298.03°C$, and $H_2/HC=2.016$.

3.4 Effect of H_2/HC ratio

The effect of the H_2/HC ratio on the total conversion, isomerization, and cracking yields was studied through reactor simulations, which results are shown in Fig. 5.

It can be observed that the total conversion and yield values are not strongly affected by the changes in H_2/HC ratio, which is in agreement with the expected behavior for three-phase hydrocracking processes reported by Kumar and Froment (2007a). In Fig. 7, the total conversion and isomerization yield slightly decreases while the cracking yield increases when the H_2/HC ratio is raised. The main explanation for such behavior lies on the fact that increasing this ratio enhances the overall mass transfer coefficient for interphase mass transfer, which intensifies the evaporation of light paraffinic compounds and increases the rate of secondary cracking explaining the increase in the cracking yield (Kumar & Froment, 2007a). Also, the high value of H_2/HC ratio indicates that more hydrogen is fed to the system compared to n-hexadecane and the concentration of hydrogen in liquid phase is expected to be higher under such conditions, which justifies the decrease in total conversion. The combination of these two factors also explains the drop in the isomerization yield as the cracking steps are going to consume the C_{16} isomers faster as well as the higher concentration of hydrogen in liquid phase slow down the isomerization rates.

Therefore, the H_2/HC ratio does not strongly affect the reactor performance as long as the trickle flow conditions are found

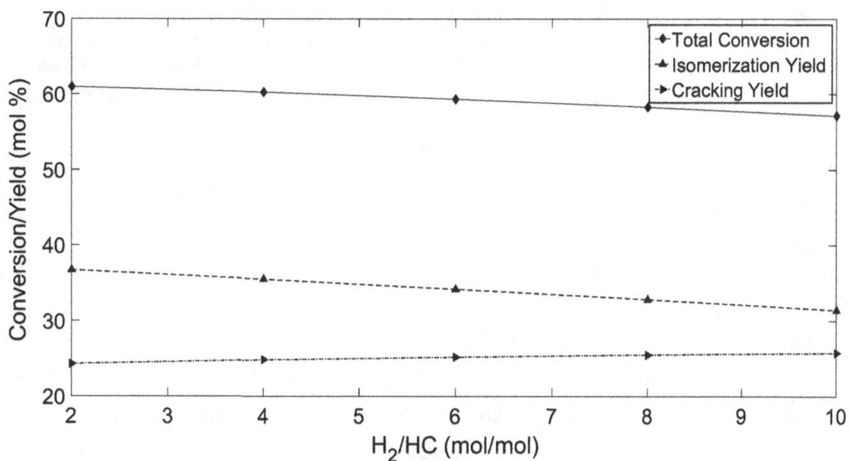

Fig. 7 Effect of the H_2/HC ratio on total conversion, isomerization, and cracking yields for n-hexadecane feed at LHSV=0.4 h^{-1}, T=298.03°C, and P=69.96 bar.

during the catalytic dewaxing operation. If the operation is carried out in gas phase, hydrogen becomes a reactant fed in excess, and this parameter is going to significantly affect the reactions rates (Kumar, 2004). It can be mainly used to control the yield losses as it governs the rate of secondary cracking due the evaporation of light paraffinic compounds.

3.5 Effect of the liquid hourly space velocity (LHSV)

The LHSV is a parameter associated with the residence time that reaction mixture spends within the catalytic bed. To study its effect on the reactor performance, several reactors simulations were carried varying its value and keeping the other operating conditions such as pressure, H_2/HC ratio, and temperature constant. The results are shown in Fig. 8.

As the residence time is represented by the inverse of the LHSV value, it is expected that the total conversion decreases as the LHSV increases as shown in Fig. 8. Lower values of LHSV indicates that the reactants are going to spend more time reacting inside the catalytic bed, and thereby, more branched isomers are formed to intensify the cracking reaction rates. It explains the higher value of cracking yield compared to the isomerization one at low LHSV. As the LHSV increases, the isomerization yield initially increases until a maximum value as the cracking reaction rates slow down but still dominating the conversion process. Therefore, less monobranched and multibranched C_{16} isomers are cracked

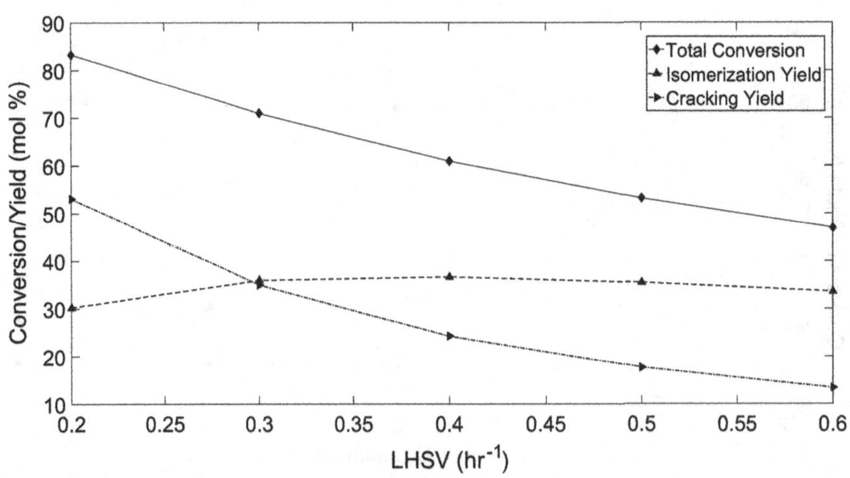

Fig. 8 Effect of the LHSV on total conversion, isomerization, and cracking yields for n-hexadecane feed at $H_2/HC = 2.016$, $T = 298.03°C$, and $P = 69.96$ bar.

down. Once the cracking yield is significantly lowered, the isomerization rates also slow down due to the low residence time and starts decreasing following the trend for total conversion, which is mainly governed by the PCP-isomerization steps at such conditions.

Therefore, the reactor simulations indicate that LHSV is also an important parameter when designing catalytic dewaxing reactors, which value must be carefully chosen so that the isomerization yield is maximized while keeping acceptable values of cracking yield and total conversion for the profitable operation.

3.6 Effects of feed carbon number

In order to verify the effect of heavy paraffinic feedstocks on the reactor performance, simulations were carried out considering additional feeds constituted by pure n-C22 and n-C24, and the results were compared to those obtained for n-hexadecane using the same operating conditions, which are shown in Fig. 9.

It can be observed that heavier paraffinic feedstocks lead to higher values of the total conversion, isomerization, and cracking yields. This can be explained by the fact that large hydrocarbons have higher reactivity once they undergo more isomerization and cracking elementary steps compared to the small ones. Similar results were observed by Kumar and Froment (2007a) for conventional hydrocracking processes. Therefore, depending on the carbon distribution of the diesel feed, heavy paraffinic molecules are

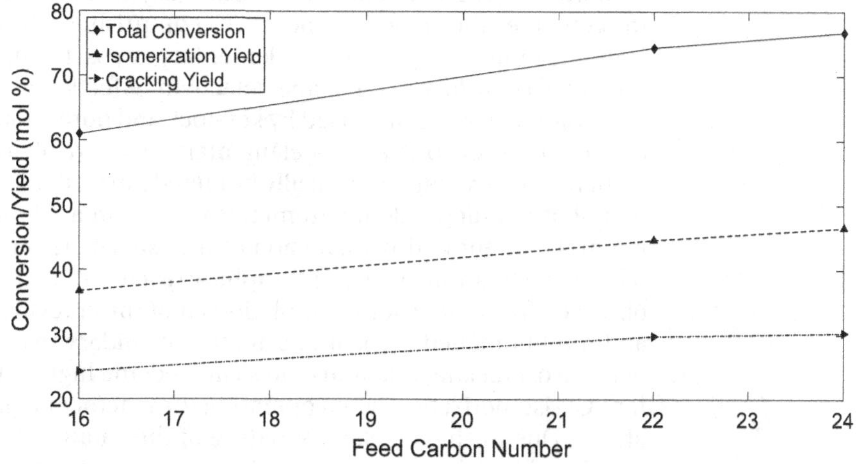

Fig. 9 Effect of the feedstock carbon number on total conversion, isomerization, and cracking yields for LHSV = $0.4\,h^{-1}$, $H_2/HC = 8.68$, $T = 298.03°C$, and $P = 69.96$ bar.

going to be converted in higher extension than the smaller ones, which is in agreement with what is observed in industrial catalytic dewaxing operations.

3.7 Effect of shape selectivity

In order to analyze the shape-selectivity feature of the Pt/ZSM-5 catalyst, the Langmuir equilibrium constants were calculated at 249.32°C and compared to those calculated using the parameters estimated by Kumar (2004) for wide-pore Pt/US-Y zeolite catalyst used in conventional hydrocracking as shown in Table 7.

It can be seen that the physisorption of normal paraffins is higher for Pt/ZSM-5 than for wide pore catalysts. The monobranched species have comparable sorption features, while dibranched and tribranched paraffinic compounds are weakly sorbed to the catalyst cages, which is in agreement with shape-selectivity theory once highly branched paraffins are sterically hindered into the catalyst pores.

4 Conclusions

In this work, the single-event kinetic modeling has been applied to estimate the kinetic parameters for the commercial Pt/ZSM-5 catalyst used in catalytic dewaxing reactors, which are independent on the feedstock type. A computer algorithm has been used to generate the reaction network comprising all the PCP-isomerization and β-scission elementary steps taking place at the catalyst surface. The shape-selectivity effects were incorporated based on rules proposed by key-lock and pore-mouth mechanisms in order to suppress elementary steps involving tertiary carbenium ions that are sterically hindered into the catalyst pores. A total of 14 independent parameters were estimated from experimental data for hydroconversion of n-hexadecane on Pt/ZSM-5 taken from literature using a multiphase reactor model. It has been observed from the reactor simulations that the cracking products and multibranched C_{16} isomers are the secondary reaction products. Also, cracking reactions rates can become higher than those for PCP-isomerization when operating the reactor at high temperatures. Due to shape-selective nature of the catalyst, most of the reactions involving multibranched carbenium ions are suppressed and small amount of multibranched C_{16} isomers are formed across the reactor. Finally, further reactor simulations confirmed that temperature, LHSV, and pressure are the main parameters affecting the catalytic dewaxing operation while the H_2/HC ratio only controls the rate at which the lighter compounds evaporates.

The kinetic parameters obtained in this work proved to be suitable to build a reactor model to design and optimize catalytic dewaxing reactors to improve the cold flow properties of diesel fuels, which related product specification such as the cloud point temperature can be easily estimated from the estimated outlet molar flow rates.

References

American Petroleum Institute (Ed.). (1997). *Technical data book-petroleum refining* (6th ed.). American Petroleum Institute.

Ancheyta, J., Sánchez, S., & Rodríguez, M. A. (2005). Kinetic modeling of hydrocracking of heavy oil fractions: A review. *Catalysis Today, 109*, 76–92.

Arroyo, J. A. M. (2000). Hydrocracking and hydroisomerization of n-paraffin mixtures and a hydrotreated gasoil on Pt/ZSM-22: Confirmation of pore mouth and key-lock catalysis in liquid phase. *Applied Catalysis, 192*, 9–22.

Baltanas, M. A., & Froment, G. F. (1985). Computer generation of reaction networks and calculation of product distributions in the hydroisomerization and hydrocracking of paraffins on Pt-containing bifunctional catalysts. *Computers and Chemical Engineering, 9*, 71–81.

Baltanas, M. A., Van Raemdonck, K. K., Froment, G. F., & Mohedas, S. R. (1989). Fundamental kinetic modeling of hydroisomerization and hydrocracking on noble metal-loaded Faujasites. 1. Rate parameters for hydroisomerization. *Industrial and Engineering Chemistry Research, 28*, 899–910.

Callejas, M. A., & Martínez, M. T. (1969). Hydrocracking of a maya residue. Kinetics and product yield distributions. *Industrial and Engineering Chemistry Research, 38*, 3285–3289.

Clymans, P. J., & Froment, G. F. (1984). Computer generation of reaction paths and rate equations in thermal cracking of normal and branched paraffins. *Computers and Chemical Engineering, 8*, 137–142.

Dewachtere, N. V., Santaella, F., & Froment, G. F. (1999). Application of a single-event kinetic model in the simulation of an industrial riser reactor for the catalytic cracking of vacuum gas oil. *Chemical Engineering Science, 54*, 3653–3660.

Domalski, E. S., & Hearing, E. D. (1988). Estimation of the thermodynamic properties of hydrocarbons at 298.15 K. *Journal of Physical and Chemical Reference Data, 17*, 1638–1678.

Dougherty, R. C., Mazzone, D. N., Socha, R. F., & Timken, H. K. C. (2001). *Production of high viscosity lubricating oil stock with improved ZSM-5 catalyst.* U.S. Patent 6,294,077 B1, Sep. 25.

Froment, G. F., Wild, J., & Bischoff, K. B. (2011). *Chemical reactor analysis and design* (3rd Ed.). New York: Wiley.

Gerasimov, D. N., Fadeev, V. V., Loginova, A. N., & Lysenko, S. V. (2015). Hydroisomerization of long-chain paraffins: Mechanisms and catalysts. Part 1. *Catalysis in Industry, 7*, 128–154.

Govindhakannan, J. (May 2003). *Modeling of a hydrogenated vacuum gas oil hydrocracker.* Ph.D. Thesis. Texas Tech. University.

Kumar, H. (August 2004). *Single event kinetic modeling of the hydrocracking of paraffins.* M.Sc. Thesis, Texas A&M University.

Kumar, H., & Froment, G. F. (2007a). A generalized mechanistic kinetic model for the hydroisomerization of long-chain paraffins. *Industrial and Engineering Chemistry Research, 46*, 4075–4090.

Laxmi Narasimhan, C. S., Thybaut, J. W., Marin, G. B., Jacobs, P. A., Martens, J. A., Denayer, J. F., & Baron, G. V. (2003). Kinetic modeling of pore mouth catalysis in the hydroconversion of n-octane on Pt-H-ZSM-22. *Journal of Catalysis, 220*, 399–413.

Laxminarasimhan, C. S., Verma, R. P., & Ramachandran, P. A. (1996). Continuous lumping model for simulation of hydrocracking. *AIChE, 42*, 2645–2653.

Martinis, J. M., & Froment, G. F. (2006a). Alkylation on solid acids. Part 1. Experimental investigation of catalyst deactivation. *Industrial and Engineering Chemistry Research, 45*, 940–953.

Martinis, J. M., & Froment, G. F. (2006b). Alkylation on solid acids. Part 2. Single-event kinetic modeling. *Industrial and Engineering Chemistry Research, 45*, 945–967.

Park, T., & Froment, G. F. (2001a). Kinetic modeling of the methanol to olefins process. 1. Model formulation. *Industrial and Engineering Chemistry Research, 40*, 4172–4186.

Park, T., & Froment, G. F. (2001b). Kinetic modeling of the methanol to olefins process. 2. Experimental results, model discrimination, and parameter estimation. *Industrial and Engineering Chemistry Research, 40*, 4187–4196.

Peng, D., & Robinson, D. B. (1976). A new two-constant equation of state. *Industrial and Engineering Chemistry Fundamentals, 15*, 59–64.

Qader, S. A., & Hill, G. R. (1969). Hydrocracking of gas oil. *Industrial and Engineering Chemistry Process Design and Development, 8*, 98–105.

Rakoczy, R. A., & Morse, P. M. (2013). Consider catalytic dewaxing as a tool to improve diesel cold-flow properties. *Hydrocarbon Processing*, 67–69.

Reiss, L. P. (1967). Cocurrent gas-liquid contacting in packed columns. *Industrial & Engineering Chemistry, Process Design and Development, 6*, 486–499.

Sánchez, S., Rodríguez, M. A., & Ancheyta, J. (2005). Kinetic model for moderate hydrocracking of heavy oils. *Industrial and Engineering Chemistry Research, 44*, 9409–9413.

Sato, Y., Hirose, H., Takahashi, F., & Toda, M. (1972). *First pacific chemical engineering congress* (pp. 187–195).

Schweitzer, J. M., Galtier, P., & Schweich, D. (1999). A single events kinetic model for the hydrocracking of paraffins in a three-phase reactor. *Chemical Engineering Science*, 2441–2452.

Skander, S., & Chitour, C. E. (2007). Group-contribution estimation of the critical properties of hydrocarbons. *Oil & Gas Science and Technology, 3*, 391–398.

Song, J. (February 2004). *Building robust chemical reaction mechanisms: Next generation of automatic model construction software.* Ph.D. Thesis. Massachusetts Institute of Technology.

Souverijns, W., Martens, J. A., Froment, G. F., & Jacobs, P. A. (1998). Hydrocracking of isoheptadecanes on Pt/H-ZSM-22: an example of pore mouth catalysis. *Journal of Catalysis, 174*, 177–184.

Svoboda, G. D., Vynckier, E., Debrabandere, B., & Froment, G. F. (1995). Single-event rate parameters for paraffin hydrocracking on a Pt/US-Y zeolite. *Industrial and Engineering Chemistry Research, 34*, 3793–3800.

Twu, C. H. (1985). Internally consistent correlation for predicting liquid viscosity of petroleum fractions. *Industrial & Engineering Chemistry, Process Design and Development, 24*, 1287–1293.

van Veen, J. A. R., Minderhoud, J. K., Huve, L. G., & Stork, W. H. J. (2008). Hydrocracking and catalytic dewaxing. In *Vol. 13. Handbook of heterogeneous catalysis* (pp. 2778–2808). New York: VCH.

Vynckier, E., & Froment, G. F. (1991). Modeling of the kinetics of complex processes based upon elementary steps. In G. Astarita, & S. I. Sandler (Eds.), *Kinetic and thermodynamic lumping of multicomponent mixtures* (pp. 131–161). Elsevier.

6

Modeling and simulation of batch and continuous crystallization processes

Yan Zhang
Process Engineering, Memorial University of Newfoundland, St John's, NL, Canada

1 Introduction to solution crystallization

Solution crystallization (hereinafter referred to as crystallization) is an important separation and purification technique widely used in pharmaceutical, food, and fine chemical industries. Crystallization by definition is a phase transformation process in which crystalline or amorphous (noncrystalline) solids are obtained from a liquid solution (Schall & Myerson, 2019). It is governed by thermodynamics of phase separation, mass and heat transfer, fluid flow and is not explicitly covered in any of the existing core chemical engineering textbooks. Unlike many other separation processes (e.g., distillation, liquid-liquid extraction, membrane separation) where product purity is the most important quality control measure, the quality control measures for final products of crystallization are multiple, including purity, crystal morphology, crystal polymorph, and crystal size distribution (Lewis et al., 2015). These product properties relate to the selected type of crystallization process as well as to the specific crystallization mode and type of hardware used for production, which makes crystallization much more than just a simple separation process. Profound understanding of the nucleation and crystal growth mechanisms is essential for a better control of the crystal properties.

Over the last two decades, significant progress has been made toward interpreting the crystallization mechanism due to the rapid technology advancements in process monitoring/control approaches and molecular simulation methods (Kee et al., 2009;

Modeling of Chemical Process Systems. https://doi.org/10.1016/B978-0-12-823869-1.00007-7

Wu et al., 2012). Process analytical technology (PAT) proposed by the FDA enables in situ monitoring of the crystallization process and thus ensures the final product quality by the timely monitoring, analyzing, and controlling of process parameters. For example, in situ attenuated total reflectance Fourier-transform infrared (ATR-FTIR) spectroscopy enables accurate monitoring and control of the solution concentration, focused beam reflectance measurement (FBRM) can track particle counts and size distribution continuously, whereas particle vision measurement (PVM) provides the real-time visualization of nucleation, crystal growth, polymorphic transformation, and crystal agglomeration. Molecular dynamic simulation is another important tool for the exploration of crystallization mechanism at the molecular level. Based on classical molecular force fields and molecular mechanics, molecular simulation, mainly including the Monte Carlo (MC) and molecular dynamics (MD) simulation, is capable of obtaining the trajectories of molecules and predicting their structures and properties by solving the Newton equation of motion. Such simulations promise molecular level insights that will rationalize experimental observations, predict phase stability and transition kinetics, and enable the fine-tuned crystal structure with desired physical and chemical properties (Anwar & Zahn, 2017; Desiraju, 2007).

The application of PAT to batch or continuous crystallization processes requires a multidisciplinary approach combining tools from different areas of science. Process modeling, simulation, and control play an important role in quantitative assessment of the PAT measurements. Only through a combination of the process simulation with PAT measurements, a true understanding of the effects of major process variables, such as agglomeration, breakage, additives, and impurities on the quality of the crystalline products, can be achieved (Gao et al., 2017). This chapter will be mainly focused on the modeling and simulation strategies for the batch and continuous crystallization processes as well as their applications in the manufacture of various crystalline (and/or amorphous) pharmaceuticals.

2 Supersaturation and metastable limit

Generally, in a crystallization process, solid particles are formed from a homogeneous liquid solution by reducing the solubility of the solute in a saturated solution. In other words, a solution must be "supersaturated" to make crystallization take place. As such, the mass transfer driving force for crystallization as well as the related thermodynamic concepts needs to be introduced first.

2.1 Solubility

The solid-liquid equilibrium data (solubility) are a crucial part of the design, development, and operation of a crystallization process (Schall & Myerson, 2019). Thermodynamic solubility represents the saturation (i.e., maximum) concentration of a compound in equilibrium with an excess of undissolved solid phase in a dissolution process, which means that solubility results from the simultaneous and opposing processes of dissolution and crystallization. Obviously solubility is a characteristic property of a specific solute-solvent combination, and different substances have greatly differing solubility. In addition, solubility highly depends on temperature. In the majority of cases, solubility of a solute in a solvent increases with temperature, although the rate of the increase varies widely from compound to compound.

So far, there is no simple reliable method to accurately predict solubility, even though thermodynamic models (NRTL, UNIFAC, and UNIQUAC) are available to estimate solubility of nonelectrolytes (Reid et al., 1987; Walas, 2013). The accurate solubility data of most chemical compounds need to be determined experimentally by either isothermal or nonisothermal method. The isothermal solubility measurement relies on the equilibrium between a solution and "excess solid," the solution is sampled, and the solution composition is determined separately by HPLC, NMR, UV-vis spectroscopy, or other techniques. The isothermal method tends to be more accurate but generally consumes more amount of compounds and longer time. In nonisothermal measurements, excess solute is mixed with a solvent, and the resulting solution is continuously stirred and slowly heated until all solids dissolve. Typically, the dissolution of a solid during nonisothermal solubility experiments is assessed by measuring the turbidity of the solution. Because heating in nonisothermal method occurs as a series of alternating heating and holding steps, the accuracy of nonisothermal solubility measurements is a function of hold time between temperature ramp steps and the ramp step size (Schall & Myerson, 2019).

2.2 Supersaturation

Supersaturation is the driving force for crystallization, and the state of supersaturation is an essential requirement for all crystallization operations (Mullin, 2001). Supersaturation is generally created by cooling a hot concentrated solution to prepare solutions containing more dissolved solid than that represented by equilibrium saturation. Supersaturation can also be achieved by removing some of the solvent from the solution by evaporation.

Mathematically, supersaturation could be expressed in a number of ways. The most common expressions of supersaturation are the concentration driving force, Δc (kg m^{-3}); the supersaturation ratio, S; and the relative supersaturation or percentage supersaturation, σ. These quantities are defined by:

$$\Delta c = c - c^* \tag{1}$$

$$S = \frac{c}{c^*} \tag{2}$$

$$\sigma = \frac{\Delta c}{c^*} = S - 1 \tag{3}$$

where c is the solution concentration and c^* is the equilibrium saturation at a given temperature.

2.3 Metastable zone and metastable limit

Based on Ostwald (1897), supersaturated solutions are classified into "labile" and "metastable" ones in which spontaneous (primary) nucleation would or would not occur, respectively. The relationship between supersaturation and spontaneous crystallization led to a diagrammatic representation of the metastable zone on a solubility-supersolubility diagram (Fig. 1). Metastable zone (also called metastable zone width) is defined as the zone between the solubility (saturation) curve and the unstable boundary (or metastable limit curve). Solubility curve consists of a series of "clear points" at which suspended solid material disappears from a solution. Whereas the metastable limit curve consists of a series of "cloud points," points on the cooling temperature

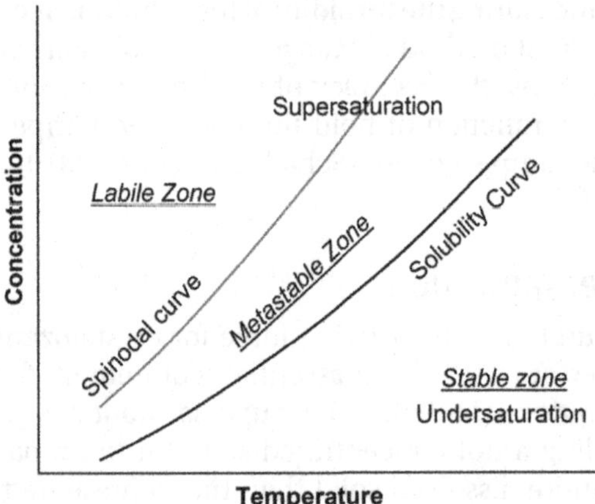

Fig. 1 Solubility and metastable zone width for a solid-liquid system.

profile at which crystal nucleation is firstly observable (Mao et al., 2010). The entire diagram (Fig. 1) is therefore divided into three zones. Zone below the solubility curve is the stable zone where crystallization is impossible, zone above the metastable limit curve is the labile zone where spontaneous crystallization takes place. Spontaneous crystallization is improbable in the metastable zone.

In contrast to the solubility (saturation limit), the metastable limit is not thermodynamically defined and strongly depends on process parameters such as cooling rate, agitation, and impurities (Nývlt et al., 1985). The determination of accurate solubility curve and metastable limit has to rely on experimental measurements with the aid of FBRM or PVM technique (Fujiwara et al., 2002).

3 Kinetics of crystallization in supersaturation

Crystallization from solution is a complex process involving several stages (Erdemir et al., 2019). The first stage is the formation of supersaturated solution, which is followed by the molecule aggregation of the dissolved solute to release of the supersaturation and move the system toward the equilibrium. The molecular aggregation process eventually leads to the formation of nuclei. The birth of the small nuclei in an initially metastable phase is called nucleation (Kashchiev & van Rosmalen, 2003). Crystal growth, which immediately follows nucleation, is governed by the diffusion of solute molecules to the surface of the existing nuclei and their incorporation into the structure of the crystal lattice (Khamskii, 1969). Crystal growth continues until all the solute in excess of saturation is consumed. Because both nucleation and crystal growth are direct functions of the level of supersaturation, a good balance between the relative nucleation and growth rates is crucial to the control of crystallization process.

3.1 Nucleation kinetics

Nucleation is a phase transition process, in which the growing nucleus of the new phase must overcome a barrier. Nucleation may occur spontaneously, or it may be induced artificially by agitation, mechanical shock, friction, and impurities present in solutions (Erdemir et al., 2019). Based on the different nucleation mechanisms, nucleation is classified into two categories, namely primary and secondary nucleation. Primary nucleation occurs in a crystal-free solution. It can be either homogeneous, which means that crystal formation occurs spontaneously in a perfectly clean solution free of impurities, or heterogeneous, where crystal formation is induced by the presence of inert foreign particles such as dust. Homogeneous nucleation rarely occurs in practice, but it forms the basis of several nucleation theories. In industry, any primary nucleation that occurs can be assumed to be heterogeneous (Erdemir et al., 2019).

Secondary nucleation is the generation of new crystals in the vicinity of parent crystals (or seeds) already present in the suspension. Secondary nucleation is thought to take place as a result of several factors, including fluid shear by agitation, contact due to collisions between crystals or between crystals and the vessel walls and agitator, and crystal attrition or breakage (Doran, 2013). Contact nucleation is considered as the most important secondary nucleation mechanism. It could originate either from the solute molecules on the surfaces of the seed crystals or from the semiordered solute clusters at the interface of seed crystals (Cui & Myerson, 2014). Contact nucleation occurs more frequently at high concentration of suspended crystals and/or high stirring speeds.

The rate at which new crystals are formed in a supersaturated solution can be represented by the empirical power-law equations, for example:

$$\text{Primary nucleation}: B = k_b \Delta c^n \tag{4}$$

$$\text{Secondary nucleation}: \ B = k_b' \Delta c^b M_T^m \tag{5}$$

where B represents the primary or the secondary nucleation rate with the unit of number per unit volume per second ($m^{-3}\,s^{-1}$), Δc denotes the supersaturation, k_b and k_b' are the rate constants with different units, M_T is the suspension density in $kg\,m^{-3}$, n and b refer to the orders of primary and secondary nucleation, and m is the empirical exponents.

3.2 Kinetics of crystal growth

Crystal growth is the series of processes by which an atom or a molecule is incorporated into the surface of a crystal, causing an increase in size. Crystal growth is typically described by the change in some dimension of a crystal with time, namely the rate of crystal growth. Because crystals are made up of a number of faces that can grow at different rates, a fundamental expression of the growth rate is the linear growth rate of a particular face (Lee et al., 2019). Normally, the growth rate of the characteristic dimension of a crystal is used to describe the growth of an entire crystal. For example, the characteristic dimension of a spherical crystal would be the diameter. For crystals with other shapes, the characteristic dimension is the second longest dimension (Lee et al., 2019).

The mechanism governing crystal growth, particularly the reason of forming a certain morphology, was not fully understood even to this day although several crystal growth theories were evolved based on considerations of the crystal surface structure. Layer growth, in which crystals are thought to grow in layer-by-layer fashion, predicts that growth takes place at relatively high supersaturation. Nonetheless, layer growth theory fails to account for observed crystal growth rates at low supersaturations (Tibury & Doherty, 2017). A spiral growth model was put forward by Frank, who postulated that dislocations in the crystal are the source of new steps and that a type of dislocation known as a screw dislocation could provide a way for the steps to grow continuously (Frank, 1949). The theory of crystal growth by spiral dislocation was further refined by Burton, Cabrera, and Frank, giving rise to what is known as the BCF theory (Burton et al., 1951). The BCF theory tells us that crystal growth rates vary from a parabolic dependence on supersaturation to a linear dependence as the supersaturation increases.

Crystal growth theories provide a theoretical basis for the correlation of experimental crystal growth data and the determination of kinetic parameters to be used in models of different crystallization processes. In general, the rate of crystal growth can be described by the following equation:

$$G = dL/dt = k_g \Delta C^g \tag{6}$$

where G (m s^{-1}) is the crystal growth rate, k_g is the temperature-dependent growth constant, the relationship of k_g and temperature (T) can be expressed by the Arrhenius equation.

$$k_g = k_{g,0} \exp(-E_G/RT) \tag{7}$$

where $k_{g,0}$ is a preexponential constant and E_G (kJ mol^{-1}) is the activation energy.

4 Crystal size distribution and population balance equations

The crystallized solid products from solution crystallization process may have different sizes and shapes, leading to different physical properties. As such, controlling both crystal shape and sizes is vital for the production of high-quality products with desirable properties. Experimental strategies for identifying crystal size distribution (CSD) and analyzing crystal structures are essential in producing optimal products. CSD can also be predicted by solving the population balance and mass balance equations simultaneously, provided the crystal nucleation and growth kinetics are available. This section starts with an analysis of CSD and how distributions are measured, followed by a general description of the population balance equation (PBE).

4.1 Crystal size distribution

The CSD is one of the important characteristics influencing the end-use applications and interacting strongly with the crystallization process itself. The CSD may be described as the number or mass (volume) of crystals per unit volume within a series of defined size intervals (Chianese, 2012). The number distribution is more directly linked to the physics of nucleation, and its determination is based on the techniques (e.g., electrical sensing zone analysis, the light-blockage) where the particles are examined one by one. The mass distribution is more related to industrial context and can be readily measured by sieving (Rasmuson, 2019).

The particle size distribution may be referred to the density distribution or cumulative distribution. The density distribution (Fig. 2) refers to the number or mass of particles of a certain size range with an average of L, whereas the cumulative distribution expresses number or mass of crystals per unit slurry volume over zero size to the size L.

The most used density distribution variable is the crystal population density, $n(L)$, which is mathematically defined as:

$$n(L) = \frac{dN_c}{dL} \tag{8}$$

where N_c stands for cumulative undersize number of particles per unit volume, and L is the linear dimension used to characterize the size of the particles, the unit of $n(L)$ is number m^{-4}.

The population density can be used to estimate the total number, N_T (number m^{-3}), the total surface area A_T (m^{-1}), and total mass, M_T (kgm^{-3}), of crystals per unit volume in the suspension by means of the following expressions:

$$N_T = \int_0^\infty n(L)dL \tag{9}$$

$$A_T = k_a \int_0^\infty L^2 n(L)dL \tag{10}$$

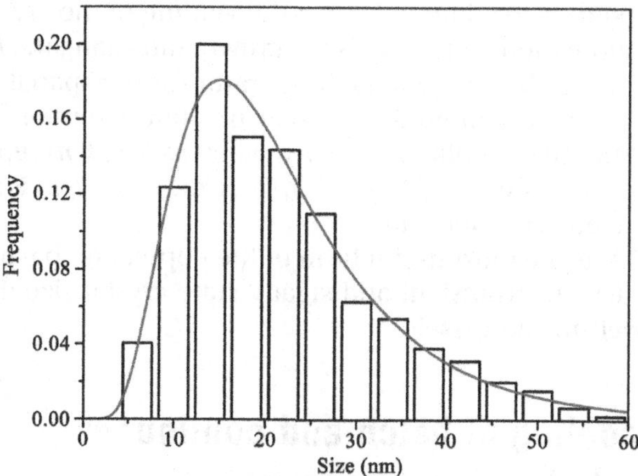

Fig. 2 Density distribution of crystal sizes.

$$M_T = \rho_c k_v \int_0^{\infty} L^3 n(L) dL \tag{11}$$

where ρ_c is the crystal mass density, k_a and k_v are the area shape factor and the volume shape factor.

4.2 The population balance equation

Broadly, the population balance equation (PBE) is an equation in the foregoing population density and may be regarded as representing a number balance on particles of a particular state (Ramkrishna, 2000). It is the capacity of PBE to address the evolutionary aspects of a dispersion that affords its distinctive value to the analysis of dispersed phase systems (solid-liquid, solid-gas, liquid-liquid etc.). In crystallization processes, the PBE is capable of describing the dynamic behavior of distribution functions with one or more internal coordinates, e.g., crystal size. A PBE neatly summarizes all of the operations that affect the number of particles with a particular set of characteristics residing within the control volume, such as nucleation, growth, particle agglomeration, and breakage.

For well-mixed and constant-volume crystallizers, general PDE can be written as:

$$\frac{\partial n}{\partial t} + \nabla n G_j + \sum_k \frac{n_k Q_k}{V} = hL, tn \tag{12}$$

where t (s) denotes time, V (m^3) is the volume of the crystallizer, n is the population density, G_j is the growth rate along the L_j internal coordinate, ∇ is the gradient operator for the spatial coordinates. Q_k and n_k represent the inlet or outlet volume flowrate ($m^3 \cdot s^{-1}$) and the population density, whereas $h(L, t, n)$ represents the net rate of crystal agglomeration/breakage, $h(L, t, n)$ is zero for homogeneous crystallization.

Eq. (12) is the most useful form of the population balance and used to describe transient and steady-state crystal size distribution in well-mixed vessels.

5 Modeling of batch and continuous crystallization processes

5.1 Batch crystallization

Batch crystallizers are extensively used in chemical and pharmaceutical industries to prepare a wide range of high-value crystal products. They are simple and flexible in operations and generally involve less process development than those for a continuous operation (Tavare, 1995). For the purpose of analysis, crystallization in any process configuration may be considered as a competitive process from the solution side and as a consecutive process from the solid side. Achievement of supersaturation may occur by cooling, evaporation, and addition of antisolvents. Supersaturation generation by evaporation is scarcely used in pharmaceutical industry because it is energy-intensive and not suitable for temperature-sensitive products. Supersaturation in batch crystallization is usually generated by cooling, or antisolvent addition, or a combination of both.

Batch crystallization is essentially an unsteady state operation with state variables (supersaturation, mass or surface areas of crystals, etc.) changing with time. Modeling of a batch crystallization process needs to consider the time-dependent batch conservation equations that include the population, mass, and energy balances (Tavare, 1987; Wey, 1985).

5.1.1 Modeling of a batch crystallization process

To better predict and appropriately control the major process variables, namely supersaturation, CSD, suspension density, and possibly product purity of a batch crystallization process, a set of conservation equations consisting of the population and material (solute) balances need to be solved simultaneously with crystal nucleation and growth rate equations.

The population balance

A batch or unseeded semi-batch crystallization process has no net inflow or outflow of the crystals. And usually no more than two internal coordinates are applied to the PBE. For two characteristic dimensions L_1 and L_2 with size-independent growth rates, the PBE takes on the following form:

$$\frac{\partial n}{\partial t}+G_1\frac{\partial n}{\partial L_1}+G_2\frac{\partial n}{\partial L_2}=B_0\delta L_1\delta L_2 \tag{13}$$

where the item $B_0\delta(L_1)\delta(L_2)$ is the statistical function for the net rate of crystal agglomeration with the unit of number·m^{-3}·s^{-1}, B_0 is the rate of nucleation of particles of zero size, and δ is the Dirac delta function.

If direct proportionality of the growth rates G_1 and G_2 with the nuclei dimensions is assumed, Eq. (13) can be further reduced to one-dimensional (1-D) PBE:

$$\frac{\partial n}{\partial t}+G\frac{\partial n}{\partial L}=B_0\delta(L) \tag{14}$$

The initial and boundary conditions of Eq. (14) are as follows:

$$n(L,0)=n_0(L) \tag{15a}$$

$$n(0,t)=\frac{B_0}{G_{L=0}} \tag{15b}$$

In Eq. (15a), the initial distribution, $n_0(L)$ is zero for an unseeded batch crystallization, for a seeded batch process, $n_0(L)$ is described as:

$$n_0(L)=\begin{cases} a_0(L_{\max}-L_s)(L_s-L_{\min}) & \text{with} \quad L_{\min}<L_s<L_{\max} \\ 0 & \text{otherwise} \end{cases} \tag{16}$$

The mass balance of solute

A mass balance equation of the crystallizing solute is described by:

$$\frac{dc}{dt}=-3\rho_c k_v G\int_0^{\infty} n(L,t)L^2 dL \tag{17}$$

with the initial condition of $c(t=0)=c_0$ (18)

where c is the solute concentration in the mother liquor, the physical definitions for other variables are the same as before.

The population and mass balance equations need to be solved together with nucleation and crystal growth rate equations for a specific method to generate supersaturation (such as controlled cooling and/or antisolvent addition) during the batch crystallization. Mathematically, finite difference and finite volume methods can be used to numerically solve the parabolic partial differential equation (PDE) of Eq. (14) and ordinary differential equation (ODE) of Eq. (17) simultaneously. Recently, a high-resolution finite volume method was developed to provide high-accuracy numerical solutions, which is capable of avoiding the numerical smearing and numerical dispersion (nonphysical oscillations) associated with finite difference and other finite volume methods (Gunawan et al., 2004).

5.1.2 Case study I: Seeded batch crystallization of (R)-mandelic acid in the presence opposite enantiomer

Over the past two decades, direct crystallization was used as a separation technique to prepare pure enantiomers from the racemic solutions. The (virtually) pure enantiomers were obtained by a seeded crystallization through the proper selection of the starting composition of the enantiomeric-enriched solution and precise monitoring/control of the enantiomeric excess (*ee*) of the desired enantiomers in the solutions. Zhang et al. utilized a seeded batch cooling crystallization process to produce pure (*R*)-mandelic acid ((*R*)-MA) solid (*ee*>96.4%) from the solution containing quite significant amount of (*S*)-MA (the opposite enantiomer) (Zhang et al., 2010).

Mathematical model without considering crystal agglomeration and breakage in PBE was used to simulate the seeded batch crystallization of (*R*)-MA with all the other equations being the same as Eqs. (15)–(18). Pure (*R*)-MA seeds with crystal size falling within the range of 212–300 μm were used in the batch crystallization. The operating conditions and some model parameters are given in Table 1. The solubility data as well as the kinetic parameters of secondary nucleation and crystal growth of (*R*)-MA can be found elsewhere (Zhang et al., 2010).

Fig. 3 demonstrates the comparison of the model predicted and experimental measured concentration profile for two crystallization batches. The magnitudes of the values for the predicted concentration profiles are in extremely good agreement with those obtained experimentally. The average deviations between the predicted and measured concentrations of (*R*)-MA are less

Table 1 Operating parameters of seeded batch crystallization of (*R*)-MA from enantiomeric-enriched solutions.

Parameter	Symbol	Run1	Run2	Unit
Starting temperature	T_{start}	25	28	°C
Seeding temperature	T_{seed}	23	26	°C
Ending temperature	T_{end}	19	22	°C
Cooling rate after seeding	r_c	0.02	0.02	$°C\,min^{-1}$
Mass of solvent	W_{sol}	150	150	g
Initial total concentration[a]	C_0	0.104	0.209	$g\,g^{-1}\,H_2O$
Initial enantiomeric excess[b]	ee_0	100	60	%
Seed amount	W_{seed}	0.31	0.31	g
Coefficient for seeding $\times 10^3$	$a_0 \times 10^3$	6.6	6.6	(–)
Enantiomeric excess of seeds	ee_s	100	100	%
Crystal density of (*R*)-MA	ρ_c	1349.0		$kg\,m^{-3}$
Shape factor of crystals	k_v	0.12		(–)

Data from Zhang, Y., Mao, S., Ray, A. K., & Rohani, S. (2010). Nucleation and growth kinetics of (R)-mandelic acid from aqueous solution in the presence of the opposite enantiomer. Crystal Growth & Design 10, *2879–2887.*

than 2.0%. The dynamic model could also predict the yield of (*R*)-MA crystals with an average error of $5.78 \pm 1.3\%$. To compare the model predicted CSD of (*R*)-MA crystals with measured data, the model predicted population densities of (*R*)-MA were solved, and the results are illustrated in Fig. 4. It was found that the model predicted CSD is narrower than that of the measured one (Zhang et al., 2010). Finally, the impact of the presence of (*S*)-MA on the optical purity of final crystal product of (*R*)-MA was studied. The optical purity (*ee*) of the final crystal product from Run 2 was found to be 98.2%, which meets the purity requirement of the commercial product of (*R*)-MA.

5.2 Continuous crystallization

Most crystallization processes in the pharmaceutical and fine chemical industry are operated batch-wise. Nevertheless, there is a growing interest in the pharmaceutical industry to move from traditional batch-wise processing to continuous crystallization processes (Alvarez & Myerson, 2010; Yang & Nagy, 2015). Continuous processes have considerable advantages over batch processes in terms of higher reproducibility, higher process efficiency, and lower

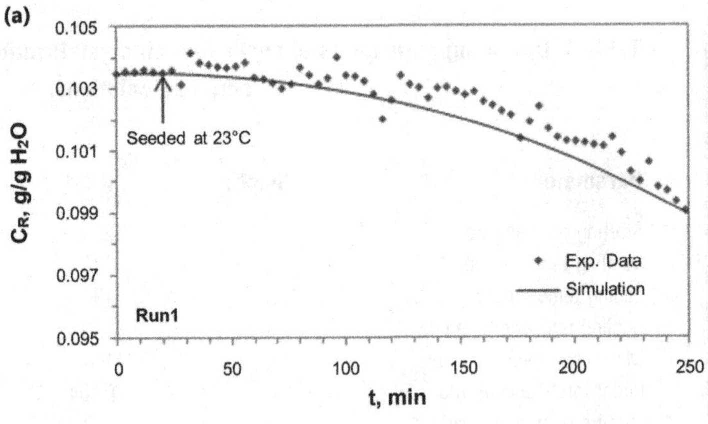

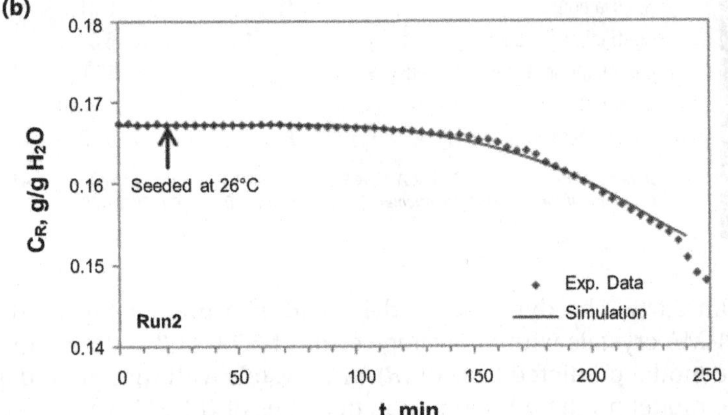

Fig. 3 Measured and calculated concentration profiles of (*R*)-MA for (A) Run 1 and (B) Run 2. Reprinted with permission from Zhang, Y., Mao, S., Ray, A. J., & Rohani, S. (2010). *Crystal Growth & Design, 10,* 2883. Copyright 2010, American Chemical Society.

operating costs when productivity if higher than 20,000 tons per year (Kramer & Lakerveld, 2019).

The two simple and ideal continuous crystallizer models are the plug flow and the complete mixing flow models, which represent two extremes. The mixed suspension mixed product removal (MSMPR) crystallizer introduced by Randolph and Larson represents the perfect mixing flow configuration (Randolph & Larson, 1988). The MSMPR crystallizers have been initially used in bulk chemical industry to produce large quantity of solid products (e.g., KCl and K_2CO_3). Most recently, the MSMPR system has also been used in pharmaceutical industry to produce active pharmaceutical ingredients (APIs).

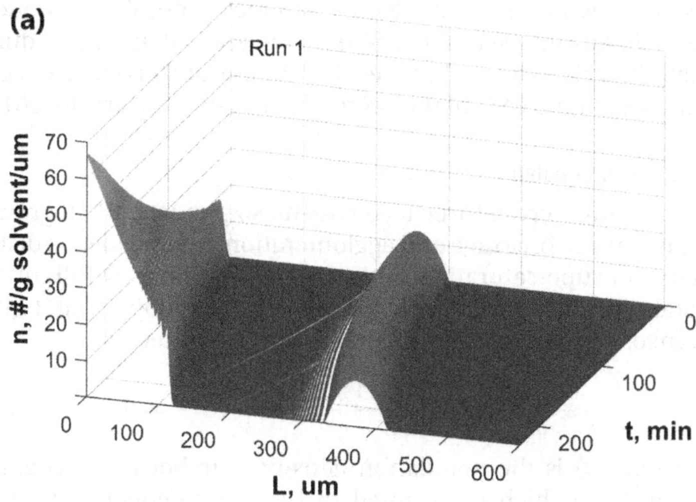

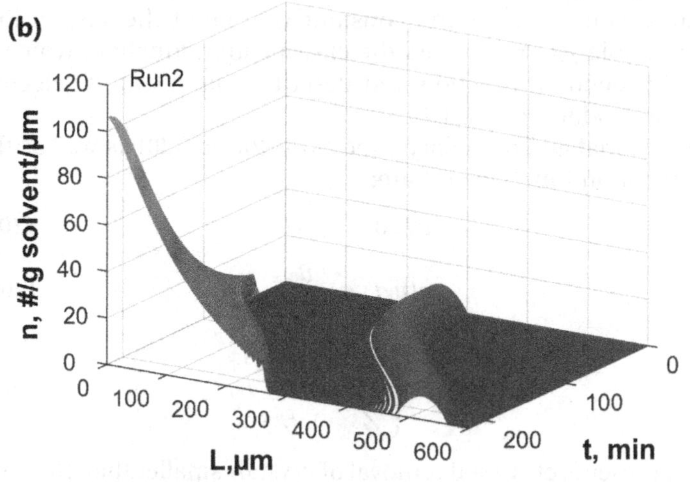

Fig. 4 Model predicted population density for (A) Run 1 and (B) Run 2. Reprinted with permission from Zhang, Y., Mao, S., Ray, A. J., & Rohani, S. (2010). *Crystal Growth & Design, 10,* 2883. Copyright 2010, American Chemical Society.

5.2.1 Modeling of an MSMPR crystallizer

The most important assumption regarding the operation of MSMPR crystallizer is the perfect mixing of the suspension without any spatial variations of attributes within the whole contents. The feed stream of the MSMPR crystallizer is crystal-free, and the product composition is equal to the crystallizer content. Additional concepts with respect to the simulation and operation of an MSMPR crystallizer include fines destruction (R model), clear liquor overflow, selective product removal (Z model), and a crystallizer with fines and selective product removal (R-Z model). The main advantage of the MSMPR concept is that it gives quick insight into the effects of process conditions and crystallizer configuration on the CSD of the product (Kramer & Lakerveld, 2019).

The population balance

Assuming a crystal-free feed stream, size-independent growth rate, no crystal breakage and agglomeration, homogeneous distributions of supersaturation in the crystallizer, the 1-D PBE of the crystals in an MSMPR with fine dissolution loop, clear liquor advance, and product classification can be written as:

$$\frac{\partial n(L,t)}{\partial t} + G\frac{\partial n(L,t)}{\partial L} + \frac{Q_p \gamma(L) n(L,t)}{V} = 0 \tag{19}$$

where $n(L, t)$ is the population density (number m^{-4}), G is the growth rate (which is assumed to be size independent), Q_p is the product flow rate (m^3 s^{-1}), L is the characteristic size of the crystals, and V (m^3) is the constant volume of the suspension. The variable $\gamma(L)$ represents the classification function, which is a size-dependent function and describes the effect of selective fines or product removal.

The initial and boundary conditions for Eq. (20) as well as the classification function $\gamma(L)$ are:

$$n(L,0) = n_0(L) \tag{20a}$$

$$n(0,t) = \frac{B_0}{G_{L=0}} \tag{20b}$$

$$\gamma(L) = \begin{cases} R & 0 \le L \le L_f \\ 1 & L_f < L \le L_p \\ Z & L > L_p \end{cases} \tag{20c}$$

$\gamma(L)$ represents classified removal of crystals smaller than fine cut-off size, L_f, at a rate R times that of an MSMPR feed rate and crystal larger than L_p at a rate Z times the feed flow rate.

The mass balance of solute

The PBE needs to be coupled with the solute mass balance in an MSMPR crystallizer:

$$\frac{dc}{dt} = \frac{Q_i c_i - \left(Q_f + Q_p + Q_{cla}\right)c + Q_f c_f}{V} - 3\rho_c k_v G \int_0^\infty n(L,t) L^2 dL \tag{21}$$

where Q_i is the feed flow rate, Q_p, Q_f, and Q_{cla} denote the flow rates of product, fine removal, and clear liquor overflow streams respectively; c_i, c, and c_f are the solute concentrations in the feed, crystallizer, and fine dissolution loop.

5.2.2 Case study II: The MSMPR crystallization of ciprofloxacin from crude APIs

Antisolvent crystallization of ciprofloxacin, an antibiotic for the treatment of infections caused by bacteria, in the presence of unknown impurities was carried out by Capellades et al. using a single-stage MSMPR crystallizer illustrated in Fig. 5 (Capellades et al., 2019). A large amount of unknown impurities was introduced during multistep syntheses, and those impurities could have significant impact on the crystallization thermodynamics and kinetics of ciprofloxacin. This case study deals with the development of the final crystallization step in manufacturing, from which the drug substance (ciprofloxacin HCl monohydrate) is isolated by continuous antisolvent crystallization.

Continuous crystallization experiments were conducted, exploring four different crystallization temperatures (5, 10, 15, and 25°C) for both purified feed and crude feed. All the MSMPR experiments were conducted with a residence time ($\tau = V/Q$) of 60 min and a constant feed to antisolvent volume ratio of 1:4. The agitation speed was also maintained at 200 rpm. In this study, the mathematical model of MSMPR crystallizer was first used to derive the kinetic parameters for crystallization processes using both purified feed and crude feed streams under various

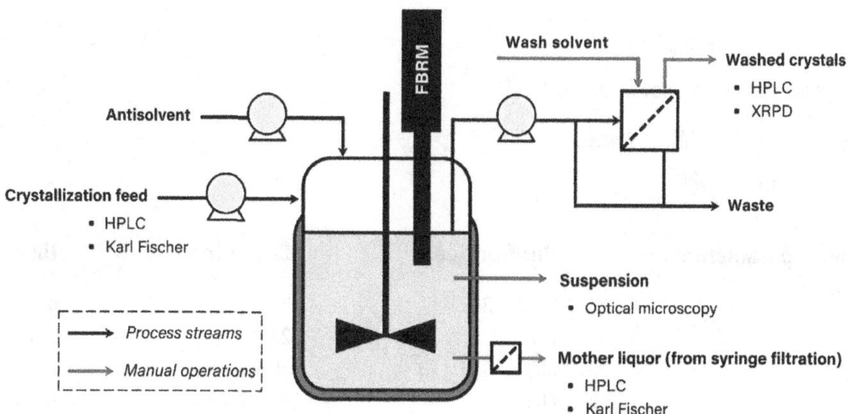

Fig. 5 Schematic diagram of the MSMPR crystallization setup. Reprinted with permission of the publisher from Capellades, G., Wiemeyer, H., & Myerson, A. S. (2019). *Crystal Growth & Design, 19,* 4010. Copyright 2019, American Chemical Society.

crystallization temperatures. Then the obtained kinetic parameters and the mathematical model were employed to predict the performance (yield, CSD, L_{43} etc.) under different operating conditions. The detailed mathematical modeling and the derived crystallization kinetic parameters are listed in Table 2.

Inhibition of crystal growth rates by considerable impurities was noticed in this study, which is reflected from the significant drop in the in the preexponential growth factor, $k_{g,0}$. It was also observed that faster growth kinetics resulted in a higher-impurity incorporation into the crystal products (Capellades et al., 2019). Due to the complexity of the crude feed system, it is impossible to approximate the number of impurities contributing to this effect or the mechanism behind these observations. Nonetheless, the MSMPR mathematical model with the derived kinetic parameter is capable of predicting crystallization yield, crystal size distribution, suspension density, and mother liquor concentration with good accuracy (Table 3 and Fig. 6).

Table 2 Mathematical modeling and kinetic parameters of the MSMPR crystallization of ciprofloxacin HCL monohydrate.

Population balance equation

$$\frac{\partial n(L,t)}{\partial t} + G\frac{\partial n(L,t)}{\partial L} + \frac{n(L,t)Q}{V} = 0$$

Mass balance equation

$$\frac{dc}{dt} = \frac{Q(c_i - c)}{V} - 3\rho_c k_v G \int_0^\infty n(L,t)L^2 dL$$

Nucleation and crystal growth kinetics

$$B = k_{b,0}\exp\left(-\frac{E_B}{RT}\right)\sigma^b M_T \quad \text{where } \sigma = \ln\left(\frac{c}{c^*}\right)$$

$$G = k_{g,0}\exp\left(-\frac{E_G}{RT}\right)\sigma^g$$

Kinetic parameters	Purified feed	Crude feed	Unit
$k_{b,0}$	2157 ± 344	2241 ± 161	$g^{-1}\,s^{-1}$
b	2.0	2.0	(–)
$k_{g,0}$	$(8.0 \pm 10.6) \times 10^6$	55.4 ± 57.9	$\mu m\,s^{-1}$
E_G	46.9 ± 3.1	17.9 ± 2.5	$kJ\,mol^{-1}$
g	1.0	1.0	(–)

Data from Capellades, G., Wiemeyer, H., Myerson, A.S. (2019). Mixed-suspension, mixed-product removal studies of ciprofloxacin from pure and crude active pharmaceutical ingredients: The role of impurities on solubility and kinetics. Crystal Growth & Design, 19, *4008–4018.*

Table 3 Experimental validation for the model prediction of steady state yields, suspension densities, and mother liquor concentrations on independent experiments.

Run	Feed source	Yield (%)		M_T (mg g^{-1})		c (g L^{-1})	
		Experimental	*Predicted*	*Experimental*	*Predicted*	*Experimental*	*Predicted*
1	Purified	89.6±2.8	88.5	19.3±0.3	19.0	1.91±0.01	2.10
2	Crude	76.5±3.4	78.2	14.0±0.4	14.3	3.66±0.06	3.39

Data from Capellades, G., Wiemeyer, H., Myerson, A.S. (2019). Mixed-suspension, mixed-product removal studies of ciprofloxacin from pure and crude active pharmaceutical ingredients: The role of impurities on solubility and kinetics. Crystal Growth & Design, 19, *4008–4018.*

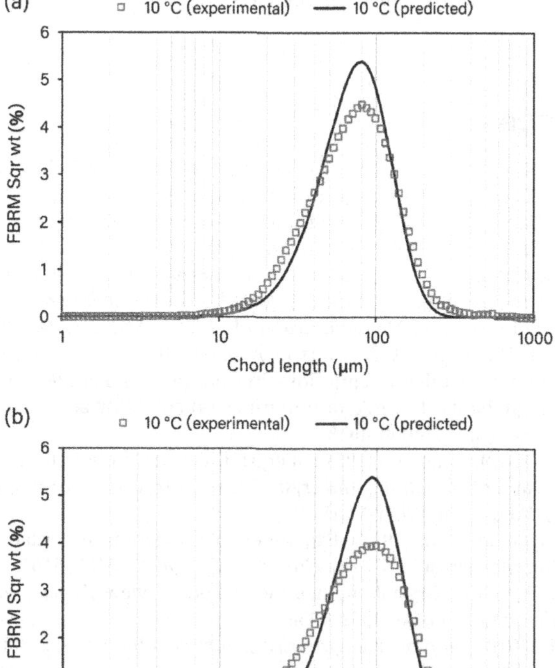

Fig. 6 Experimental and predicted steady-state chord length distributions, (A) Run 1 with purified feed; and (B) Run 2 with crude feed. Reprinted with permission of the publisher from Capellades, G., Wiemeyer, H., & Myerson, A. S. (2019). *Crystal Growth & Design, 19*, 4013. Copyright 2019, American Chemical Society.

Summary

Solution crystallization plays an important role in manufacturing fine chemicals, pharmaceuticals, and many other value-added products. The quality control measures of final products of crystallization, such as purity, crystal morphology, crystal polymorph, and crystal size distribution, highly depend on many process factors (crystallization mode, type of crystallizer, etc.), particularly the way and the extent of the release of supersaturation. Process modeling and simulation not only provide quantitative assessment of the process but also enable the product quality control with the aid of in situ monitoring technology. This chapter will be mainly focused on the introduction of essential thermodynamics and kinetics of solution crystallization process, followed by the mathematical models and simulation case studies of batch and continuous crystallization processes.

References

Alvarez, A. J., & Myerson, A. S. (2010). Continuous plug flow crystallization of pharmaceutical compounds. *Crystal Growth & Design, 10*, 2219–2228.

Anwar, J., & Zahn, D. (2017). Polymorphic phase transitions: Macroscopic theory and molecular simulation. *Advanced Drug Delivery Reviews, 117*, 47–70.

Burton, W. K., Cabrera, N., & Frank, F. C. (1951). The growth of crystals and the equilibrium structure of their surfaces. *Philosophical Transactions of the Royal Society of London. Series A, Mathematical and Physical Sciences, 243*, 299–358.

Capellades, G., Wiemeyer, H., & Myerson, A. S. (2019). Mixed-suspension, mixed-product removal studies of ciprofloxacin from pure and crude active pharmaceutical ingredients: The role of impurities on solubility and kinetics. *Crystal Growth & Design, 19*, 4008–4018.

Chianese, A. (2012). Characterization of crystal size distribution. In A. Chianese, & H. J. M. Kramer (Eds.), *Industrial crystallization process monitoring and control* (pp. 1–6). Weinheim: Wiley-VCH.

Cui, Y., & Myerson, A. S. (2014). Experimental evaluation of contact secondary nucleation mechanisms. *Crystal Growth & Design, 14*, 5152–5157.

Desiraju, G. R. (2007). Crystal engineering: A holistic view. *Angewandte Chemie, International Edition, 46*, 8342–8356.

Doran, P. M. (2013). *Bioprocess engineering principles* (2nd ed., pp. 538–562). Waltham, MA: Elsevier.

Erdemir, D., Lee, A. Y., & Myerson, A. S. (2019). Crystal nucleation. In A. S. Myerson, D. Erdemir, & A. Y. Lee (Eds.), *Handbook of industrial crystallization* (pp. 76–88). New York: Cambridge University Press.

Frank, F. C. (1949). The influence of dislocations on crystal growth. *Discussions of the Faraday Society, 5*, 48–54.

Fujiwara, M., Chow, P. S., Ma, D. L., & Braatz, R. D. (2002). Paracetamol crystallization using laser backscattering and ATR-FTIR spectroscopy: Metastability, agglomeration, and control. *Crystal Growth & Design, 2*, 363–370.

Gao, Z., Rohani, S., Gong, J., & Wang, J. (2017). Recent developments in the crystallization process: Toward the pharmaceutical industry. *Engineering, 3*, 343–353.
Gunawan, R., Fusman, I., & Braatz, R. D. (2004). High resolution algorithms for multidimensional population balance equations. *AICHE Journal, 50*, 2738–2749.
Kashchiev, D., & van Rosmalen, G. M. (2003). Review: Nucleation in solutions revisited. *Crystal Research and Technology, 38*, 555–574.
Kee, N. C. S., Tan, R. B. H., & Braatz, R. D. (2009). Selective crystallization of the metastable α-form of L-glutamic acid using concentration feedback control. *Crystal Growth & Design, 9*, 3044–3051.
Khamskii, E. (1969). *Crystallization from solutions and melts* (pp. 1–13). New York: Springer.
Kramer, H. J. M., & Lakerveld, R. (2019). Selection and design of industrial crystallizers. In A. S. Myerson, D. Erdemir, & A. Y. Lee (Eds.), *Handbook of industrial crystallization* (pp. 198–200). New York: Cambridge University Press.
Lee, A. Y., Erdemir, D., & Myerson, A. S. (2019). Crystals and crystal growth. In A. S. Myerson, D. Erdemir, & A. Y. Lee (Eds.), *Handbook of industrial crystallization* (pp. 41–51). New York: Cambridge University Press.
Lewis, A. E., Seckler, M. M., Kramer, H., & Van Rosmalen, G. (2015). *Industrial crystallization: Fundamentals and applications* (pp. xxiii–xxiv). New York: Cambridge University Press.
Mao, S., Zhang, Y., Rohani, S., & Ray, A. K. (2010). Kinetics of (*R,S*)- and (*R*)-mandelic acid in an unseeded cooling batch crystallizer. *Journal of Crystal Growth, 312*, 3340–3348.
Mullin, J. W. (2001). *Crystallization* (4th ed., pp. 123–132). Oxford: Butterworth-Heinemann.
Nývlt, J., Söhnel, O., Matachová, M., & Broul, M. (1985). *The kinetics of industrial crystallization*. Amsterdam: Elsevier.
Ostwald, W. (1897). Studien über die bildung und umvandlung fester köper. *Zeitschrift für Physikalische Chemie, 22*, 289–330.
Ramkrishna, R. (2000). *Population balances: Theory and applications to particulate systems in engineering* (pp. 4–20). New York: Academic Press.
Randolph, A. D., & Larson, M. A. (1988). *Theory of particulate processes: Analysis and techniques of continuous crystallization* (2nd ed., pp. 64–100). San Diego: Academic Press.
Rasmuson, A. C. (2019). Crystallization process analysis by population balance modeling. In A. S. Myerson, D. Erdemir, & A. Y. Lee (Eds.), *Handbook of industrial crystallization* (pp. 172–193). New York: Cambridge University Press.
Reid, R. C., Prausnitz, J. M., & Poling, B. E. (1987). *The properties of gases and liquids* (4th ed.). New York: McGraw-Hill.
Schall, J. M., & Myerson, A. S. (2019). Solutions and solution properties. In A. S. Myerson, D. Erdemir, & A. Y. Lee (Eds.), *Handbook of industrial crystallization* (pp. 1–23). New York: Cambridge University Press.
Tavare, N. S. (1987). Batch crystallizers: A review. *Chemical Engineering Communications, 61*, 259–318.
Tavare, N. S. (1995). *Industrial crystallization process simulation analysis and design* (pp. 93–100). New York: Springer.
Tibury, C. J., & Doherty, M. F. (2017). Modeling layered crystal growth at increasing supersaturation by connecting growth regimes. *AICHE Journal, 64*, 1338–1352.
Walas, S. M. (2013). *Phase equilibria in chemical engineering* (pp. 395–432). Stoneham: Butterworth-Heinemann.
Wey, J. S. (1985). Analysis of batch crystallization processes. *Chemical Engineering Communications, 35*, 231–252.

Wu, J. X., Xia, D., van den Berg, F., Amigo, J. M., Rades, T., Yang, M., & Rantanen, J. (2012). A novel image analysis methodology for online monitoring of nucleation and crystal growth during solid state phase transformations. *International Journal of Pharmaceutics, 433*, 60–70.

Yang, Y., & Nagy, Z. K. (2015). Advanced control approaches for combined cooling/antisolvent crystallization in continuous mixed suspension mixed product removal cascade crystallizers. *Chemical Engineering Science, 127*, 362–373.

Zhang, Y., Mao, S., Ray, A. K., & Rohani, S. (2010). Nucleation and growth kinetics of (*R*)-mandelic acid from aqueous solution in the presence of the opposite enantiomer. *Crystal Growth & Design, 10*, 2879–2887.

PART III

Macro scale modeling of process systems

7

Fuel processing systems

Prakash V. Ponugoti and Vinod M. Janardhanan
Department of Chemical Engineering, Indian Institute of Technology Hyderabad, Hyderabad, Telangana, India

1 The need for fuel processing units

Fuel cells in general deliver the best performance when they are operated on hydrogen. The H_2-rich mixtures that are produced from steam reforming of natural gas or coal gasification generally contains other species such as CO, CO_2, CH_4, and N_2 (You et al., 2012). As per ISO FDIS 1467 specification, the H_2 should be 99.97% pure, and the maximum allowable CO concentration is 0.2 ppm for low-temperature polymer electrolyte membrane (LTPEM) fuel cells (U.S. Department of Energy et al., 2016). ISO FDIS 1467 specification also puts upper limit on other contaminants such as He, CO_2, NH_3, HCOOH, HCHO, and H_2S. The impurities present in the primary fuel source, if not removed before processing, enter into the cell stack and cause cell performance degradation (Cheng et al., 2007). Today, more than 90% of H_2 is produced by the steam reforming of natural gas (Cheng et al., 2007; Rajalakshmi et al., 2003). Reforming of oxygenated hydrocarbons (alcohols) and partial oxidation of hydrocarbons are other methods for the production of H_2 (Bharadwaj & Schmidt, 1995; Pan et al., 2005). The reformate gas produced by these reforming methods contains CO as the major impurity. The CO tolerance of LTPEM fuel cells is very poor due to the use of Pt as the electrode catalyst. At low temperatures, the CO that binds strongly on active Pt sites does not desorb. Thus, the active sites eventually get completely covered by CO, preventing the H_2 from being adsorbed on the Pt surface at normal operating potentials (Borup et al., 1997; Farrell et al., 2007). The level of performance drop as a result of CO contamination depends on the concentration of CO, exposure time, operating temperature, and the electrode catalyst used (Cheng et al., 2007). Therefore, the CO

Modeling of Chemical Process Systems. https://doi.org/10.1016/B978-0-12-823869-1.00008-9

concentration needs to be further reduced by subjecting the syngas to shift reactions or pressure swing adsorption (PSA) for use in PEM fuel cells (You et al., 2012). Although, PSA can produce 99.999% pure H_2 (Ertl et al., 2008), it is not an option for portable fuel cell systems.

In addition to CO, the reforming process generally results in the production of CO_2. The steam reforming of natural gas results in close to 25% CO_2 in the product gas mixture (Cheng et al., 2007; Rajalakshmi et al., 2003). There are claims that the CO_2 reacts with H_2 according to reverse water gas shift reaction (rWGSR) internally within the anode leading to the formation of CO, which eventually impedes the cell operation (Gu et al., 2005). However, since rWGSR is endothermic, the kinetics is not really favored at the operating temperature of LTPEM fuel cell, and the CO_2 is more likely to affect the cell performance due to fuel dilution (U.S. Department of Energy et al., 2016), which brings down the Nernst potential and current density. Methanol (CH_3OH) is another potential fuel for portable fuel cell applications, and the steam reforming of CH_3OH, mainly produces CO, CH_4, HCHO, and HCOOH as impurities (Narusawa et al., 2003). The CO content can be brought down to 50–100 ppm by shift reaction, however, can lead to the formation of CH_4, as methanation is a side reaction that may occur alongside water gas shift. The concentration of CH_4 that is formed as a result of shift reaction or CH_3OH reforming does not lead to cell degradation, rather its effect on cell performance is due to fuel dilution. Unlike CH_4, HCHO does affect the performance of the cell. However, its poisoning effect is not as severe as CO. The other impurity, HCOOH does not affect the cell performance even at 1000 ppm. Both CO and HCHO poisoning of LTPEM fuel cells are reversible in nature, i.e., the cell performance can be recovered by removing the impurities from the fuel stream (Narusawa et al., 2003). NH_3 is another impurity for PEM fuel cell systems, which enters the feed stream if hydrogen-rich reformate is produced from the reforming of distributed natural gas where NH_3 is used as a tracer (Cheng et al., 2007). Traces of NH_3 can also be produced from the partial oxidation reactions that uses air (Rajalakshmi et al., 2003). NH_3 concentrations of more than 10 ppm can cause substantial loss in cell performance (Rajalakshmi et al., 2003).

Similar to the quality of fuel, the quality of air is an equally important factor, which decides the performance and durability of the fuel cells. Although, pure O_2 can be used in laboratory research, for all practical applications, air is used at the cathode side, and the contaminants such as NO_x and SO_x present in air can severely affect the cell performance. The concentration of NO_x and SO_x present in atmospheric air is location-specific

(Zamel & Li, 2011). According to some reports, the cell performance loss due to NO_x is reversible under short exposure times. However, long-term exposure leads to irreversible loss in activity (Zamel & Li, 2011).

The high-temperature polymer electrolyte membrane fuel cells (HTPEMs) operate in the temperature range of 120–180°C and are more tolerant to CO (Li et al., 2003). All contaminants that affect the performance of LTPEM fuel cell also affect HTPEM fuel cell performance. However, owing to the higher operating temperature, the tolerance of HTPEM fuel cells toward these impurities is better than that of LTPEM fuel cells. The desorption kinetics of CO on Pt becomes more favorable as the operating temperature increases (Unnikrishnan et al., 2018, 2019). Therefore, the CO tolerance increases with increase in operating temperature. For instance, when a CO concentration of 1% leads to significant performance losses in an HTPEM fuel cell at 125°C, the same cell can be operated with almost same performance as that of neat H_2 at 175°C (Li et al., 2003). HTPEM fuel cells can tolerate up to 3% CO with minimal performance loss (Zuliani & Taccani, 2012). There are also reports wherein the reformate fuel from CH_3OH reforming is directly fed to HTPEM fuel cell without any additional purification steps involved (Pan et al., 2005).

For both HTPEM and LTPEM fuel cells, H_2S is a poison, which leads to irreversible losses in the electrode activity (Prass et al., 2019). Very low concentrations of sulfur compounds in the fuel stream can lead to significant deactivation of electrode catalysts. In addition to H_2S, these compounds include carbonyl sulfide (COS), carbon disulfide (CS_2) and methanethiol (CH_3SH) (U.S. Department of Energy et al., 2016). Like CO poisoning, performance degradation due to sulfur compounds is also cumulative in nature. The degradation increases with increase in exposure time. However, its exact effect depends on the catalyst loading and relative humidity. HTPEM fuel cells operating with 5 ppm H_2S with degradation rate similar to that of pure H_2 over 3000 h have been reported (Schmidt & Baurmeister, 2006). Reports are not available on the tolerance of HTPEM fuel cells toward HCHO. The effect of fuel dilution on the performance of PEMFC was mentioned earlier, and any fuels that cause fuel dilution effect in LTPEM fuel cell will have the same effect in HTPEM fuel cell as well.

Unlike the PEM fuel cells, the high-temperature fuel cells such as solid oxide fuel cells (SOFCs) and molten carbonate fuel cells (MCFCs) are fuel flexible owing to their high operating temperature. The stringent fuel quality requirements that are necessary for the operation of PEM fuel cells are not required for high-temperature fuel cells. In fact, as the operating temperature

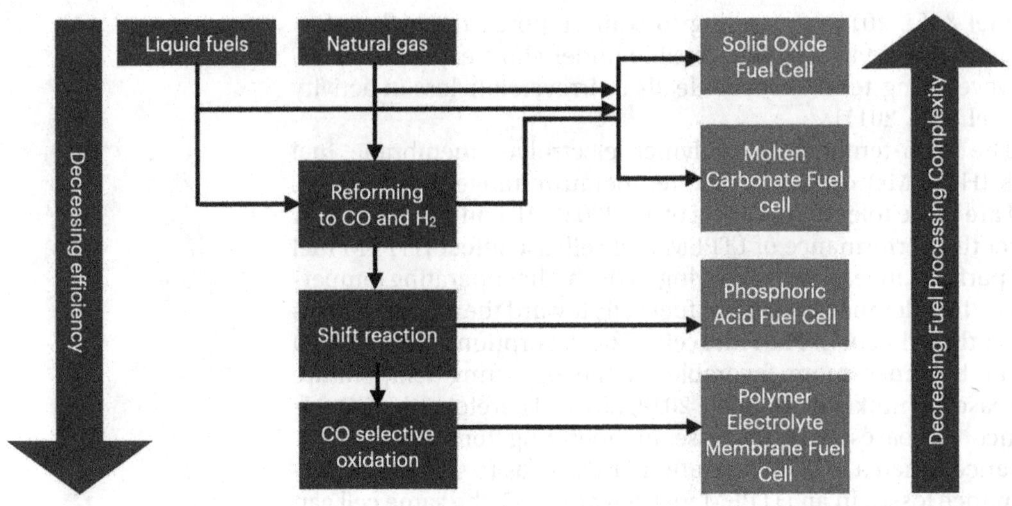

Fig. 1 Fuel processing complexity and fuel cell operating temperature.

increases, the complexity associated with fuel processing decreases as shown in Fig. 1.

The CO, which is a poison in the PEM fuel, can be electrochemically oxidized in high-temperature fuel cells. Both SOFCs and MCFCs are capable of electrochemical oxidation of syngas, which can be produced internally within the stack from natural gas/CH_4 (Steele, 1999). These are called internal reforming cells. Although, it is possible to internally reform CH_4 in SOFC, the Ni, which is typically used as active electro-catalyst in SOFC, is prone to carbon deposition when enough steam is not present in the anode (Takeguchi et al., 2003). Biogas is another potential fuel for SOFC, which is produced by the anaerobic digestion of organic matter and is rich in CH_4. The typical composition of biogas is 50%–75% CH_4, 25%–50% CO_2, 0%–10% N_2, and 0%–3% H_2S (Appari et al., 2014). Due to the high CH_4 content, SOFCs can be operated directly on biogas with modified anode materials if it does not contain H_2S (Shiratori et al., 2008). These anode materials (certain perovskites) must be capable of suppressing the coke formation.

H_2S is one of the major impurities that can severely poison of Ni-based anodes. However, the performance drop in the presence of H_2S depends on the anode materials used and the operating conditions such as temperature and current density. A concentration of 50 ppm H_2S in H_2 can lead to 17% drop in current density at 0.6 V. If operated on biogas, even 1 ppm H_2S can lead to 9% drop in voltage at a current density of 200 mA/cm^2 (Shiratori et al., 2008). However, at low concentrations of H_2S, its poisoning effect is reversible at temperatures above 700°C (Zha et al., 2007). The formation of Ni sulfide also leads to loss in catalytic activity. However, its formation from H_2S occurs only when the concentration of H_2S in the fuel stream is more than 8000ppm at temperatures around 1000°C. Both of these conditions are very unlikely during SOFC operation. Other sulfur compounds such as COS and CH_3SH also lead to cell performance drop. Internally at SOFC operating temperatures, these impurities get converted to H_2S (Haga et al., 2008). So, the mechanism of electrode poisoning due to other sulfur compounds is finally due to H_2S poisoning.

Siloxanes are another class of species that leads to cell performance loss. If the biogas is generated from wastewater treatment plants, it often contains siloxanes or silica-containing compounds. Siloxane concentrations ranging from 2 to 400 mg/Nm^3 have been reported in the literature (Madi et al., 2015). Cyclic siloxanes containing four or five Si atoms are most abundant in biogas generated from wastewater. Presence of even 1 ppm siloxane containing four Si atoms leads to significant increase in the polarization resistance of the cell and causes faster degradation at higher concentrations (Madi et al., 2015).

NH_3, which is a poison for PEM fuel cell, is a fuel when it comes to SOFC. NH_3 is a rich carrier of H_2 and has established distribution infrastructure. As a potential H_2 carrier, NH_3 can solve the storage and distribution challenges associated with H_2. Although slightly lower than H_2, stable performance has been reported by using NH_3 (Ma et al., 2006). Cl is another contaminant that is known to degrade the cell performance when present in considerable quantities. The presence of Cl can lead to the formation of $NiCl_2$ even with 10ppm at 800°C (Haga et al., 2008).

Most of the works reported in the literature study the effect of one impurity at a time. However, for a comprehensive understanding on the effect of contaminants on real-time operation of fuel cells, it is important to consider multiple contaminants simultaneously (U.S. Department of Energy et al., 2016). The levels of impurities that can be tolerated by different types of fuel cells are given in Table 1.

Thus, a fuel processing unit is essential to ensure that the reformate fuel sent to the fuel cell is free from impurities or contains only tolerable levels so that it does not damage the fuel cell components.

2 Fundamentals of fuel processing

Fuel processing deals with the conversion of primary fuels such as natural gas, petroleum, biogas, and coal into a hydrogen-rich mixture that can be used in fuel cells. Steam reforming, dry reforming, partial oxidation, and autothermal reforming are the common

Table 1 Fuel cell type and tolerance to impurities.

Fuel cell type	Impurities	Tolerance level
LTPEM	CO	0.2 ppm
	CO_2	2 ppm
	H_2S	0.004 ppm
	NH_3	0.1 ppm
	HCOH	1 ppm
	HCOOH	2 ppm
HTPEM	CO	3 ppm
	H_2S	5 ppm
SOFC	H_2S[a]	–
	COS[a]	–
	CH_3SH[a]	–
	Siloxane	Less than 1 ppm

[a]No standards available.

reforming techniques in use today. Water electrolysis in PEM and solid oxide electrolyzer is another route for the production of H_2. This section will introduce the reader to the basic principles of common reforming techniques. The choice of the fuel processing method depends on the operating requirements of the fuel cell stack.

2.1 Steam reforming

Steam reforming (SR) is a technique where the primary hydrocarbon or alcohol fuels are converted in the presence of a catalyst using steam (H_2O) into a mixture of H_2, CO, CO_2, CH_4, and unconverted steam. Primarily one is interested in a mixture of H_2 and CO from steam reforming reaction, and the reforming of hydrocarbons results in an H_2/CO ratio more than 1. Since water and primary fuels are stable molecules, the reaction is highly endothermic. For instance, CH_4 steam reforming has a reaction enthalpy of 206 kJ/mol (Ashrafi et al., 2008). For large-scale operation, the reaction is typically carried out in tubular reactors in the temperature range of 400–750°C (Xu & Froment, 1989). Similar to temperature, pressure also affects the reaction equilibrium according to *Le* Chatelier's principle. Low pressure favors steam reforming reaction in the forward direction due to the higher number of molecules on the product side. However, it is generally carried out at supply pressure of the product gas. Industrial steam reformers operate at pressures above 20 bar due to the pressure requirements of the downstream process. Unlike these industrial reformers, the fuel processor for fuel cell applications operates at atmospheric-pressure conditions since fuel cells operate at atmospheric-pressure conditions.

The general stoichiometric equation for the steam reforming of hydrocarbons is

$$C_nH_m + nH_2O \leftrightarrow (n + m/2)H_2 + nCO, \quad (1)$$

and for alcohol fuels, the stoichiometric equation can be written as

$$C_nH_mOH + (2n - 1)H_2O \leftrightarrow nCO_2 + \frac{(4n + m - 1)}{2}H_2 \quad (2)$$

Although, this reaction can be performed on a number of catalysts, industrially Ni is the preferred one due to its lower cost compared with the Pt group metals. One of the problems associated with Ni is its propensity to form carbon when enough steam is not present. Generally, steam reforming is carried out at a steam to carbon (S/C) ratio of 3 to avoid carbon deposition, which affects the catalyst activity. At temperatures above 600–650°C, carbon deposition may occur due to thermal cracking (Ertl et al., 2008). Since there is always excess steam present during steam reforming of hydrocarbon fuel, the reaction is always accompanied by water gas shift reaction (WGS).

$$CO + H_2O \leftrightarrow H_2 + CO_2 \quad (3)$$

However, the same is not true in the case of reforming of alcohol fuels, due to the absence of CO in the product mixture. Although, the overall reaction can be represented by the stoichiometry of Eq. (1) or Eq. (2), in reality the reaction proceeds through a number of elementary steps.

2.2 Dry reforming

In dry reforming (DR), the hydrocarbon fuel is converted to synthesis gas using CO_2 as a reactant. Therefore, dry reforming is also known as CO_2 reforming. In today's context of changing global climate, CO_2 reforming is a very attractive process as it can fix CO_2 back into the carbon cycle. Thermodynamically this reaction is more endothermic than steam reforming reaction with enthalpy of reaction being 247 kJ/mol for CH_4 dry reforming (Ashrafi et al., 2008). It is also more challenging to practice it industrially due to the carbon formation reactions that accompany. The carbon may result from cracking reactions of CH_4 or from Boudouard reaction (Pawar et al., 2015). Similar to steam reforming, Ni is the preferred catalyst for dry reforming as well. The dehydrogenation reactions of hydrocarbons leading to the formation of carbon occur at high temperatures, while Boudouard reaction is generally favored at lower temperatures (Haag et al., 2007). The stoichiometric equation for the dry reforming of hydrocarbons is

$$C_nH_m + nCO_2 \leftrightarrow 2n\,CO + \frac{m}{2}H_2 \tag{4}$$

The reaction results in an H_2/CO ratio of 1 for CH_4. For application in fuel cells, particularly for PEM fuel cells, the off gases from dry reforming need to be subjected to water gas shift reaction or pressure swing adsorption. Most of the experimental work reported in the literature deals with the reforming of stoichiometric mixtures of CO_2 and CH_4. The carbon formation tendency of dry reforming reaction increases with increasing CH_4 concentration. Studies that focus on non-stoichiometric CH_4-rich mixtures are important from the context of utilization of biogas as source for syngas since the CH_4-to-CO_2 ratio in biogas is more than 1. Ru and Rh are proven to be promising catalysts due to their high dry reforming rate and low carbon deposition (Ferreira-Aparicio et al., 2000). However, these materials are prohibitively expensive for industrial-scale production of syngas from dry reforming.

Several recent research on dry reforming is focused on the identification of new catalyst materials that can resist carbon formation and hence to avoid catalyst deactivation. Ormerod and coworkers have reported Ni-doped $SrZrO_3$ catalyst that was nearly carbon resistant for non-stoichiometric CH_4-rich mixture (Evans, Good, et al., 2014; Evans, Staniforth, et al., 2014).

2.3 Partial oxidation

Catalytic partial oxidation (CPOx) is another important reaction to covert hydrocarbons and alcohols into synthesis gas. The general stoichiometric equation for partial oxidation of hydrocarbon fuels can be written as

$$C_nH_m + \frac{n}{2}O_2 \leftrightarrow nCO + \frac{m}{2}H_2 \tag{5}$$

For alcohol fuels the equation becomes

$$C_nH_mOH + \frac{(n-1)}{2}O_2 \leftrightarrow nCO + \frac{(m+1)}{2}H_2 \tag{6}$$

CPOx produces a H_2/CO ratio close to 2 and is particularly attractive for fuel cell applications due to fast reaction times. However, again for application in PEM fuel cells, the product gas needs to be subjected to shift reactions to bring down the CO content. Pt and Rh are excellent catalysts for CPOx (Schwiedernoch et al., 2003). The reaction is slightly exothermic, with a reaction enthalpy of 36 kJ/mol for CH_4. Unlike steam and dry reforming, external energy supply is not required for CPOx. Once the reaction mixture is lit off, no further heat supply to the reactor is required (Schwiedernoch et al., 2003). Another advantage of CPOx reaction over steam reforming is the short contact times. For steam reforming, contact times of the order of 1 s are required to achieve appreciable conversion of CH_4. Whereas CPOx only requires milliseconds (Hickman & Schmidt, 1993). This is really promising for fuel cell systems, as CPOx reactors can quickly respond to load changes. Due to the availability of O_2 in the system, total oxidation of HC can compete with POx. If total oxidation occurs alongside POx, then it is possible that the steam, which is produced as a result of total oxidation, can undergo steam reforming with the HC fuel. The work of Horn et al. suggests that H_2 and CO are formed as a result of steam reforming once the O_2 is completely consumed near the reactor inlet (Horn et al., 2006).

2.4 Water gas shift reaction

Water gas shift reaction is required to reduce the CO concentration in the product gases from SR or DR. The reaction stoichiometry is given by Eq. (3). The WGS reaction is a slightly exothermic reaction with a reaction enthalpy equal to 41 kJ/mol and is reversible. Low temperatures are required for high-equilibrium conversions; however, the kinetics are not favored at low temperatures. The reaction is carried out in two steps, a high-temperature shift (HTS) at 500°C using Fe_2O_3/Cr_2O_3 catalyst followed by a low-temperature shift (LTS) reaction using $Cu/Zn/Al_2O_3$ in the temperature range of 210–230°C (Wheeler et al., 2004). Practically it is not possible to convert all the CO by WGS. Therefore, the product stream from WGS is subjected to preferential oxidation reaction to further reduce the CO to ppm levels. The preferential oxidation of CO is generally carried out using Pt-Fe/Al_2O_3 (Choi & Stenger, 2005). The Fe and Cu catalysts that are conventionally used in WGS suffer from coke formation by Fe in the presence of excess fuel and the irreversible oxidation of Cu if exposed to O_2 (Twigg, 1996). Both of these are possible scenarios with respect to incomplete conversion of fuel and oxidant in the reforming and partial oxidation reactions, respectively. Although the reaction kinetics are not favorable at low temperature, practically one requires high conversion near reaction equilibrium, and high conversions are possible only at lower temperatures. This led to the exploration of other catalysts such as noble metals and ceria-based systems as these catalysts could overcome the issues associated with Fe and Cu catalysts (Whittington et al., 1995). For applications in fuel cell system, the WGS reaction should be carried out in short residence time reactors so that the fuel processing unit can respond quickly to the load requirements of the fuel cell. Schmidt and coworkers reported on the activity of various catalyst for water gas shift reaction. According to their analysis, the activity follows the order Ni > Ru > Rh > Pt > Pd with Ni giving highest conversion of 80%. With the addition of Ce, they could increase the activity of Pt catalyst (Wheeler et al., 2004).

2.5 Autothermal reforming

In autothermal reforming (ATR), the fuel is converted in the presence of H_2O/CO_2 and O_2. The SR reaction is highly endothermic, whereas oxidation reaction is exothermic. ATR is a thermoneutral process, where the heat requirement for SR is supplied by the partial oxidation reactions. The general stoichiometric equation for ATR is written as

$$C_nH_M + nH_2O + \frac{n}{2}O_2 \leftrightarrow nCO_2 + \left(\frac{m}{2} + n\right)H_2 \quad \Delta H \leq 0 \qquad (7)$$

The advantage of autothermal reforming compared with other processes is that the reformer system becomes more compact as no external heat supply is required. Furthermore, it produces far less CO compared with steam reforming or partial oxidation and is free from carbon formation. For ATR of CH_4 under adiabatic conditions, thermodynamic analysis dictates an air-to-fuel (A/F) ratio of 2.1 and steam-to-fuel (S/F) ratio of 1.1; however, to compensate for the heat loss, the reaction is generally carried out at higher A/F than the theoretical requirement (Hoang & Chan, 2004). Ni and Rh are the commonly used catalysts for autothermal reforming of CH_4.

3 Recent developments in the reforming of common fuels

The aforementioned reforming technologies can be applied to convert a number of primary fuels to H_2-rich mixture for use in fuel cell systems.

SOFCs are capable of operating on CH_4 rich fuels; however, for reasons of cell longevity and stable operation, the primary fuels are reformed in a fuel processing unit and the gas mixture resulting from the fuel processing unit is sent to the cell stack. Fig. 2 shows the various sources of primary fuel and the treatment methods for use in various types of fuel cells. The last part of the figure shows electrolysis of water for the production of H_2 using nuclear and renewable sources such as solar and wind. Reverse operation of fuel cells (regenerative fuel cells) to produce H_2 from H_2O electrolysis has gained significant interest in the recent past, and therefore, it is discussed as a separate section (Section 4) of this chapter.

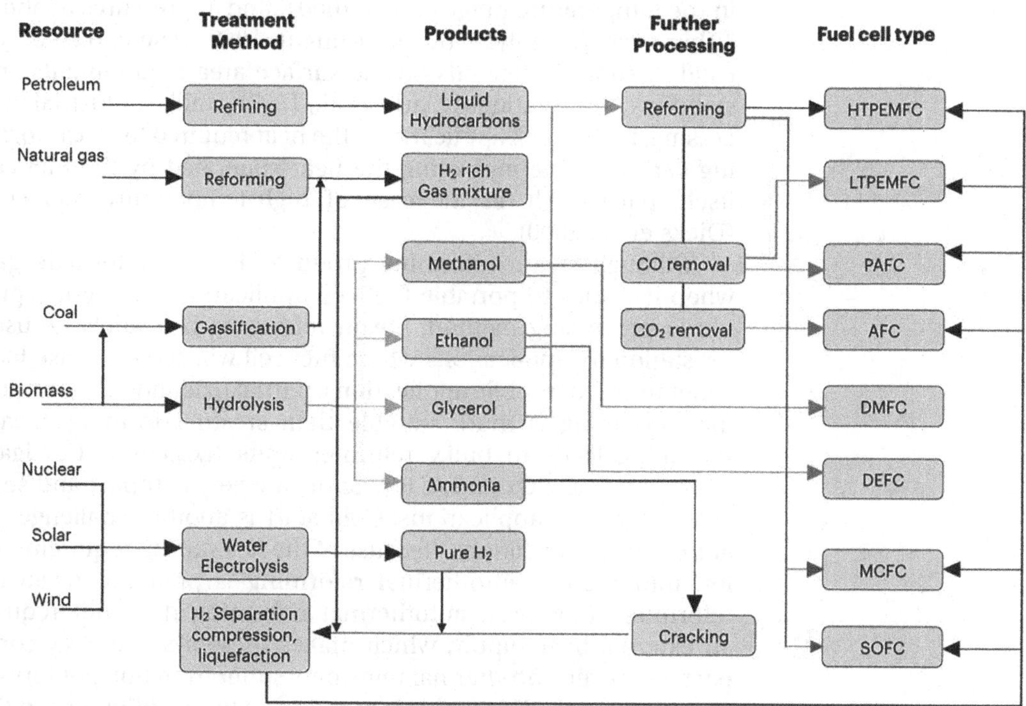

Fig. 2 Fuel sources and reforming methods. Different fuel sources and treatment methods for application in different for fuel cell technologies.

Reforming of some of the common fuels and recent developments in their processing are described below. It should be kept in mind that the operating conditions of a fuel cell system depend on the type of the cell (LTPEM, HTPEM, SOFC, etc.) used, and therefore, the reformer design and the degree of its integration with the stack can lead to systems of varying complexity and cost.

3.1 Reforming of hydrocarbons

Most of the hydrogen today is produced by the steam reforming of natural gas, which predominantly contains CH_4. The reforming of CH_4 is a highly endothermic reaction with the following stoichiometry.

$$CH_4 + H_2O \leftrightarrow 3H_2 + CO \quad \Delta H_R = 206.1\ \text{kJ/mol} \tag{8}$$

The syngas produced is H_2-rich with H_2/CO ratio of 3. Commercially these reactions are carried out in tubular gas-fired furnaces in the temperature range of 600–1000 K and at pressures of about 30 bar over Ni catalysts (Xu & Froment, 1989). The choice of the catalyst support depends on the surface area requirements and stability considerations (Clarke et al., 1997). Unlike industrial processing, for fuel cell applications, the heat required for the reforming can partly be met using the heat generated by the fuel cell itself, particularly in the case of high-temperature fuel cells (Dicks et al., 2000).

Although steam reforming produces H_2-rich reformate gas, when it comes to portable fuel cell applications, it is not a preferred processing method. Steam reforming can safely be used for stationary applications where fuel cell will serve as base load generators. For mobile applications, partial oxidation or autothermal reforming is more suitable than steam reforming. Steam reforming leads to bulky reformer units (Lesieur & Corrigan, 2001), whereas POx and ATR reactors are very compact and suitable for mobile applications. Cold start is another challenge for automotive applications. Because of the fast startup requirements for automobiles, autothermal reforming is preferred to steam reforming. Moreover, autothermal reforming does not require an external heat supply, which makes these systems very compact. However, autothermal reformers suffer from hotspot formation (Kang et al., 2006). This is mainly due to the difference in the rates of partial oxidation and steam reforming. During the startup of autothermal reactors, only partial oxidation is carried out to bring the reactor temperature to the light-off temperature at which the autothermal reaction becomes self-sustaining (Hoang & Chan, 2004). Micro-structured reactors can be used to thermally couple steam reforming with exothermic catalytic combustion reactions. This approach is different from that of ATR. In this configuration, the heat integration is through the external wall, which separates steam reforming from the combustion reactions (Chen et al., 2017; Lakhete & Janardhanan, 2014). Micro-structured reactors with integrated membrane separation are also suitable for portable fuel cell applications.

It is also worth mentioning that, there are contradictory reports on the efficiency of SR compared with PO_x or ATR. Some reports claim that SR is more efficient than PO_x or ATR for fuel cell applications, as SR can effectively utilize the waste anode gas to produce heat required for the heating purpose. However, PO_x or ATR cannot fully utilize the waste anode gas as they are thermally neutral (Qi et al., 2007).

Hydrocarbons can be converted to H_2-rich reformate by pyrolysis. Propane is an ideal fuel for pyrolysis. The energy requirement for propane pyrolysis is much lower compared with that of CH_4. CH_4 is the most stable hydrocarbon, and therefore, the energy requirement for the pyrolysis of CH_4 is 37.8 kJ/mol of H_2, whereas for propane the energy requirement is 26 kJ/mol of H_2. The higher energy requirement for CH_4 pyrolysis is due to the higher C—H bond energy of CH_4 molecule (440 kJ/mol) compared with that of propane (402 kJ/mol). Typically, high-surface-area carbonaceous materials such as carbon black and activated carbon are used as catalysts for pyrolysis reactions. Muradov performed the pyrolysis of propane, CH_4, and gasoline in the temperature range of 850–950°C with a residence time of 20–50 s (Muradov, 2003). He obtained higher H_2 concentration with propane fuel compared with CH_4 and gasoline. One of the problems with pyrolysis reactor is the incomplete conversion. For instance, in his experiments using CH_4 as feedstock, Muradov obtained a product gas containing 80% H_2 with CH_4 as the balance. He also found small amounts of C_2 hydrocarbons. Such a reformer needs further processing before sending the fuel to the fuel cell stack. Muradov proposed a methanator before sending the reformate gas to the fuel cell stack (Muradov, 2003).

Researchers have also explored the possibility of using gasoline and diesel as a primary fuel for mobile applications. These fuels can be reformed in an onboard reformer for application in fuel cells. The challenges that are associated with the use of gasoline or diesel are coke formation and sulfur poisoning. Therefore, several catalysts have been developed for ATR of gasoline and diesel. Since gasoline and diesel are hydrocarbon mixtures of alkanes, alkenes, and aromatics, their reforming is usually studied using surrogate fuels. According to Kang et al. the C—C bond cleavage occurs one bond at a time, and the higher hydrocarbons are eventually converted to C1 species. The larger hydrocarbons have lower C—C bond energy compared with lower hydrocarbons. The higher hydrocarbons may also be converted to aromatics, and they decrease the overall rate of the reaction (Kang et al., 2006). The reformate from ATR of C_8H_{18} (which is considered as surrogate fuel for gasoline) under conditions of $H_2O/C=1.25$ and $O_2/C=0.5$ contains CO, CO_2, CH_4, and C_2H_6. Obviously, the reformate gas from the ATR of gasoline or diesel cannot be fed directly to the fuel cell stack without further processing. A comparison between the reforming of surrogate fuels and commercial gasoline and diesel carried out in a fixed-bed reactor shows that the product yields are different (Kang et al., 2006). Maintaining the exit temperature of the ATR below 750°C will help reduce CH_4 formation (Springmann et al., 2004).

A high-temperature shift reactor can be used to bring down the CO content further (Springmann et al., 2004). This requires the reformate gas to be cooled down to the operating temperature of the HTS reactor using a heat exchanger. Otherwise, the reformate will lead to coke formation on the shift catalyst due to Boudouard reaction. Further CO reduction can be achieved by employing an LTS reactor, and final removal of CO is achieved using a preferential oxidizer (PROX) although, this step leads to small loss in H_2 concentration (Lindström et al., 2009). For use in PEM fuel cell stack, the gases from the shift reactor further need to be cooled down to the operating temperature of PEM fuel cell, which is 80°C for LTPEM and 120–200°C for HTPEM.

Different approaches are used for heating the ATR to the autoignition temperature. The reactor can be heated by direct combustion of the fuel that leads to compact designs; however, in this case, there is a risk of contamination by particulate matter. In an alternate approach, air can be heated in a burner where the fuel is burnt and the hot air can then be used for preheating the reactor. This leads to increased startup time; however, the system will be more durable (Lindström et al., 2009). A schematic of the diesel ATR by PowerCell is shown in Fig. 3.

In this design, a low-temperature partial oxidation zone is created in front of the catalyst. This zone creates a completely homogeneous mixture before the catalytic combustion (Lindström et al., 2009).

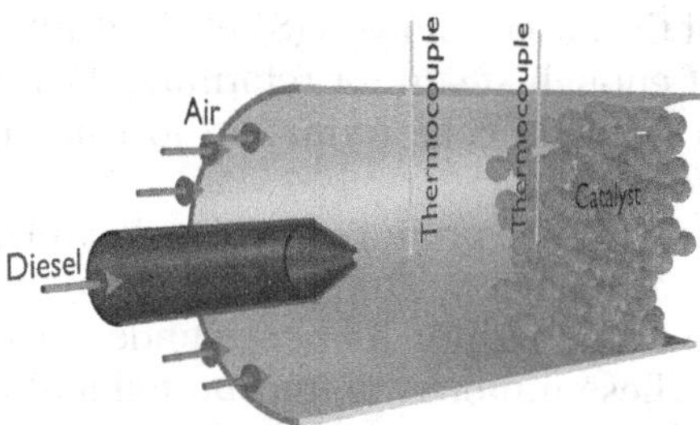

Fig. 3 Schematic of autothermal reformer (ATR) design of M4 Power Cell.

3.2 Alcohol reforming

Methanol is the simplest alcohol and a rich carrier of H_2. Since it is in the liquid form at room temperature, the transportation and handling are easy. Its high-energy density and easy availability make CH_3OH an ideal source of H_2 for transport applications. The reforming of CH_3OH results in an H_2-rich mixture. The composition of the resulting mixture from the steam reforming of CH_3OH depends on the S/C ratio, operating temperature, and pressure. At lower temperature, the steam reforming of CH_3OH leads to abundant formation of CH_4. As the reforming temperature increases, the CH_4 content in the product mixture decreases and the H_2 content increases. Due to the presence of hydroxyl group and the absence of C—C bond, the activation energy for methanol steam reforming is low (Chougule & Sonde, 2019). The three major reactions that occur concurrently during methanol steam reforming reaction are (i) reforming reaction, (ii) decomposition reaction, and (iii) water gas shift reaction. In the absence of sufficient steam, CH_3OH undergoes decomposition leading to unwanted by-products. As the steam content increases, the by-product formation decreases, H_2 selectivity increases, and CO selectivity decreases. The by-products of CH_3OH decomposition include CH_4, CO_2, methyl formate, and dimethyl ether (DME) (Choi & Stenger, 2005). As the steam content increases the by-products, CH_4, DME, and methyl formate start to disappear.

CH_3OH reformers can be integrated into the cell assembly in case of HTPEM fuel cells. In these internal methanol reforming cells, the reformer is in contact with the membrane electrode assembly of the fuel cell. The reformer needs to be operated in the temperature range of 433–473 K, which is the upper operating temperature of HTPEM fuel cell. At these temperatures, $CuO/ZnO/Al_2O_3$ catalysts are very active and selective for the reforming of CH_3OH (Ji et al., 2019). However, Cu-based catalysts suffer from thermal sintering, and high concentrations of CO enhance the sintering of Cu-based catalysts (Sá et al., 2010). Coke forms in the absence of enough steam for reforming. However, oxidative reforming can mitigate coke formation just like in the case of hydrocarbon reforming.

Hydrogen can also be produced by the electrolysis of CH_3OH. The process is carried out in a fuel cell. In a PEM fuel cell, CH_3OH is passed to the anode of the cell where it undergoes oxidation and forms proton. These protons are transported to the cathode and instead of passing air to the cathode, N_2 is passed. Since N_2 is an inert, it does not react with H_2. The operation of the cell is similar to that of a concentration cell. In a PEM fuel cell, the flux of H^+ through the membrane is maintained by removing the hydrogen by combining with O_2. In the CH_3OH electrolysis case, H_2 is removed by flowing N_2, thereby maintaining the concentration gradient required for the flux of protons. Although in literature, it is known as CH_3OH electrolysis, the process is in fact that of a fuel cell. Zakaria et al. reported the electrolysis of CH_3OH in a tandem cell, which is a combination of two PEM cells. The anode of one cell is in contact with the cathode of the other cell, and CH_3OH is passed to the anode. Through one cathode, air is passed and through the other cathode, N_2 is passed. The part of the cell through which air is passed acts as a fuel cell, and the power generated by the fuel cell is sent to the other cell due to the anode cathode contact. Thus, the fuel cell generated power is used to oxidize CH_3OH passed though the other cell (Zakaria et al., 2020). A schematic of such a cell is shown in Fig. 4.

Ethanol (EtOH) is another feedstock that can be reformed to produce H_2. The advantages of H_2 production from EtOH include (i) renewable nature of EtOH, (ii) free from sulfur compounds that

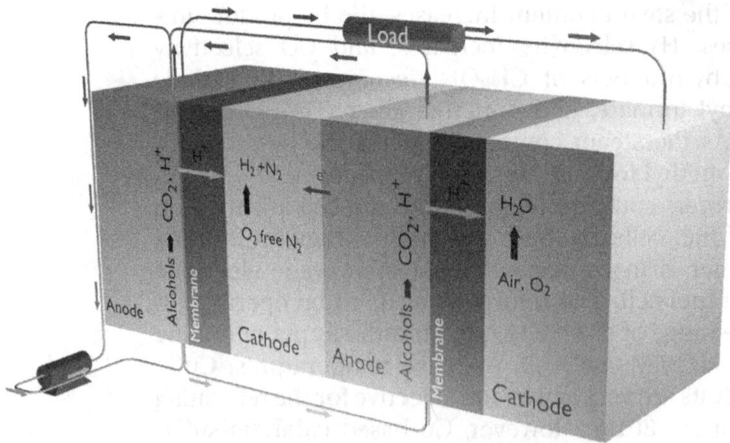

Fig. 4 Schematic representation of tandem cell (components are not drawn to scale).

poisons the catalyst, (iii) easy transportation, low toxicity, and biodegradability. There are several pathways that can occur during the steam reforming of EtOH, which depends on the metal catalyst employed. The dehydration of EtOH leads to the formation of C_2H_4, which polymerizes into coke. The decomposition of EtOH leads to the formation of CH_4, which further undergoes steam reforming leading to the formation of H_2 and CO. Dehydrogenation of EtOH forms acetaldehyde, which further undergoes decarbonylation or steam reforming. Decarbonylation of acetaldehyde forms CH_4 and CO. In addition to these, several other reactions such as decomposition, water gas shift, methanation, and coke formation can occur simultaneously (Haryanto et al., 2005; Huang et al., 2008). The formation of ethylene and further coke formation by polymerization reaction can be avoided by using high steam-to-EtOH ratio for the reforming reaction (Punase et al., 2019). The higher steam content also reduces the CO concentration in the reaction products (Haryanto et al., 2005). Several catalyst systems have been studied for the steam reforming of EtOH. Among oxide catalysts, ZnO shows excellent selectivity toward H_2. Ni/La_2O_3-Al_2O_3 is an equally good catalyst. However, without the addition of La_2O_3, the catalyst immediately deactivates. Rh is another good catalyst for EtOH steam reforming when supported on CeO_2 or Al_2O_3.

EtOH may also be subjected to autothermal reforming using Ni, Rh, or Pt catalyst. Huang et al. performed the autothermal reforming of EtOH over iron-promoted Ni catalyst. The time on stream analysis showed good catalyst stability and high H_2 selectivity. However, without iron promotion, the Ni catalyst stability was poor and the selectivity to H_2, CO_2 and CO dropped over time (Huang et al., 2008).

4 Electrochemical H_2 production

Electrochemical methods for the production of H_2 have gained significant attention recently. Unlike the production of H_2 from hydrocarbons or alcohols, the H_2 that is produced by the electrochemical splitting of H_2O is pure. The electricity production from renewable sources had caused a renewed interest in water electrolysis for the production of H_2. The excess electricity that is produced during the off-peak period from wind and solar needs to be locally stored, and this is generally achieved using batteries. Energy storage in the form of H_2 is an alternative to batteries. The excess electricity can be used to electrolyze water by reverse operation of a fuel cell. The same H_2 fuel can then be used in a fuel cell to produce electricity during the intermittent periods or can be used for other applications.

For the electrolysis of water, one may use a PEM electrolyzer, alkaline electrolyzer, or a solid oxide electrolyzer (SOE). SOE is more ideal for large-scale applications (Pettersson et al., 2006). Alkaline electrolyzers are highly matured technology and enjoy commercial-scale operation. However, they do have several disadvantages such as low operating pressure, low operating current density, and gas crossover through the diaphragm (Carmo et al., 2013).

4.1 PEM electrolyzer

The PEM electrolyzer is similar to a PEM fuel cell in architecture. It contains an electrolyte membrane, typically Nafion, and two electrodes each consisting of a catalyst layer and a gas diffusion layer. The gas diffusion layer is generally made of carbon cloths, and this poses challenges to the stability of membrane electrode assembly (MEA) in PEM electrolyzers. The high potentials used during the operation of the electrolyzer cause corrosion of carbon cloth on the oxygen electrode side. The electrode on which the oxidation or splitting of water occurs is called the anode. At the anode, water is oxidized according to

$$H_2O \leftrightarrow 2H^+ + 2e^- + 0.5O_2, \tag{9}$$

and the H^+ ions are transported to the cathode (H_2 electrode) through the electrolyte membrane and underdo reduction according to

$$2H^+ + 2e^- \leftrightarrow H_2. \tag{10}$$

The overall cell reaction becomes

$$H_2O \leftrightarrow H_2 + 0.5O_2. \tag{11}$$

The process requires an energy input of 285.8 kJ/mol (Carmo et al., 2013). One of the advantages of PEM electrolyzers is the low gas crossover through the solid polymer membrane and hence operability under wide ranges of power input (Carmo et al., 2013). PEM electrolyzers are capable of operating at current densities as high as 2 A/cm^2. High-current operation generally reduces the overall electrolysis cost. The major challenge for improved performance of PEM electrolyzers is the development of electrocatalysts having appreciable rate of hydrogen evolution reaction (HER) and oxygen evolution reaction (OER) and resistant to the oxidative environment of the PEM electrolyzer. For instance, RuO_2 is a very good electrocatalyst that leads to low activation losses for OER compared with other metals such as Pd and Pt. However, RuO_2 corrodes by forming RuO_4 under high-current-density operation. A schematic representation of the PEM electrolyzer is shown in Fig. 5.

4.2 Solid oxide electrolyzer

In solid oxide electrolysis cell (SOEC) operation, the H_2O is reduced at the cathode according to

$$H_2O + 2e^- \leftrightarrow H_2 + O^{2-}. \tag{12}$$

The O^{2-} ions are then transported toward the anode where it is oxidized according to

$$O^{2-} \leftrightarrow \frac{1}{2}O_2 + 2e^-. \tag{13}$$

The overall cell reaction is

$$H_2O \leftrightarrow H_2 + \frac{1}{2}O_2. \tag{14}$$

Comparison of the anodic and cathodic reactions for PEM and SOEC electrolyzer reveals that H_2O is oxidized at the anode in the case of PEM, whereas in the case of SOEC, H_2O is reduced at the cathode. In both cases, H_2 is produced at the cathode, and O_2 is produced at the anode. The SOECs operate at much elevated temperature compared with PEM electrolyzers. As the operating temperature increases, the electrical energy demand of the system decreases (Menon et al., 2014; Zheng et al., 2017). For

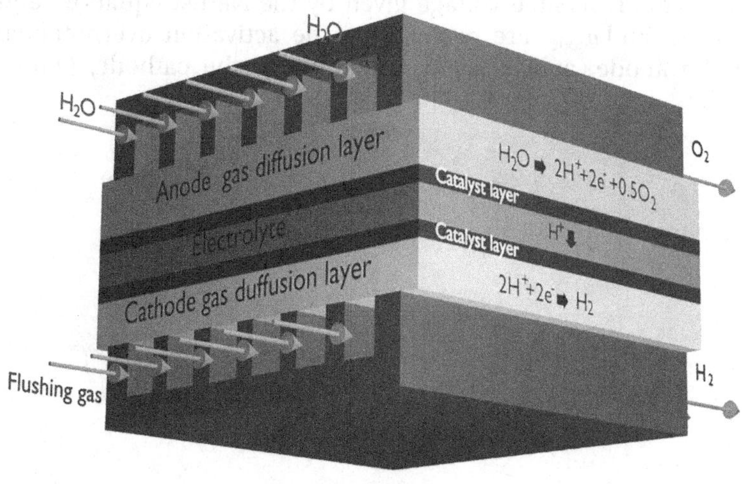

Fig. 5 Schematic of the operation of PEM electrolyzer (cell components are not drawn to scale).

instance, Schiller et al. reported a cell voltage of 1.4 V at −1 A/cm^2 and 800°C. However, when the operating temperature was increased to 850°C, the cell voltage reduced to 1.28 V. The high operating temperature also leads to faster kinetics and high energy efficiency (Schiller et al., 2009). Degradation in the presence of fuel impurities is also reported for SOEC operation. The degradation is more severe while operating at high electrolysis current density (Sun et al., 2019). Some reports claim very good reversibility of SOFC and the use of the same electrode materials for operation in the SOEC mode. For such cases, Ni-YSZ, YSZ, and LSCF are respectively the typical cathode, electrolyte, and anode materials (Sun et al., 2019). Ni agglomeration/migration leading to loss of percolation and oxygen electrode delamination are the main degradation mechanisms reported for SOEC operation. There are also reports that question the redox stability of Ni-YSZ electrode. Therefore, a number of perovskite materials have been tested for their suitability as SOEC cathode (Deka et al., 2019). A schematic representation of water electrolysis in SOEC is shown in Fig. 6.

4.3 Overpotential loses

The open-circuit voltage for H_2-O_2 redox couple at normal conditions is 1.23 V. Therefore, theoretically the electrolysis water should be achievable at any voltage higher than 1.23 V. However, due to the various overpotential losses, higher potentials are required to achieve the same. The electrolysis voltage of the cell E_{cell} is given by

$$E_{cell} = E_{rev} + \eta_a + \eta_c + \eta_\Omega + \eta_{conc} \tag{15}$$

E_{rev} is the reversible voltage given by the Nernst equation; and $\eta_a, \eta_c, \eta_\Omega$, and η_{conc} are respectively the activation overpotential at the anode, activation overpotential at the cathode, Ohmic

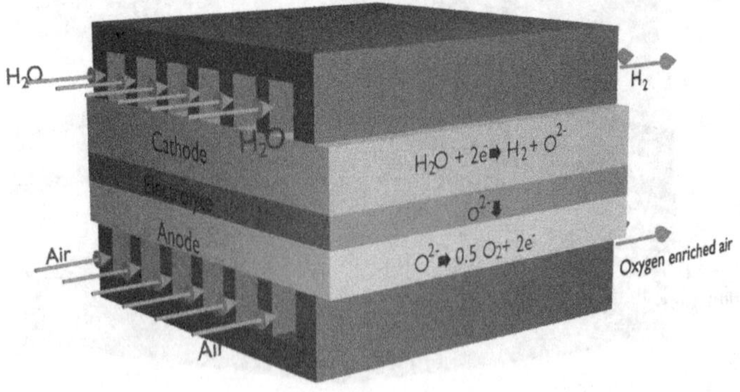

Fig. 6 Schematic of the operation of SOEC (cell components are not drawn to scale).

overpotential, and concentration overpotential. All overpotentials are voltage losses. The more the voltage loss, less efficient the cell operation becomes. The activation overpotentials are due to the activation barrier of the electrochemical reactions. It represents the extra voltage that needs to be sacrificed in order to drive the electrode reactions in the respective directions. The ohmic overpotential is due to the resistance offered by the cell components toward the transport of electrons and ions. However, the ionic conductivity of the electrolyte membrane is much lower than the electrical conductivity of the electron conducting phases. Therefore, in calculating the Ohmic overpotential, generally only the ionic contribution is accounted for. The concentration overpotential is due to the diffusion limitation at high current densities. High current density represents high faradaic reaction rates, and the diffusion flux may not be able to supply reactants to the reaction site at the rate required by the faradaic reactions, leading to large concentration overpotential η_{conc}.

5 Kinetic models for reforming

As discussed in the previous sections, a number of primary fuels can be reformed to produce syngas, which is further processed in shift reactors to reduce the CO concentration. Accurate kinetic models will aid in the design of fuel reformers and finding the optimum operating conditions for a particular fuel cell application. Hydrogen for fuel cell applications is mostly produced from CH_4, CH_3OH, or Ethanol. Therefore, the discussion on kinetic models is limited to the reforming of these fuels.

5.1 Methane reforming

There are several kinetic models reported in the literature for steam reforming of CH_4. One of the popular kinetic models for steam reforming of CH_4 is the one proposed by Xu and Froment (Table 2). They proposed a mechanistic kinetic model, which is derived by considering elementary step mechanisms. They have considered several reaction schemes for the development of the model. The kinetic model is valid for reforming of CH_4 performed over Ni-based catalysts. They estimated the activation energy for steam reforming reaction as 240 kJ/mol and that of WGSR as 67.13 kJ/mol (Xu & Froment, 1989).

In SOFCs, it is possible to reform the fuel internally due to the high operating temperature. Arguing that the steam reforming experiments performed on Ni/Al_2O_3 bare little relevance to the

Table 2 Kinetic models for reforming of CH_4.

Sl no.	Model equations	Reforming type	Remarks	Ref
1	$r_1 = \frac{\left(k_1/p_{H_2}^{2.5}\right)\left(p_{CH_4}p_{H_2O} - \frac{p_{H_2}^3 p_{CO}}{K_1}\right)}{D^2}$ $r_2 = \frac{\left(k_2/p_{H_2}\right)\left(p_{CO}p_{H_2O} - \frac{p_{H_2}p_{CO_2}}{K_2}\right)}{D^2}$ $r_3 = \frac{\left(k_3/p_{H_2}^{3.5}\right)\left(p_{CH_4}p_{H_2O}^2 - \frac{p_{H_2}^4 p_{CO_2}}{K_3}\right)}{D^2}$ $D = 1 + K_{CO}p_{CO} + K_{H_2}p_{H_2} + K_{CH_4}p_{CH_4} + K_{H_2O}p_{H_2O}/p_{H_2}$	SR	Rate equations are developed based on the following rate limiting steps $CHO* + * \leftrightarrow CO* + H*$ $CO* + O* \leftrightarrow CO_2* + *$ $CHO* + O* \leftrightarrow CO_2* + H*$ $k_1 = 1.84 \times 10^{-4}$, $k_2 = 7.55$, $k_3 = 2.19 \times 10^{-5}$ $E_1 = 240$, $E_2 = 67.13$, $E_3 = 243.9$ $k_1 = k_3 = (bar^{1/2}/kg.\ cat.\ h)$, $k_2 = kmol/kg.\ cat.\ h.\ bar$ $E = kJ/mol$	Xu and Froment (1989)
2	$r = \frac{k(T)p_{CH_4}}{\left(1 + K_H(T)p_{H_2}^{1/2} + K_S(T)p_{H_2O}/p_{H_2}\right)^2}$	SR	Based on experiments performed on Ni/YSZ $E_a = 135$ kJ/mol $k = 21$ mol/s.cm^2.bar	Dicks et al. (2000)
3	$r_{CO} = k_{CO}p_{CH_4}^{n_{CH_4}}p_{H_2O}^{n_{H_2O}}\left[1 - \frac{p_{CO}p_{H_2}^3}{K_{p,CO}p_{CH_4}p_{H_2O}}\right]$ $r_{CO_2} = k_{CO_2}p_{CH_4}^{m_{CH_4}}p_{H_2O}^{m_{H_2O}}\left[1 - \frac{p_{CO_2}p_{H_2}^4}{K_{p,CO_2}p_{CH_4}p_{H_2O}^2}\right]$	SR	Based on experiments performed on micro-channel reactors. These are power law models and is not based on any mechanistic analysis $k_{CO} = 6.13 \times 10^9$ $k_{CO_2} = 8.33 \times 10^3$ $n_{CH_4} = 0.343$ $n_{CO_2} = -0.053$ $m_{CH_4} = 1.127$ $m_{CO_2} = 1.681$ $E_{CO} = 198.8$ $E_{CO_2} = 152.9$ $E = kJ/mol$	Wang et al. (2013)

4	$r_1 = \frac{k_1 p_{CH_4} p_{H_2O}^{0.5} / p_{H_2}^{1.25} \left(1 - \left(p_{CO} p_{H_2}^3 / K_1 p_{CH_4} p_{H_2O}\right)\right)}{D^2}$ $r_2 = \frac{k_2 p_{CO} p_{H_2O}^{0.5} / p_{H_2}^{0.5} \left(1 - \left(p_{CO_2} p_{H_2} / K_2 p_{CO} p_{H_2O}\right)\right)}{D^2}$ $r_3 = \frac{k_3 p_{CH_4} p_{H_2O} / p_{H_2}^{1.75} \left(1 - \left(p_{CO_2} p_{H_2}^4 / K_3 p_{CH_4} p_{H_2O}^2\right)\right)}{D^2}$ $D = 1 + K_{CO} p_{CO} + K_H p_H^{0.5} + K_{H_2O}(p_{H_2O}/p_{H_2})$	SR	The kinetics constants k_1, k_2 and k_3 are found to be temperature dependent. Please refer to Hou and Hughes (2001) for parameter values	Hou and Hughes (2001)
5	$r_{CH_4} = \frac{k p_{CH_4} p_{O_2}}{D}$ $r_{CO2} = \frac{k p_{CO} p_{O_2}}{(1 + K p_{O_2}) D}$ $r_{CO2} = \frac{k p_{H_2}^{1/2} p_{O_2}}{(1 + K p_{O_2}) D}$ $D = k_4 K p_{O_2} + k_4 K_5 p_{O_2} p_{CO} + k_1 p_{CH_4} + k_1 K p_{CH_4} p_{O_2}$	POx	k values are estimated based on Arrhenius equation and K values are estimated based on van't Hoff equation	Elmasides et al. (2000)
6	$r_{CH_4} = \frac{k C_{CH_4}}{\left(1 + f(k) C_{H_2^{1/2}}\right)\left(1 + f(k) C_{H_2} + f(k) C_{CO}\right)} (1 - \eta)$ $r_{WGS} = \frac{f(k) C_{CH_2O}}{\left(1 + f(k) C_{H_2} + f(k) C_{CO}\right)^2} (1 - \eta)$	SR/DR	The k values in the denominator are functions of rate constants appearing in the microkinetic model. η is defined as the equilibrium factor	Maestri et al. (2008)

operation of SOFC, Dicks et al. examined CH_4 reforming on Ni/YSZ. They estimated the activation energy for the reaction to be 135 kJ/mol, first order with respect to CH_4 and negative order with respect to H_2O (Dicks et al., 2000). Their LHHW rate expression is given in Table 2.

In another work, Wang et al. adopted a power law model (Table 2). They proposed a parallel mechanism based on the steam reforming of CH_4 performed on microchannel reactors. According to the parallel mechanism, H_2O reacts with CH_4 to form both CO and CO_2 (Wang et al., 2013). The product distribution has to be made out of the rates of CO and CO_2 formation.

Hou and Hughes performed experiments using commercially prepared Ni/α-Al_2O_3 catalyst in an integral flow reactor. Like Zu and Froment, they also did mechanism discrimination to arrive at the final rate expressions, which are given in Table 2 (Hou & Hughes, 2001). They also estimated the activation energy of steam reforming to be 209 kJ/mol, a value comparable with that estimated by Xu and Froment.

Microkinetic models for steam reforming of CH_4 over Ni-based catalyst are also reported in the literature. These models do not assume any rate-limiting steps. The mechanisms that are developed based on the experiments performed on Ni/Al_2O_3 catalyst were able to reproduce experiments performed on Ni/YSZ. The group of Prof. Deutschmann at the Karlsruhe Institute of Technology has developed a microkinetic model consisting of 42 reactions among six gas-phase species and 12 surface adsorbed species based on experiments performed on Ni/Al_2O_3, and the mechanism was excellent in reproducing the experiments performed on Ni/YSZ (Hecht et al., 2005; Janardhanan & Deutschmann, 2006). Although, the implementation of the microkinetic models is more complex compared with mechanistic kinetic models or power law models, the microkinetic models are capable of predicting the state of the surface. The activation energy for the elementary reactions present in the microkinetic models is calculated either based on DFT or based on UBI-QEP method (Shustorovich, 1990). The pre-exponential (frequency) factors for the elementary steps can be calculated based on transition state theory. However, in many instances, these values fail to reproduce the experimental observations. Therefore, the pre-exponential factors are generally fitted to reproduce the experimental data. One of the major challenges in developing microkinetic models that are thermodynamically consistent is the lack of thermochemical properties of the surface adsorbed species. Given the forward reaction rate parameters, the reverse reaction parameters should be calculated based on the equilibrium constant estimated from thermodynamics properties. However, the lack of

Table 3 Kinetic models for steam reforming of CH_3OH and C_2H_5OH.

Sl no.	Model equations	Remarks	Ref
1	$-r_1 = k_1\left(p_{CH_3OH} - \frac{p_{H_2}^2 p_{CO}}{K_1}\right)$ $-r_2 = k_2 p_{CH_3OH}^2$ $-r_3 = k_3\left(p_{CO}p_{H_2O} - \frac{p_{H_2}p_{CO_2}}{K_3}\right)$ $-r_4 = k_4 p_{CH_3OH}^2$ $-r_5 = k_5 p_{CH_3OCHO}$	r_1 corresponds to CH_3OH decomposition r_2 Corresponds to DME formation, r_3 corresponds to WGS reaction, r_4 corresponds to MF formation and r_5 corresponds to CH4 formation	Choi and Stenger (2005)
2	$r = k_0 \exp(-E/RT) p_{MeOH}^m p_{H_2O}{}^w$	$E = 105.1$ kJ/mol, $m = 0.263$, $w = 0.0325$, $k_0 = 5.307 \times 10^{12}$ (mol-kg-s-Pa)	Jiang et al. (1993a)
3	$r = \frac{kK_2K_3^{-0.5} p_{CH_3OH} p_{H_2}^{-0.5}}{\left(1 + K_1K_3^{-0.5} p_{CH_3OH} p_{H_2}^{-0.5} + K_3^{-0.5} p_{H_2}^{-0.5}\right)^2}$	The parameters are temperature dependent	Jiang et al. (1993b)
4	$r = -A_0 \exp(-E_a/RT) \frac{p_{CH_3OH}^{0.18} p_{O_2}^{0.18}}{p_{H_2O}^{0.14}}$	E_a 115 kJ/mol and A_0 is 6×10^8 mol/min-g-kPa$^{0.22}$	Chan and Wang (2004)
5	$r_{EtOH} = \frac{k_0 \exp(-E/RT)\left(C_{EtOH} - \left(C_{CO_2}^2 C_{H_2}^6 / K_p C_{H_2O}^3\right)\right)}{\left(1 + (K_A C_{EtOH}) + \left(K_G C_{CO_2} C_{H_2}^3 / C_{H_2O}^2\right) + \left(K_E C_{CO_2}^2 C_{H_2}^6 / C_{H_2O}^3\right)\right)}$	E_a is 4.43×10^6 kJ/mol, $k_0 = 8.91 \times 10^2$, $K_A = 3.83 \times 10^7$, $K_G = 0$, $K_F = 0$	Akande et al. (2006)
6	$r_A = \frac{k_0 \exp(-E/RT)\left(C_A - C_C^2 C_D^6 / K_p C_B^3\right)}{\left(1 + K_1C_A + K_2C_B + K_3C_C + K_4C_D + K_5\frac{C_C C_D^{1/2}}{C_B} + K_6\frac{C_B}{C_D} + K_7\frac{C_C C_D^{3/2}}{C_B^2} + K_8 C_D^{1/2}\right)^2}$ $r_A = \frac{k_0 \exp(-E/RT)\left(C_B^3 - C_C^2 C_D^6 / K_p C_A\right)}{\left(1 + K_1C_A + K_2\frac{C_C^{2/3} C_D^2}{C_D^{1/3}} + K_3C_C + K_4C_D + K_5 C_C^{1/3} C_D^{1/2} + K_6\frac{C_C^{2/3} C_D}{C_A^{2/3}} + K_7\frac{C_A^{2/3}}{C_C^{1/3} C_D^{1/2}} + K_8 C_D^{1/2}\right)^3}$ $r_A = \frac{k_0 \exp(-E/RT)\left(C_A - C_C^2 C_D^6 / K_p C_B^3\right)}{\left(1 + K_1\frac{C_C^2 C_D^6}{C_B^3} + K_2\frac{C_C C_D^2}{C_B} + \frac{C_C C_D^4}{C_B^2} + K_4 C_C\right)}$ $r_A = \frac{k_0 \exp(-E/RT)\left(C_A - C_C^2 C_D^6 / K_p C_B^3\right)}{\left(1 + K_1C_A + K_2\frac{C_C C_D^2}{C_B} + \frac{C_C C_D^4}{C_B^2} + K_4 C_C\right)}$	A- EtOH B- Water C- CO_2 D- H_2 The first two are based on LHHW approach and the last two are based on ER mechanism	Akpan et al. (2007)

Continued

Table 3 Kinetic models for steam reforming of CH_3OH and C_2H_5OH—Cont'd

Sl no.	Model equations	Remarks	Ref
7	$r_{SRE} = \frac{kK_{EtOH}\left(p_{EtOH}/p_{H_2}^{1/2}\right)\left(1-\frac{p_{CO_2}^2 p_{H_2}^4}{K_r p_{EtOH} p_{H_2O}^2}\right)C_S^2}{D_1+D_2}$ $r_{WGS} = \frac{k_w K_{HCOO} p_{CO_2}\left(1-\frac{p_{H_2O} p_{CO}}{K_W p_{H_2} p_{CO_2}}\right)C_S^2}{D_1+D_2}$ $r_{ED} = \frac{k_D K_{CH_3CHO}\left(\frac{p_{CO_2}^2 p_{H_2}^3}{p_{H_2O}^2}\right)\left(1-\frac{p_{H_2O}^2 p_{CH_4} p_{CO}}{K_D p_{CO_2}^2 p_{H_2}^3}\right)}{D_1+D_2}$ $D_1 = 1+K_{CO_2}p_{CO_2}+K_{CO}p_{CO}+K_{CH_4}p_{CH_4}+K_{HCOO}p_{CO_2}p_{H_2}^{1/2}+K_{H_2}p_{H_2}^{1/2}$ $D_2 = \frac{K_{CH_3CHO} p_{CO_2}^2 p_{H_2}^5}{p_{H_2O}^3} + \frac{K_{EtOH} p_{EtOH}}{p_{H_2}^{1/2}} + \frac{K_{OH} p_{H_2O}}{p_{H_2}^{1/2}}$	Cs is the total concentration of surface sites $E_{SER}=82.7$ kJ/mol, $E_{WGS}=43.6$ kJ/mol $E_{ED}=71.3$ kJ/mol $k_{SER}=1.16\times10^{20}\,m^2/mol\text{-}s$ $k_{EGS}=4.64\times10^{16}\,m^2/mol\text{-}s$ $k_{ED}=4.46\times10^{19}\,m^2/mol\text{-}s$	Sahoo et al. (2007)
8	$r_{ED} = k_{ED}\frac{y_{EtOH}}{y_{CH_4} y_{H_2}^{1/2}}\frac{1}{D^2}$ $r_{ER} = k_{ER}\frac{y_{EtOH} y_{H_2O}}{y_{CH_4} y_{H_2}}\frac{1}{D^2}$ $r_{WGS} = k_{WGS}\left(\frac{y_{CO} y_{H_2O}}{y_{H_2}^{1/2}} - \frac{y_{CO_2} y_{H_2}^{1/2}}{K}\right)\frac{1}{D^2}$ $r_{SRM} = k_{SRM}\left(\frac{y_{CH_4} y_{H_2O}}{y_{H_2}} - \frac{y_{CO} y_{H_2}^2}{K}\right)\frac{1}{D^2}$ $D = 1 + K_1 y_{EtOH} + K_2\frac{y_{H_2O}}{y_{H_2}^{1/2}} + K_3 y_{CO_2}$	$E_{ED}=85.9$ kJ/mol $E_{ER}=418$ kJ/mol $E_{SRM}=151$ kJ/mol $E_{WGS}=107$ kJ/mol	Graschinsky et al. (2010)

thermochemical properties does not allow the calculation of equilibrium constant and hence, the reverse reaction parameters. Fitting the reverse reaction parameters arbitrarily can lead a reaction mechanism that is thermodynamically inconsistent. Therefore, the pre-exponential factors for elementary heterogeneous mechanisms must be estimated by imposing thermodynamic constraints (Appari et al., 2014). Prof. Deutschmann's group has also developed elementary heterogeneous kinetic models for the partial oxidation of CH_4 on Rh surface (Schwiedernoch et al., 2003). This mechanism consists of 38 reactions between 6 gas-phase species and 11 surface adsorbed species. A detailed kinetic model consisting of 19 steps for the POx of CH_4 on Rh and Pt was proposed by the group of Prof. Schmidt (Elmasides et al., 2000). Several heterogeneous elementary reaction mechanisms can be downloaded from the website (www.detchem.com).

Mechanistic kinetic models for CH_4 partial oxidation have also been reported in the literature. Elmasides et al. proposed a kinetic model for POx of CH_4 over Ru dispersed over TiO_2 doped with Ca^{2+}. They assumed dissociative and irreversible adsorption of CH_4 on the catalytically active sites. The O_2 adsorption is assumed as a two-step equilibrium process with the first step leading to the oxidation of the active site and the second step leading to the formation of active adsorbed oxygen species. CO is formed as a result of reaction between the carbon and surface adsorbed oxygen, and H_2 is formed as a result of recombination reactions. The rate equation proposed by the authors is given in Table 2.

Maestri et al. also reported the steam and dry reforming of CH_4 on Rh. They concluded that the conversion of CH_4 mainly proceeds by a pyrolysis pathway and carbon oxidation by surface adsorbed OH radicals, which are produced by the co-reactant H_2O (for steam reforming) or CO_2 (dry reforming) (Maestri et al., 2008). They have reported hierarchy of kinetic models, a full microkinetic model, a reduced microkinetic model, and a two-step rate model. The two-step rate model involves one rate expression for SR and another for WGS. Similarly, the rate expression for DR involves one for DR and another for WGS. For both cases, the rate expressions are identical. The only difference is in the reaction ratio. The two-step rate model proposed by these authors is presented in Table 2.

5.2 Methanol reforming

There is little agreement over the mechanism of methanol reforming. Initially it was thought that CH_3OH undergoes decomposition according to

$$CH_3OH \leftrightarrow CO + 2H_2 \quad (16)$$

and then CO undergoes WGS reaction: called a decomposition shift mechanism. However, this mechanism was ruled out for being lower than equilibrium composition of CO in experimental measurements. The Cu-based ($Cu/ZnO/Al_2O_3$) catalyst used for methanol reforming is very active for WGS. However, in the presence of methanol, the WGS reaction did not occur. This prompted Jiang et al. to propose an alternate route involving methyl formate. According to this mechanism, methanol initially decomposes to methyl formate, which then reacts with water to form formic acid. The formic acid then decomposes to give CO_2, which further participates in rWGSR to produce CO (Jiang et al., 1993a).

$$2CH_3OH \rightarrow HCOOCH_3 + 2H_2 \quad (17)$$

$$HCOOCH_3 + H_2O \rightarrow CH_3OH + HCOOH \quad (18)$$

$$HCOOH \rightarrow H_2 + CO_2 \quad (19)$$

Jiang et al. performed independent experiments to validate these reactions and proposed a multi-step reaction mechanism consisting of seven reactions. Based on the analysis of these seven reactions, they proposed an LHHW-type reaction equation (Jiang et al., 1993a). The activation energy was found to be 110 kJ/mol. Prior to that they proposed a power law model based on the overall methanol steam reforming reaction, and they found the reaction rate to be independent of water partial pressure (Jiang et al., 1993b).

For methanol reforming, the decomposition, steam reforming, DME formation, CH_4 formation, methyl formate formation, and water gas shift reactions occur concurrently. Choi et al. assumed the decomposition and water gas shift reactions to be reversible and all other reactions to be irreversible and proposed rate laws for each of them (Choi & Stenger, 2005). They are presented in Table 3. All the five rate expressions were required to reproduce the experimental observations. However, it should be kept in mind that these are power law models, and the parameters are determined by fitting against experimental data. The power law models do not provide any insight into the mechanistic details of the process. For the autothermal reforming of CH_3OH, Chan et al. derived a power law model. They found the reaction to be positive order with respect to oxygen and negative order with respect to water (Chan & Wang, 2004).

5.3 Ethanol reforming

Hydrogen production from ethanol (EtOH) fits very well into the ecosystem of biorefinery. One of the advantages of hydrogen production from EtOH compared with methanol is that EtOH is non-toxic and can be produced in a distributed manner under the concept of biorefinery. The reaction is strongly endothermic with a reaction enthalpy of 174 kJ/mol (Akpan et al., 2007). The steam reforming of EtOH produces a number of products that include C_2H_4, CH_3CHO, CH_4 CO, CO_2, coke, and H_2 (Punase et al., 2019). The side products C_2H_4 and CH_3OH are produced as a result of EtOH decomposition, and the C_2H_4 polymerizes to form coke. However, a high steam-to-EtOH ratio is known to suppress the formation of side products. Furthermore, these intermediates are stable only at temperatures below 473 K, and at temperatures above 673 K, they can be reformed to a mixture of CH_4, CO_2 and H_2 (Akpan et al., 2007).

In order to derive the rate kinetics of crude EtOH reforming, Akande et al. performed experiments in a packed-bed reactor using Ni/Al_2O_3 catalyst in the temperature range of 593–793 K. They have considered the following four steps for the derivation of the kinetic model (Akande et al., 2006).

$$C_2H_6O + (*) \leftrightharpoons C_2H_6O(*) \quad (20)$$

$$C_2H_6O(*) + (*) \leftrightharpoons CH_4O(*) + CH_2(*) \quad (21)$$

$$CH_4O(*) + H_2O(g) \leftrightharpoons CO_2 + 3H_2 + (*) \quad (22)$$

$$CH_2(*) + 2H_2O(g) \leftrightharpoons CO_2 + 3H_2 + (*) \quad (23)$$

Assuming each one of the above reactions to be rate-limiting, they have derived rate expression for EtOH reforming using Eley-Rideal (ER) mechanism. It should be noticed that, although the rate expressions are in terms of C_2H_6O, the experiments are performed using crude EtOH, which has a different molecular composition. However, the authors claim that there is no loss in accuracy by expressing the rate expression in terms of C_2H_6O. They found the model that is derived by considering Eq. (21) as rate-limiting leads to better comparison with experimental data compared with other models. The rate expression is given in Table 3.

In the literature, there is no consensus on the rate-limiting step or even the elementary steps involved in the steam reforming of EtOH. Several authors claim that the rate-determining step involves the formation of CH_4 from CH_2-CH_2O, which is contrary to the assumption of Akande et al. (2006).

Akpan et al. also reported mechanistic kinetic model for reforming of crude ethanol in the temperature range of 673–863 K (Akpan et al., 2007). They have derived rate expression using LHHW and ER approaches. For LHHW model development, they have considered an elementary reaction mechanism consisting of six steps, whereas the elementary steps considered for the development of ER mechanism involved five steps. They further performed a model discrimination by calculating the average absolute deviation between the model predicted values and experimental measurements. Based on this procedure, they finally shortlisted four models that are capable of predicting the steam reforming of crude EtOH on commercial Ni catalyst. These rate expressions are presented in Table 3.

Sahoo et al. reported the steam reforming of EtOH on Co/Al_2O_3 catalyst in the temperature range of 673–973 K (Sahoo et al., 2007). The experiments were performed in a fixed-bed reactor, and the data were used to validate the kinetic models they derived. They have considered a 17-step reaction mechanism, based on which they derived the rates of EtOH steam reforming, water gas shift, and EtOH decomposition. The rate expressions are provided in Table 3.

Graschinsky et al. proposed a kinetic model for the steam reforming of EtOH over $RhMgAl_2O_4/Al_2O_3$ catalyst (Graschinsky et al., 2010). They proposed a 14-step reaction mechanism, and of the 14 steps, those steps involving two active sites were considered as RDS for the derivation of kinetic models-based LHHW approach. Based on model discrimination, they concluded that the dissociation of formyl radical and surface reaction of hydroxyl radical with CH_3 and CO as rate-limiting ones. The final form of the expressions they derived is given in Table 3.

6 Reactor choice

The industrial hydrogen production technology cannot be scaled down to meet the requirements of a small-scale fuel processor. The small-scale fuel processor must be compact and highly efficient, so that it can be tightly integrated into a fuel cell system. A number of basic reactor types are available to choose from when it comes to fuel processing. The choice of the reactor depends on a number of factors including the operational cost and the capital expenses. The discussion in this section will be limited to gas-solid reactors, where the gas phase is composed of gaseous reactants including inerts, and the catalyst makes up the solid phase.

Two major challenges associated with the reforming of light hydrocarbons are the catalyst deactivation due to coke formation and sulfur poisoning. The light hydrocarbons can be converted to CH_4 at low temperatures using SR. SR at low temperature suppresses coke formation; however, if sulfur species are present in the fuel, it can severely deactivate the catalyst at low temperature as a result of metal sulfide formation. Sulfur enters the fuel as an ordering agent. For instance, natural gas might contain 4–6 ppm sulfur. At high operating temperature, the sulfur poisoning can be mitigated; however, the high temperature leads to cracking of higher hydrocarbons (Löffler et al., 2003).

6.1 Reactor designs based on packed bed

Packed-bed reactors are the most common reactor type employed for fuel processing applications. These reactors contain a straight tube filled with catalyst particles. When steam reforming or dry reforming is carried out in a packed-bed reactor, heat input needs to supplied through the reactor walls. When exothermic reactions are carried out, the heat needs to be removed through the reactor walls. This can be achieved by having a jacket around the reactor through which a cooling liquid is passed. The reaction temperature can be controlled by controlling the flow rate of the cooling medium. One of the biggest advantages of using packed-bed reactors is the ease of bed replacement. All catalysts lose activity over time. Once the catalyst activity drops below acceptable limits, the catalyst bed can be replaced with a new one. Even the catalyst itself can be changed.

A detailed design of a packed-bed reactor therefore involves the optimization of several factors such as particle size, reactor length, flow rate, pressure drop considerations, and heat removal/addition option.

Although steam reforming is not suitable for portable fuel cell application, it is ideal for stationary applications as the reformate is hydrogen-rich, and the reactors enjoy high efficiency. Several designs of packed-bed reactors are reported in the literature for steam reforming. IdaTech reported a fuel reformer for the processing of light hydrocarbons intended for stationary applications. The entire unit consists of a desulfurization unit, a pre-reformer unit operating at low temperature, which converts hydrocarbons heavier than CH_4 to CH_4 followed by a high-temperature reformer to convert CH_4. The reformate is then heat-exchanged and cleaned using a hydrogen permeable membrane. For fuel cell applications, the packed-bed reactor may consist of a bundle of tubes packed with the catalyst bed. These reactors contain small-diameter tubes, and the heat required for the endothermic steam reforming reaction is provided by the combustion of feed gas or raffinate from the membrane (Löffler et al., 2003). In IdaTech's design, the heat required is supplied by burning the raffinate from the membrane separation unit. The schematic of a compact heat exchanger reactor is shown in Fig. 7.

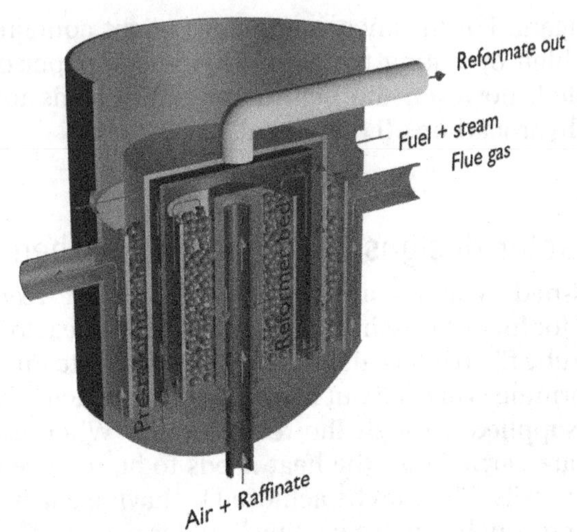

Fig. 7 Schematic of a compact heat exchanger reactor.

High-heat integration is necessary to ensure high efficiency of the fuel reforming unit. The heat value of the residual gases from the membrane separation unit can be used to provide heat of reforming and the heat required to vaporize water. Heat can also be recovered from the product hydrogen that must be supplied at the fuel cell operating temperature (Löffler et al., 2003). The unconverted hydrogen from the fuel cell anode can also be burned to recover the heat.

Ahmed et al. designed an ATR for the production of H_2 for fuel cell applications. They utilized the heat generated by POx to heat O_2 or air and water, which are used in the reformer chamber. The air-to-fuel and steam-to-fuel ratios are optimized in such a way that the H_2 production is optimized and the use of conventional material for fuel processing such as SS is facilitated. The cylindrically configured reactor contains an inner reforming zone and an outer reforming zone, containing catalyst that is capable of promoting POx and SR. The gases from the outer reforming zone enter a sulfur removal zone containing sulfur-removing agents such as ZnO. A water gas shift zone follows the sulfur removal zone (Ahmed et al., 2004).

There are other inventions with minor variations to the aforementioned designs. In short, all the designs based on packed bed contain pellet catalyst loaded as catalyst bed in reactor tubes. Major design variations are in the way heat integration is achieved in the reactors and the downstream process for the reduction of CO concentration.

6.2 Reactor design based on monoliths

Monolith reactors are class of reactors having a large number of parallel channels. The monoliths can be either metallic or ceramic (cordierite). They are very commonly used in exhaust gas treatment systems. They are characterized in terms of number of channels per square inch (cpsi). The higher the cpsi value, higher is the number of channels per square inch. The walls of the monolith can be coated with catalyst of choice. It's common to washcoat the monolith walls to increase the surface area. The washcoat creates a porous structure on the walls. The typical thickness of these washcoat layers is 10μm. Once the washcoat is applied, the square channels become circular in shape. The thickness of the channels at the corners will be more than that at the edges. One of the biggest advantages of monolith reactors is the low pressure drop at high flow rates. Moreover, due to the small thickness of the washcoat layer, the pore diffusion paths are shorter and the diffusion resistance will be lower. However, the interfacial mass transfer will persist. Unlike the fixed-bed reactors, the flow regime in monolith reactors is laminar, which does not eliminate the external mass transfer resistance. Their operation is close to ideal plug flow reactors, and therefore, they are best suited for kinetic studies. One of the drawbacks of the monolith reactors is the difficulty in replacing the catalyst. Once deactivated, the entire unit needs to be replaced. Furthermore, washcoating the monolith is also a challenge particularly when the cpsi values are high. One needs to ensure uniform thickness of the washcoat layer throughout the length of the monolith reactor. Another drawback of cordierite monolith is the poor thermal conductivity. Due to the poor thermal conductivity, the heat transfer from external wall (radial heat transfer) becomes a challenge in the case of large-diameter reactors.

There are reformer designs that are based on the characteristics of monolith. Lesieur et al. in a US patent designed a fuel reformer that consists of repeat units of reformers separated by flat plates. Each of the reformer units consists of reformer passages sandwiched between burner and heat exchanger sections. All these sections contain monolithic open cell foam components. These surfaces can be washcoated and catalyzed as desired. The monolith foam components provide diffuse flow path between the heat transfer walls. This design results in a lightweight compact fuel reformer (Lesieur & Corrigan, 2001). In general, the use of monolith reactors results in compact reformer configurations (Lesieur & Corrigan, 2001).

The use of these engineered catalyst supports offers tremendous scope for process intensification. Using a catalytic plate reactor, a close heat integration between exothermic combustion and endothermic steam reforming can be achieved. In the catalytic plate reactors, one side of the wall is coated with reforming catalyst and the other side is coated with oxidation catalyst (Janardhanan et al., 2011; Lakhete & Janardhanan, 2013; Zanfir & Gavriilidis, 2002, 2004). Since the channel dimensions are very small (20–100 μm), the heat transfer efficiency is more than 80% (Qi et al., 2007).

6.3 Membrane reactors

Membrane reactors are another class of reactors with significant potential for simultaneous reaction and separation. Chen et al. proposed an oxygen ion transport membrane (ITM) reactor. The system essentially contains planar membrane wafers, which are stacked to form a module that allows the passage of natural gas through the flow channels. A schematic of the ITM configuration proposed by Chen et al. is shown in Fig. 8 (Chen, 2004; Chen et al., 2004). The membrane material must be chemically and mechanically stable and should have sufficient mixed ion conductivity. Typically, perovskite materials are used as membrane materials. Catalysts may be incorporated into the pores of the membrane layer. O_2 from the low-pressure air flowing through one side of the membrane is transported to the other side as O^{2-} ions, where it combines with CH_4. The consumption of O_2 within the reactor maintains the driving force for O^{2-} ion transfer from the low-pressure air side to the reaction side. Overall the reaction is that of autothermal reforming. However, since pure O_2 is reacting instead of air, the resulting syngas is not diluted with N_2. This removes any additional N_2 purification steps required at the downstream. Although N_2 is not harmful for the fuel cell, its presence in the fuel can reduce the open-circuit potential as a result of fuel

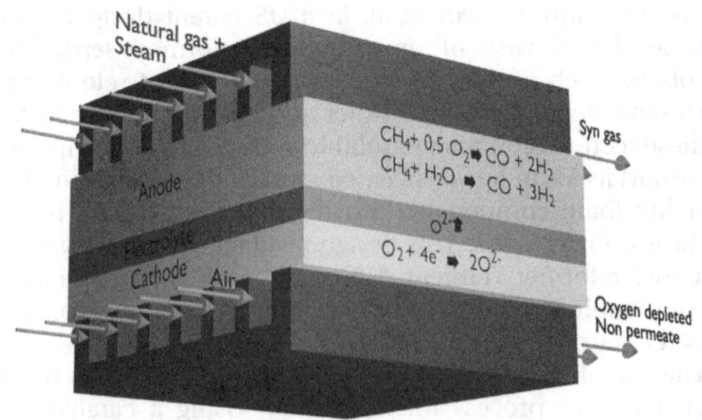

Fig. 8 Schematic of the ITM membrane reactor (components are not drawn to scale).

dilution effect (Chen, 2004; Chen et al., 2004). The cost-saving potential of ITM-assisted natural gas reforming is 30%.

A hydrogen transport membrane (HTM) can be combined with ITM syngas reactor to produce pure H_2 (Schwartz et al., 2004). The HTM is a selective membrane, which allows only the transport of atomic hydrogen. The syngas from the ITM reactor is sent to the HTM reactor, which operates around 400°C, where CO reacts with the steam in the syngas to produce more H_2. The H_2 is then separated using the selective membrane. The advantage of simultaneous H_2 removal is that the reaction never becomes thermodynamically limited (Schwartz et al., 2004). HTM operation if not based proton transport mechanism is a pressure-driven one, and therefore, requires high pressure within reforming unit (Qi et al., 2007). Payzant et al. using computation approach identified materials that transport protons in the temperature range of 300–800°C (Payzant et al., 2004). This membrane does not require high-pressure operation.

7 Reactor modeling

Reactor models are important in the validation of kinetic models. The reactor models can range from simple 0-D models to complex 3D models. Appropriate choice of reactor model is important in ensuring the accuracy of the estimated parameters. The complex 3D models are generally simulated using commercial CFD codes, where the species transport equations are solved along with the Navier-Stokes equations. The solution of the governing equation can be taxing for systems with large number of chemical species. Furthermore, such models are not suitable for parameter estimation using evolutionary techniques such as genetic algorithm due to the significantly high computational time required. On the other hand, 0-D or 1D models are very fast in execution and can handle large number of chemical species with ease. However, in order for the system to be represented using a 0D or 1D model, the experiments must be designed accordingly.

7.1 Zero-dimensional stirred tank model

The 0D models are ideal for representing a stirred tank reactor configuration. For such a system, the species balance equation is given by:

$$\rho\frac{dY_k}{dt}=\frac{1}{V}(\dot{Q}_{\text{in}}\rho_{\text{in}}Y_{k,\text{in}}-\dot{Q}_{\text{out}}\rho Y_k)+\dot{\omega}_k M_k A_{\text{sv}}+\dot{g}_k M_k \tag{24}$$

Here Y_k is the mass fraction of the species k, ρ is the density of the gas mixture, $Y_{k,\text{in}}$ is the mass fraction of species k at the inlet, ρ_{in} is the density of the gas mixture at the reactor inlet, V is the reactor volume, $\dot{Q}$ is the volumetric flow rate, M_k is the molecular weight of species k, A_{sv} is the surface area per unit volume, t is the time, $\dot{\omega}_k$ is the molar production rate of gas-phase species due to surface reactions, and $\dot{g}_k$ is the molar production rate due to gas-phase reactions. The form of the surface reaction source term may vary depending on the units of $\dot{\omega}_k$. In heterogeneous reactions, it is common to express the rate of the reaction in terms of available surface area or in terms of the mass of the catalyst.

The outflow rate is calculated from

$$\dot{Q}_{\text{out}} = \frac{T_{\text{out}}\overline{M_{\text{in}}}}{T_{\text{in}}\overline{M_{\text{out}}}}\dot{Q}_{\text{in}} \tag{25}$$

In terms of enthalpy, the energy balance equation may be written as

$$\frac{dH}{dt} = \sum_k \left(h_{k,\text{in}}\dot{Q}_{\text{in}}\rho_{\text{in}}Y_{k,\text{in}} - h_k\dot{Q}_{\text{out}}\rho Y_k \right) \tag{26}$$

Here h_k is the specific enthalpy of the species. While using this equation, the temperature needs to be calculated from the enthalpy. A root finding algorithm such as Newton-Raphson can be used for this purpose. Instead, the energy balance equation may also be written directly in terms of the rate of reaction

$$\rho C_{\text{p}}\frac{dT}{dt} = \sum_i (\Delta H_{\text{R}})_i r_{i,\text{s}} A_{\text{sv}} + \sum_i (\Delta H_{\text{R}})_j r_{j,\text{g}} \tag{27}$$

In the above equation, C_{p} is the specific enthalpy, $(\Delta H_{\text{R}})_i$ and $r_{i,\text{s}}$ are respectively the molar reaction enthalpy and reaction rate for ith surface reaction. $(\Delta H_{\text{R}})_j$ and $r_{i,\text{g}}$ are respectively the molar reaction enthalpy and reaction rate for jth gas-phase reaction. Generally, the reaction enthalpy for the surface reactions is not known. Therefore, one more alternative way exists for the formulation of energy balance equation in terms of the molar production rates of the species.

$$\rho C_{\text{p}}\frac{dT}{dt} = \sum_{k=1}^{K_{\text{g}}} \left(\dot{\omega}_k h_k A_{\text{sv}} + \dot{g}_k h_k \right) \tag{28}$$

The reader may keep in mind that these energy balance equations are written for adiabatic conditions. A packed-bed reactor with $L \approx D$ may be represented using a stirred tank reactor model, provided the reactor volume is small, and the reactor is operated under differential conditions, i.e., small conversions. The major assumption involved in the CSTR model is that the properties such as temperature, pressure, and compositions are uniform within the reactor volume.

7.2 One-dimensional plug flow model

For a reactor that is operated in the integral mode, 1D models are more suitable than 0D model. The major assumption involved in the plug flow model is that there is no axial dispersion, i.e., upstream properties are not influenced by the downstream. The properties are uniform along the radial direction due to infinite mixing. The species transport equation for a 1D plug flow model is given by

$$\rho u A_c \frac{dY_k}{dx} = M_k (A_s \dot{\omega}_k + A_c \dot{g}_k). \tag{29}$$

And the total continuity equation is given by

$$\frac{d(\rho u A_c)}{dx} = A_s \sum_{k=1}^{K_g} \dot{\omega}_k M_k. \tag{30}$$

In the above equations, A_c is the area of cross section, A_s is the surface area of the reactor, $\dot{g}_k$ is the gas-phase reaction rate, and K_g is the number of gas-phase species. The right-hand side of the above equation will vanish in the event of no mass addition or depletion in the reactor. This term becomes important when there is significant carbon deposition in the reactor during reforming reactions. The energy balance equation is written as:

$$\begin{aligned} \rho u A_c C_p \frac{dT}{dx} &+ \sum_{k=1}^{K_g} h_k M_k (\dot{\omega}_k A_s + \dot{g}_k A_c) \\ &= U A_s (T_w - T) \end{aligned} \tag{31}$$

The above equation is written for non-adiabatic conditions. Under adiabatic condition, the last term will be zero. T_w is the temperature of the reactor wall, and T is the temperature within the reactor. The energy balance equation can also be written in terms of enthalpy balance.

$$\rho u \frac{dH}{dx} = U \frac{A_s}{A_c} (T_w - T) \tag{32}$$

Here H is the specific enthalpy of the gas mixture calculated as

$$H = \sum_k X_k h_k M_k \tag{33}$$

where X_k is the mole fraction of species k, and h_k is the specific enthalpy of species k. The formulation of the energy balance equation using H results in a differential algebraic equation (DAE) system rather than ODE. If the wall temperature is assumed to be in equilibrium with the gas temperature, one can introduce the heat loss to the outside environment, which may be required in the case of exothermic reactors. Experiments that are carried out in honey-comb monolith reactors are ideal for representation using plug flow models. In these reactors, the catalysts are coated on the reactor wall. The L/D ratio of the reactor must be ≈15 for the use of plug flow models.

The 1D plug flow model equations can be used to model catalytic plate reactors with appropriate temperature boundary conditions.

7.3 One-dimensional packed-bed model

In a packed-bed reactor, the catalyst is supported on pellets that are porous. The packed-bed reactor may also be modeled using the 1D plug flow equations with an additional equation for solving the pressure drop. A more complicated heterogeneous axial dispersion model may also be developed for packed bed. The species transport equation with axial dispersion is written as:

$$\rho u \frac{dY_k}{dx} = \frac{d}{dx}\left(D_{\mathrm{km}}\,\rho \frac{dY_k}{dx}\right) + \dot{g}_k M_k + \dot{\omega}_k A_{\mathrm{sv}} M_k \tag{34}$$

The total continuity equation is

$$\frac{d(\rho u)}{dx} = A_{\mathrm{sv}} \sum_k \dot{\omega}_k M_k \tag{35}$$

In the absence of mass addition or depletion, the right-hand side of the above equation vanishes. For making the species balance equation heterogeneous, one must solve an additional algebraic equation:

$$\dot{\omega}_k M_k = k_{\mathrm{m}}(\rho_{\mathrm{s}} Y_{k,\mathrm{s}} - \rho Y_k) \tag{36}$$

and the reaction source term $\dot{\omega}_k$ should be calculated using the surface mass fractions $Y_{k,\,\mathrm{s}}$ instead of Y_k. The mass transfer coefficient k_{m} may be calculated from Sherwood number. A number of correlations are available for Sherwood number calculation. For the solid phase, the energy balance equation is

$$\frac{d}{dx}\left(k \frac{dT_{\mathrm{s}}}{dx}\right) = -hA_{\mathrm{sv}}(T_{\mathrm{s}} - T) + \sum_k h_k M_k \dot{\omega}_k A_{\mathrm{sv}} \tag{37}$$

Here T_{s} is the temperature of the solid, k is the thermal conductivity of the solid, h is the heat transfer coefficient for heat transfer from solid phase to gas phase. The reaction rate $\dot{\omega}_k$ should be calculated using the surface temperature T_s rather than the gas-phase temperature T. For the gas phase, the energy balance equation becomes

$$\rho u C_p \frac{dT}{dx} = \frac{d}{dx}\left(k_f \frac{dT}{dx}\right) + hA_{sv}(T_s - T) + \sum_k h_k \dot{g}_k M_k \qquad (38)$$

8 Sizing of reactor for applications in fuel cells

Finding the optimum reactor size is critical for the design of fuel cell systems. In general, the reforming system will consist of three reactors in series with in-between heat exchangers. The first one is a reforming reactor, which converts the primary fuel into a H_2-rich reformate gas. The second is a shift reactor, which reduces the CO concentration, and the third one is the preferential oxidation reactor, which further reduces the CO concentration. A shift reactor and PROX are not necessary in the case of high-temperature fuel cells. The right amount of water is always advantageous for the reformer, shift reactor, and PROX reactor (Choi & Stenger, 2005). A larger shift reactor results in a smaller PROX reactor. A number of parameters need to optimized for a compact fuel processor design. In the case of a methanol reformer, it is possible to operate the reformer at the operating temperature of the shift reactor, due to the low activation energy requirement for methanol reforming. However, the same is not true in the case of CH_4 reforming. In that case, a heat exchanger is required to bring down the temperature of the reformate gas to the operating temperature of the shift reactor. Another important factor that needs to be considered the cost of the system. The reactor cost mainly depends on the catalyst loading in the reactor, and the cost of the fuel processing unit increases with fuel purity (pure H_2) requirement.

With conventional steam reforming process, larger the scale of production, lower is the operation cost. Therefore, it is not possible to scale down the current industrial steam reforming process for fuel cell applications. Instead one needs to look for new designs and construction materials that lead to a lower cost per kW of hydrogen produced. Several compact reformer designs have been proposed for fuel cell applications. One such design is catalytic plate reformer where one side of the plate is coated with reforming catalyst and the other side is coated with oxidation catalyst. The anode off gases from the fuel cell are burnt in the oxidation channel, which produces the heat required for the endothermic reforming reaction. In comparison with reforming reactor, POx reactors are more compact. However, the H_2:CO ratio in POx reaction is 2. This means more CO needs to be converted in shift reactor. Thus, the purification system needed for a POx reactor will be more expensive compared with SR.

9 Summary

Fuel processing is indispensable even in internally reforming high-temperature fuel cells as any impurity beyond the tolerable limits can lead to severe performance loss and finally failure of the fuel cell. All fuel cells work best with pure hydrogen; however, the hydrogen produced from primary sources such as natural gas or oxygenated hydrocarbons contains CO, which is a poison for PEM fuel cells. Although, CO is the most widely studied impurity for PEM fuel cell systems, there are several other species such as sulfur-containing compounds, NO_x SO_X, and NH_3, which affect the fuel cell performance. Therefore, it is important to perform fuel processing before feeding the fuel to the fuel cell stack. The most commonly used primary fuels for fuel cell applications are natural gas, methanol, and ethanol. These fuels can be converted to H_2_rich mixture using a number of reforming techniques. In order to design an optimal fuel processing system for a fuel cell application, one needs kinetic models for the reforming reaction. With the help of kinetic models and numerical reactor models, one can arrive at the best possible reactor configuration and operating conditions.

References

Ahmed, S., Lee, S. H. D., Carter, J. D., & Krumpelt, M. (2004). *Method for generating hydrogen for fuel cells.* US 6713040B2.

Akande, A., Aboudheir, A., Idem, R., & Dalai, A. (2006). Kinetic modeling of hydrogen production by the catalytic reforming of crude ethanol over a co-precipitated Ni - Al2 O3 catalyst in a packed bed tubular reactor. *International Journal of Hydrogen Energy, 31*, 1707–1715. https://doi.org/10.1016/j.ijhydene.2006.01.001.

Akpan, E., Akande, A., Aboudheir, A., Ibrahim, H., & Idem, R. (2007). Experimental, kinetic and 2-D reactor modeling for simulation of the production of hydrogen by the catalytic reforming of concentrated crude ethanol (CRCCE) over a Ni-based commercial catalyst in a packed-bed tubular reactor. *Chemical Engineering Science, 62*, 3112–3126. https://doi.org/10.1016/j.ces.2007.03.006.

Appari, S., Janardhanan, V. M., Bauri, R., Jayanti, S., & Deutschmann, O. (2014). A detailed kinetic model for biogas steam reforming on Ni and catalyst deactivation due to sulfur poisoning. *Applied Catalysis A: General, 471*, 118–125. https://doi.org/10.1016/j.apcata.2013.12.002.

Ashrafi, M., Pfeifer, C., Pro, T., & Hofbauer, H. (2008). Experimental study of model biogas catalytic steam reforming : 2. Impact of sulfur on the deactivation and regeneration of Ni-based catalysts. *Energy & Fuels, 22,* 4190–4195.

Bharadwaj, S. S., & Schmidt, L. D. (1995). Catalytic partial oxidation of natural gas to syngas. *Fuel Processing Technology, 42,* 109–127.

Borup, R. L., Sauer, D. E., & Stuve, E. M. (1997). Electrolyte interactions with vapor dosed and solution dosed carbon monoxide on platinum (111). *Surface Science, 374,* 142–150. https://doi.org/10.1016/S0039-6028(96)01223-X.

Carmo, M., Fritz, D. L., Mergel, J., & Stolten, D. (2013). A comprehensive review on PEM water electrolysis. *International Journal of Hydrogen Energy, 38,* 4901–4934. https://doi.org/10.1016/j.ijhydene.2013.01.151.

Chan, S. H., & Wang, H. M. (2004). Thermodynamic and kinetic modelling of an autothermal methanol reformer. *Journal of Power Sources, 126,* 8–15. https://doi.org/10.1016/j.jpowsour.2003.08.016.

Chen, C. M. (2004). *Ceramic membrane reactor systems for converting natural gas to hydrogen and synthesis gas.* ITM Syngas.

Chen, C. M., Bennett, D. L., Carolan, M. F., Foster, E. P., Schinski, W. L., & Taylor, D. M. (2004). ITM syngas ceramic membrane technology for synthesis gas production. *Studies in Surface Science and Catalysis, 147,* 55–60. https://doi.org/10.1016/s0167-2991(04)80027-2.

Chen, J., Yan, L., Song, W., & Xu, D. (2017). Methane steam reforming thermally coupled with catalytic combustion in catalytic microreactors for hydrogen production. *International Journal of Hydrogen Energy, 42,* 664–680. https://doi.org/10.1016/j.ijhydene.2016.12.114.

Cheng, X., Shi, Z., Glass, N., Zhang, L., Zhang, J., Song, D., Liu, Z.-S., Wang, H., & Shen, J. (2007). A review of PEM hydrogen fuel cell contamination: Impacts, mechanisms, and mitigation. *Journal of Power Sources, 165,* 739–756. https://doi.org/10.1016/j.jpowsour.2006.12.012.

Choi, Y., & Stenger, H. G. (2005). Kinetics, simulation and optimization of methanol steam reformer for fuel cell applications. *Journal of Power Sources, 142,* 81–91. https://doi.org/10.1016/j.jpowsour.2004.08.058.

Chougule, A., & Sonde, R. R. (2019). Modelling and experimental investigation of compact packed bed design of methanol steam reformer. *International Journal of Hydrogen Energy, 44,* 29937–29945. https://doi.org/10.1016/j.ijhydene.2019.09.166.

Clarke, S. H., Dicks, A. L., Pointon, K., Smith, T. A., & Angie, S. (1997). Catalytic aspects of the steam reforming of hydrocarbons in internal reforming fuel cells. *Catalysis Today, 38,* 411–423.

Deka, D. J., Gunduz, S., Kim, J., Fitzgerald, T., Shi, Y., Co, A. C., & Ozkan, U. S. (2019). Hydrogen production from water in a solid oxide electrolysis cell: Effect of Ni doping on lanthanum strontium ferrite perovskite cathodes. *Industrial and Engineering Chemistry Research, 58,* 22497–22505. https://doi.org/10.1021/acs.iecr.9b03731.

Dicks, A. L., Pointon, K., & Siddle, A. (2000). Intrinsic reaction kinetics of methane steam reforming on a nickel/zirconia anode. *Journal of Power Sources, 86,* 523–530.

Elmasides, C., Ioannides, T., & Verykios, X. E. (2000). Kinetic model of the partial oxidation of methane to synthesis gas over Ru/TiO2 catalyst. *AIChE Journal, 46,* 1260–1270. https://doi.org/10.1002/aic.690460618.

Ertl, G., Knözinger, H., Schüth, F., & Weitkamp, J. (Eds.). (2008). *Handbook of heterogeneous catalysis* (2nd ed.). Weinheim, Germany: Wiley-VCH Verlag GmbH & Co.

Evans, S. E., Good, O. J., Staniforth, J. Z., Ormerod, R. M., & Darton, R. J. (2014). Overcoming carbon deactivation in biogas reforming using a hydrothermally synthesised nickel perovskite catalyst †. *RSC Advances, 4*, 30816–30819. https://doi.org/10.1039/C4RA00846D.

Evans, S. E., Staniforth, J. Z., Darton, R. J., & Ormerod, R. M. (2014). A nickel doped perovskite catalyst for reforming methane rich biogas with minimal carbon deposition. *Green Chemistry, 16*, 4587–4594. https://doi.org/10.1039/c4gc00782d.

Farrell, C. G., Gardner, C. L., & Ternan, M. (2007). Experimental and modelling studies of CO poisoning in PEM fuel cells. *Journal of Power Sources, 171*, 282–293. https://doi.org/10.1016/j.jpowsour.2007.07.006.

Ferreira-Aparicio, P., Rodrıguez-Ramosa, I., Anderson, J., & Guerrero-Ruiz, A. (2000). Mechanistic aspects of the dry reforming of methane over ruthenium catalysts. *Applied Catalysis, A: General, 202*, 183–196. https://doi.org/10.1007/978-3-319-99441-3_91.

Graschinsky, C., Laborde, M., Amadeo, N., Le Valant, A., Bion, N., Epron, F., & Duprez, D. (2010). Ethanol steam reforming over Rh(1%)MgAl2O4/Al 2O3: A kinetic study. *Industrial and Engineering Chemistry Research, 49*, 12383–12389. https://doi.org/10.1021/ie101284k.

Gu, T., Lee, W.-K., & Van Zee, J. W. (2005). Quantifying the 'reverse water gas shift' reaction inside a PEM fuel cell. *Applied Catalysis, B: Environmental, 56*, 43–50. https://doi.org/10.1016/j.apcatb.2004.08.016.

Haag, S., Burgard, M., & Ernst, B. (2007). Beneficial effects of the use of a nickel membrane reactor for the dry reforming of methane: Comparison with thermodynamic predictions. *Journal of Catalysis, 252*, 190–204. https://doi.org/10.1016/j.jcat.2007.09.022.

Haga, K., Adachi, S., Shiratori, Y., Itoh, K., & Sasaki, K. (2008). Poisoning of SOFC anodes by various fuel impurities. *Solid State Ionics, 179*, 1427–1431. https://doi.org/10.1016/j.ssi.2008.02.062.

Haryanto, A., Fernando, S., Murali, N., & Adhikari, S. (2005). Current status of hydrogen production techniques by steam reforming of ethanol : A review current status of hydrogen production techniques by steam reforming of ethanol : A review. *Energy*. https://doi.org/10.1021/ef0500538.

Hecht, E. S., Gupta, G. K., Zhu, H., Dean, A. M., Kee, R. J., Maier, L., & Deutschmann, O. (2005). Methane reforming kinetics within a Ni–YSZ SOFC anode support. *Appl. Catal. A, 295*, 40–51. https://doi.org/10.1016/j.apcata.2005.08.003.

Hickman, D. A., & Schmidt, L. D. (1993). Production of syngas by direct catalytic oxidation of methane. *Science (80-.), 259*, 343–346. https://doi.org/10.1126/science.259.5093.343.

Hoang, D. L., & Chan, S. H. (2004). Modeling of a catalytic autothermal methane reformer for fuel cell applications. *Applied Catalysis, A: General, 268*, 207–216. https://doi.org/10.1016/j.apcata.2004.03.056.

Horn, R., Degenstein, N. J., Williams, K. A., & Schmidt, L. D. (2006). Spatial and temporal profiles in millisecond partial oxidation processes. *Catalysis Letters, 110*, 169–178. https://doi.org/10.1007/s10562-006-0117-8.

Hou, K., & Hughes, R. (2001). The kinetics of methane steam reforming over a Ni/α-Al2O catalyst. *Chemical Engineering Journal, 82*, 311–328. https://doi.org/10.1016/S1385-8947(00)00367-3.

Huang, L., Xie, J., Chen, R., Chu, D., Chu, W., & Hsu, A. T. (2008). Effect of iron on durability of nickel-based catalysts in auto-thermal reforming of ethanol for hydrogen production. *International Journal of Hydrogen Energy, 33*, 7448–7456. https://doi.org/10.1016/j.ijhydene.2008.09.062.

Janardhanan, V. M., Appari, S., Jayanti, S., & Deutschmann, O. (2011). Numerical study of on-board fuel reforming in a catalytic plate reactor for solid-oxide fuel cells. *Chemical Engineering Science, 66*, 490–498. https://doi.org/10.1016/j.ces.2010.11.023.

Janardhanan, V. M., & Deutschmann, O. (2006). CFD analysis of a solid oxide fuel cell with internal reforming: Coupled interactions of transport, heterogeneous catalysis and electrochemical processes. *Journal of Power Sources, 162*, 1192–1202. https://doi.org/10.1016/j.jpowsour.2006.08.017.

Ji, F., Yang, L., Li, Y., Sun, H., & Sun, G. (2019). Performance enhancement by optimizing the reformer for an internal reforming methanol fuel cell. *Energy Science & Engineering, 7*, 2814–2824. https://doi.org/10.1002/ese3.461.

Jiang, C. J., Trimm, D. L., Wainwright, M. S., & Cant, N. W. (1993a). Kinetic mechanism for the reaction between methanol and water over a Cu-ZnO-Al2O3 catalyst. *Appl. Catal. A, 97*, 145–158. https://doi.org/10.1016/0926-860X(93)80081-Z.

Jiang, C. J., Trimm, D. L., Wainwright, M. S., & Cant, N. W. (1993b). Kinetic study of steam reforming of methanol over copper-based catalysts. *Appl. Catal. A, 93*, 245–255. https://doi.org/10.1016/0926-860X(93)85197-W.

Kang, I., Bae, J., & Bae, G. (2006). Performance comparison of autothermal reforming for liquid hydrocarbons, gasoline and diesel for fuel cell applications. *Journal of Power Sources, 163*, 538–546. https://doi.org/10.1016/j.jpowsour.2006.09.035.

Lakhete, P., & Janardhanan, V. M. (2013). Modeling process intensified catalytic plate reactor for synthesis gas production. *Chemical Engineering Science*, 1–7. https://doi.org/10.1016/j.ces.2013.05.021.

Lakhete, P., & Janardhanan, V. M. (2014). Modeling process intensified catalytic plate reactor for synthesis gas production. Author' s personal copy *Chemical Engineering Science, 110*, 13–19.

Lesieur, R. R., & Corrigan, T. J. (2001). *Compact fuel gas reformer assemblage.* 6203587 B1.

Li, Q., He, R., Gao, J.-A., Jensen, J. O., & Bjerrum, N. J. (2003). The CO poisoning effect in PEMFCs operational at temperatures up to 200°C. *Journal of the Electrochemical Society, 150*, A1599. https://doi.org/10.1149/1.1619984.

Lindström, B., Karlsson, J. A. J., Ekdunge, P., De Verdier, L., Häggendal, B., Dawody, J., Nilsson, M., & Pettersson, L. J. (2009). Diesel fuel reformer for automotive fuel cell applications. *International Journal of Hydrogen Energy, 34*, 3367–3381. https://doi.org/10.1016/j.ijhydene.2009.02.013.

Löffler, D. G., Taylor, K., & Mason, D. (2003). A light hydrocarbon fuel processor producing high-purity hydrogen. *Journal of Power Sources, 117*, 84–91. https://doi.org/10.1016/S0378-7753(03)00357-4.

Ma, Q., Peng, R., Tian, L., & Meng, G. (2006). Direct utilization of ammonia in intermediate-temperature solid oxide fuel cells. *Electrochemistry Communications, 8*, 1791–1795. https://doi.org/10.1016/j.elecom.2006.08.012.

Madi, H., Lanzini, A., Diethelm, S., Papurello, D., Van Herle, J., Lualdi, M., Gutzon Larsen, J., & Santarelli, M. (2015). Solid oxide fuel cell anode degradation by the effect of siloxanes. *Journal of Power Sources, 279*, 460–471. https://doi.org/10.1016/j.jpowsour.2015.01.053.

Maestri, M., Vlachos, D. G., Beretta, A., Groppi, G., & Tronconi, E. (2008). Steam and dry reforming of methane on Rh: Microkinetic analysis and hierarchy of kinetic models. *Journal of Catalysis, 259*, 211–222. https://doi.org/10.1016/j.jcat.2008.08.008.

Menon, V., Janardhanan, V. M., & Deutschmann, O. (2014). A mathematical model to analyze solid oxide electrolyzer cells (SOECs) for hydrogen production. *Chemical Engineering Science, 110*, 83–93. https://doi.org/10.1016/j.ces.2013.10.025.

Muradov, N. (2003). Emission-free fuel reformers for mobile and portable fuel cell applications. *Journal of Power Sources, 118*, 320–324. https://doi.org/10.1016/S0378-7753(03)00078-8.

Narusawa, K., Hayashida, M., Kamiya, Y., Roppongi, H., Kurashima, D., & Wakabayashi, K. (2003). Deterioration in fuel cell performance resulting from hydrogen fuel containing impurities: Poisoning effects by CO, CH4, HCHO and HCOOH. *JSAE Review, 24*, 41–46. https://doi.org/10.1016/S0389-4304(02)00239-4.

Pan, C., He, R., Li, Q., Jensen, J. O., Bjerrum, N. J., Hjulmand, H. A., & Jensen, A. B. (2005). Integration of high temperature PEM fuel cells with a methanol reformer. *Journal of Power Sources, 145*, 392–398. https://doi.org/10.1016/j.jpowsour.2005.02.056.

Pawar, V., Ray, D., Subrahmanyam, C., & Janardhanan, V. M. (2015). Study of short-term catalyst deactivation due to carbon deposition during biogas dry reforming on supported Ni catalyst. *Energy & Fuels, 29*, 8047–8052. https://doi.org/10.1021/acs.energyfuels.5b01862.

Payzant, E. A., Scott, A., & Armstrong, T. R. (2004). *Pyrochlore/perovskite proton transport membranes*. Department of Energy.

Pettersson, J., Ramsey, B., & Harrison, D. (2006). A review of the latest developments in electrodes for unitised regenerative polymer electrolyte fuel cells. *Journal of Power Sources, 157*, 28–34. https://doi.org/10.1016/j.jpowsour.2006.01.059.

Prass, S., Friedrich, K. A., & Zamel, N. (2019). Tolerance and recovery of ultralow-loaded platinum anode electrodes upon carbon monoxide and hydrogen sulfide exposure. *Molecules, 24*, 1–14. https://doi.org/10.3390/molecules24193514.

Punase, K. D., Rao, N., & Vijay, P. (2019). A review on mechanistic kinetic models of ethanol steam reforming for hydrogen production using a fixed bed reactor. *Chemical Papers, 73*, 1027–1042. https://doi.org/10.1007/s11696-018-00678-6.

Qi, A., Peppley, B., & Karan, K. (2007). Integrated fuel processors for fuel cell application: A review. *Fuel Processing Technology, 88*, 3–22. https://doi.org/10.1016/j.fuproc.2006.05.007.

Rajalakshmi, N., Jayanth, T. T., & Dhathathreyan, K. S. (2003). Effect of carbon dioxide and ammonia on polymer electrolyte membrane fuel cell stack performance. *Fuel Cells, 3*, 177–180. https://doi.org/10.1002/fuce.200330107.

Sá, S., Silva, H., Brandão, L., Sousa, J. M., & Mendes, A. (2010). Catalysts for methanol steam reforming—A review. *Applied Catalysis, B: Environmental, 99*, 43–57. https://doi.org/10.1016/j.apcatb.2010.06.015.

Sahoo, D. R., Vajpai, S., Patel, S., & Pant, K. K. (2007). Kinetic modeling of steam reforming of ethanol for the production of hydrogen over Co/Al2O3 catalyst. *Chemical Engineering Journal, 125*, 139–147. https://doi.org/10.1016/j.cej.2006.08.011.

Schiller, G., Ansar, A., Lang, M., & Patz, O. (2009). High temperature water electrolysis using metal supported solid oxide electrolyser cells (SOEC). *Journal of Applied Electrochemistry, 39*, 293–301. https://doi.org/10.1007/s10800-008-9672-6.

Schmidt, T. J., & Baurmeister, J. (2006). Durability and reliability in high temperature reformed hydrogen PEFCs. *ECS Transactions, 3*, 861–869. https://doi.org/10.1017/CBO9781107415324.004.

Schwartz, J., Apte, P., Drnevich, R., & Anderson, A. (2004). *Integrated ceramic membrane system for hydrogen production*. Department of Energy.

Schwiedernoch, R., Tischer, S., Correa, C., & Deutschmann, O. (2003). Experimental and numerical study on the transient behavior of partial oxidation of methane in a catalytic monolith. *Chemical Engineering Science, 58*, 633–642. https://doi.org/10.1016/S0009-2509(02)00589-4.

Shiratori, Y., Oshima, T., & Sasaki, K. (2008). Feasibility of direct-biogas SOFC. *International Journal of Hydrogen Energy, 33*, 6316–6321. https://doi.org/10.1016/j.ijhydene.2008.07.101.

Shustorovich, E. (1990). The bond order conservation approach to chemisorption and heterogeneous catalysis: Applications and implications. *Advances in Catalysis, 37*, 101–163.

Springmann, S., Bohnet, M., Docter, A., Lamm, A., & Eigenberger, G. (2004). Cold start simulations of a gasoline based fuel processor for mobile fuel cell applications. *Journal of Power Sources, 128*, 13–24. https://doi.org/10.1016/j.jpowsour.2003.09.038.

Steele, B. C. H. (1999). Running on natural gas. *Nature, 400*, 619–621.

Sun, X., Hendriksen, P. V., Mogensen, M. B., & Chen, M. (2019). Degradation in solid oxide electrolysis cells during long term testing. *Fuel Cells, 19*, 740–747. https://doi.org/10.1002/fuce.201900081.

Takeguchi, T., Kikuchi, R., Yano, T., Eguchi, K., & Murata, K. (2003). Effect of precious metal addition to Ni-YSZ cermet on reforming of CH$_4$ and electrochemical activity as SOFC anode. *Catalysis Today, 84*, 217–222. https://doi.org/10.1016/S0920-5861(03)00278-5.

Twigg, M. V. (Ed.). (1996). *Catalyst handbook*. London: Manson.

U.S. Department of Energy, Ohi, J. M., Vanderborgh, N., & Voecks, G. (2016). *Hydrogen fuel quality specifications for polymer electrolyte fuel cells in road vehicles*. https://doi.org/10.1111/j.1439-0507.2006.01349.x.

Unnikrishnan, A., Rajalakshmi, N., & Janardhanan, V. M. (2018). Mechanistic modeling of electrochemical charge transfer in HT-PEM fuel cells. *Electrochimica Acta, 261*, 436–444. https://doi.org/10.1016/j.electacta.2017.12.150.

Unnikrishnan, A., Rajalakshmi, N., & Janardhanan, V. M. (2019). Kinetics of electrochemical charge transfer in HT-PEM fuel cells. *Electrochimica Acta, 293*, 128–140. https://doi.org/10.1016/j.electacta.2018.09.171.

Wang, F., Qi, B., Wang, G., & Li, L. (2013). Methane steam reforming: Kinetics and modeling over coating catalyst in micro-channel reactor. *International Journal of Hydrogen Energy, 38*, 5693–5704. https://doi.org/10.1016/j.ijhydene.2013.03.052.

Wheeler, C., Jhalani, A., Klein, E. J., Tummala, S., & Schmidt, L. D. (2004). The water-gas-shift reaction at short contact times. *Journal of Catalysis, 223*, 191–199. https://doi.org/10.1016/j.jcat.2004.01.002.

Whittington, B. I., Jiang, C. J., & Trimm, D. L. (1995). Vehicle exhaust catalysis: I. The relative importance of catalytic oxidation, steam reforming and water-gas shift reactions. *Catalysis Today, 26*, 41–45. https://doi.org/10.1016/0920-5861(95)00093-U.

Xu, J., & Froment, G. F. (1989). Methane steam reforming, methanation and water-gas shift: I. Intrinsic kinetics. *AIChE Journal, 35*, 88–96.

You, Y. W., Lee, D. G., Yoon, K. Y., Moon, D. K., Kim, S. M., & Lee, C. H. (2012). H2 PSA purifier for CO removal from hydrogen mixtures. *International Journal of Hydrogen Energy, 37*, 18175–18186. https://doi.org/10.1016/j.ijhydene.2012.09.044.

Zakaria, K., Thimmappa, R., Mamlouk, M., & Scott, K. (2020). Hydrogen generation by alcohol reforming in a tandem cell consisting of a coupled fuel cell and electrolyser. *International Journal of Hydrogen Energy, 45*, 8107–8117. https://doi.org/10.1016/j.ijhydene.2020.01.123.

Zamel, N., & Li, X. (2011). Effect of contaminants on polymer electrolyte membrane fuel cells. *Progress in Energy and Combustion Science, 37*, 292–329. https://doi.org/10.1016/j.pecs.2010.06.003.

Zanfir, M., & Gavriilidis, A. (2002). An investigation of catalytic plate reactors by means of parametric sensitivity analysis. *Chemical Engineering Science, 57*, 1653–1659.

Zanfir, M., & Gavriilidis, A. (2004). Influence of flow arrangement in catalytic plate reactors for methane steam reforming. *Chemical Engineering Research and Design, 82*, 252–258.

Zha, S., Cheng, Z., & Liu, M. (2007). Sulfur poisoning and regeneration of Ni-based anodes in solid oxide fuel cells. *Journal of the Electrochemical Society, 154,* B201–B206. https://doi.org/10.1149/1.2404779.

Zheng, Y., Wang, J., Yu, B., Zhang, W., Chen, J., Qiao, J., & Zhang, J. (2017). A review of high temperature co-electrolysis of H2O and CO2 to produce sustainable fuels using solid oxide electrolysis cells (SOECs): Advanced materials and technology. *Chemical Society Reviews, 46,* 1427–1463. https://doi.org/10.1039/c6cs00403b.

Zuliani, N., & Taccani, R. (2012). Microcogeneration system based on HTPEM fuel cell fueled with natural gas: Performance analysis. *Applied Energy, 97,* 802–808. https://doi.org/10.1016/j.apenergy.2011.12.089.

Crude to chemicals: The conventional FCC unit still relevant

Iqbal Mohammed Mujtaba[a], Yakubu Mandafiya John[a], Aminu Zakari Yusuf[b], and Raj Patel[a]
[a]*Chemical Engineering, University of Bradford, Bradford, United Kingdom.*
[b]*Nigerian Gas Marketing Limited, Abuja, Nigeria*

1 History of direct crude processing

The term "Crude Oil" is used interchangeably with "Petroleum," and it comprises a mixture of hydrocarbon products and other compounds of sulfur, nitrogen, and oxygen in different proportions that results in different chemical and physical properties (Speight, 2020a). According to various scholars, there has been a long controversy on the origin of crude oil in the past with some who believe that it probably came from outer space during the earth's formation, while others believe it could be some form of organic or inorganic substance obtained from decayed vegetation and animals (Figueirôa et al., 2019). However, crude oil is not a commodity of the modern world only; people in Africa, Asia, Europe, and America have been collecting and using crude oil that seeps from the ground, for over 70,000 years (Craig, 2020). It was used in Babylon as a building material for walls and towers, and at Mohenjo-Daro, asphalt was used to stop water leaking from tanks. It was used in ancient Persia as a medicine and source of lighting and was extracted in China in the 4th century using bamboo sticks (Murty, 2020). One of the properties of the crude oil enjoyed by the people of old is its flammability, which enabled its use for lighting or as a source of illumination. As useful as it was to a dark world, people found a need to purify crude oil because of its unpleasant odor and the huge amount of soot produced (Murty, 2020), including having minimal value of useful fuels (Speight, 2020a). These needs gave birth to the modern way of refining crude oil using techniques such as distillation and cracking.

Modeling of Chemical Process Systems. https://doi.org/10.1016/B978-0-12-823869-1.00003-X

Even though considerable effort was invested, the transition from the direct use of petroleum crude as fuel to a more efficient application in place today was not a straightforward trajectory. According to Leprince (2001), oil industry developed from carrying out separation activities to molecular reorganization processes to upgrade both quantitative and qualitative disproportions in the petroleum products. At the beginning, to carry out molecular rearrangements, oil industry relied on thermal cracking to amend and improve quantitative balances and thermal reforming to upgrade the qualitative extents. However, they were challenged by low selectivity, poor yield, extreme technological necessities resulting from high pressure and temperature operations, by inefficient production cycles and high coking (Corma et al., 2017). Therefore, the use of thermal cracking from direct crude refining to obtain more gasoline, diesel fuel, aviation fuel, and kerosene (Murty, 2020) was incredibly challenging.

Some of the early efforts to distil crude oil into fuels in the Middle Eastern countries, Russia, Europe, and Persia were in the 13th century where products such as tar for building and kerosene for fuel were produced. By the 1915, crude oil was distilled in batches of about 5–6 barrels in tubular furnaces to obtain products such as fuels and lubricants in a series of boiling steps (Murty, 2020). Prior to this, in 1795, the British hand-dug wells to extract crude oil in Yenangyaung Myanmar (Longmuir, 2000). They were also involved in direct crude refining, which was based on heating the crude to produce a few fractions.

The modern oil industry is reported to have developed crude oil distillation units, of commercial value, around 1859–1862 (Speight, 2020b). Russia, however, was distilling crude oil to produce kerosene for use in lamps in churches and monasteries around 1825 and producing about 3500 barrels per day (Murty, 2020). In general, the earliest refining process for direct crude refining as reported in the literature is the thermal cracking process technology (Corma et al., 2017, 2018; Murty, 2020).

2 Update on crude to chemical processing

Petroleum Refineries are gradually transitioning from the “fuel” type—producing gasoline, diesel, etc., to “chemical” type—producing basic chemical raw materials such as ethylene, propylene, benzene, toluene, and xylene as necessary building blocks for producing polymers and other petrochemicals (Alabdullah, Shoinkhorova, et al., 2021; Corma et al., 2017; Zhou et al., 2021, 2022). It is also believed that the use of crude oil sources for fuel, chemicals, and material production will soon be unsustainable due to environmental issues. However, with the Covid-19 pandemic and the current Russia–Ukraine war, price of petroleum crude oil and fuels such as gasoline and diesel is on the rise, endorsing the current reliance of the world's economy around the crude-oil-based economy (Neely, 2022). The last time petroleum crude oil was above $100 per barrel was in 2014, but now, the price is over $100 again due to the war in Ukraine (Liadze et al., 2022; Mbah & Wasum, 2022). Even though the world relies heavily on the global crude-oil-based economy for energy, there is a shift to the use of crude oil to manufacture chemicals that are used as feedstock to the petrochemical industries (Corma et al., 2017; Yadav et al., 2020; Zhou et al., 2022).

From literature, the usage of basic chemical materials globally rose annually by 3.6% since 2019, and market demand for petrochemicals was about 700 million tons in 2019, and this is projected to steadily rise to 1 billion tons annually by 2050 (Al-Absi et al., 2020; Zhou et al., 2021). Substantial investment in the production of chemicals such as polypropylene from propene and ethane is growing steadily (Jenkins, 2014), with a demand of over 100 million tons per year of propylene (Rodríguez-Vallejo et al., 2020). The demand for these chemicals is currently forcing refiners to crack petroleum crude directly to fuels (Al-Khattaf & Ali, 2018; Corma et al., 2017, 2018). This is because of the profitability of integrating petrochemical plants with conventional refineries to convert direct crude to chemicals (Al-Absi et al., 2020).

Oil companies and research institutions, such as Saudi Aramco and the petrochemical maker Sabic are developing processes for petrochemicals production (Al-Ghamd, 2017; Tullo, 2019). By integrating petrochemicals technology with refinery operations and appropriate catalysts, Shell makes use of steam crackers and refinery operations to convert crude oil to chemicals (Shell Global, n.d.). ExxonMobil has also processed direct crude oil via cracking in Singapore since 2014 and Chinese Petroleum Corporation (CPC) producing chemicals such as p-xylene in China (Keusenkothen et al., 2009; Kumar et al., 2022; Tullo, 2019). Sinopec is the first to successfully own an industrial crude to chemicals plant and currently processing crude oil to ethylene, propylene, and other chemical materials in China. The Sinopec technology is based on direct crude oil catalytic cracking technology, which avoids the traditional oil refining steps of going through atmospheric and vacuum distillation and other refining processes to produce light olefins and aromatics. This approach increases the productivity of high-value chemicals such as ethylene, propylene, and light aromatics, while considerably minimizing comprehensive energy consumption and carbon emissions (Echemi, 2021).

According to the Sinopec technology results, total production rate of low-carbon olefins and aromatics doubled to 500,000 tons of high-value chemicals from every 1 million tons of crude oil processed. Zhou et al. (2022) working at the China University of Petroleum (East China) used the catalytic pyrolysis method to dehydrogenate crude oil in a pilot process to produce light olefins and aromatics in the presence of Ca/ZSM-5 catalyst at 600°C, and they obtained 34.34 wt% C_2-C_4 olefins yield and 44.42 wt% aromatics yield. An interesting study at the King Abdullah University of Science and Technology (Alabdullah, Rodriguez-Gomez, et al., 2021) developed a multizone fluidized bed (MZFB) reactor that implements a new catalyst formulation to carry simultaneous complex reactions in a single reactor vessel, which tends to improve the cracking process that for many years was basically the major challenge for direct cracking of crude oil in a single vessel. They achieved 30 wt% conversion of direct crude to olefins with minimal amount of dry gas generated. King Abdullah University of Science and Technology has invested heavily in research and innovation (Alabdullah, Rodriguez-Gomez, et al., 2021) to study and produce innovative technology that can maximize chemicals production from petroleum crude oil. There are two major ways to maximize chemicals directly from crude oil, which are heavy oil hydrocracking and catalytic cracking to produce more low-carbon olefins and aromatics (Zhou et al., 2022).

The hydrocracking approach is carried out in three forms: namely maximization of naphtha, the use of steam cracking to produce ethylene and propylene, and the use of a continuous reforming process to produce Benzene, Toluene, and Xylene (BTX). This approach produces clean fuel and petrochemicals from crude oil or heavy oils, and it is considered the most versatile and efficient process (Ding et al., 2021). The heavy oil catalytic cracking approach is also used to increase high aromatic contents products as well as propylene and naphtha to add to the BTX pool (Zhang et al., 2022; Zhou et al., 2022).

Several advantages of processing entire crude oil directly into petrochemicals range from reducing the cost of production of light olefins with minimal energy consumption and carbon emissions to bypassing the atmospheric and vacuum distillation units, thereby cutting down capital and operating costs (Amghizar et al., 2017; Bryden et al., 2014; Zhou et al., 2022). Due to this, many oil and chemical companies have developed several production plants arrangements for the conversion of crude oil directly into petrochemicals, such as light olefins and aromatics (Al-Absi et al., 2020; Zhou et al., 2022). Some of the plants use diverse feedstocks (petroleum crude) such as Arab light, Arab extra light, Arab super light, Hydrotreated Arab light, Alaskan crude oil, and Agbami crude oil and processing at temperatures ranging from 400°C to 900°C to achieve olefins yield ranging from 2.5 wt% to 42.9 wt% and ethylene (18–23.2 wt%) and propylene (12.1–13.8 wt%) (Al-Absi et al., 2020; Al-Absi & Al-Khattaf, 2018; Bourane et al., 2016; Corma et al., 2018; Usman et al., 2017; Zhou et al., 2022). The catalysts used in these processes are mainly Equilibrium FCC catalysts/ZSM-5, Pentasil Zeolites (MFI), and USY Zeolites.

According to Deloitte (2019), the new ways to process crude oil directly to chemicals have the potential to revolutionize the entire production and use of chemicals by 2025. The process (Unfolding Process Technology) is compared with the traditional process technology in Fig. 1. They claimed that this process of producing chemicals directly from crude oil has yields of over 40% and will yield more when the traditional or conventional refineries are integrated with the petrochemical industries.

Deloitte (2019) observed that the process integration could take advantage of the emerging technological advancement to be run by digital technologies such as the Internet-of-Things (IoT), specialized drones, trained robots, the artificial intelligence skills, machine learning, cloud computing, and blockchain. Other

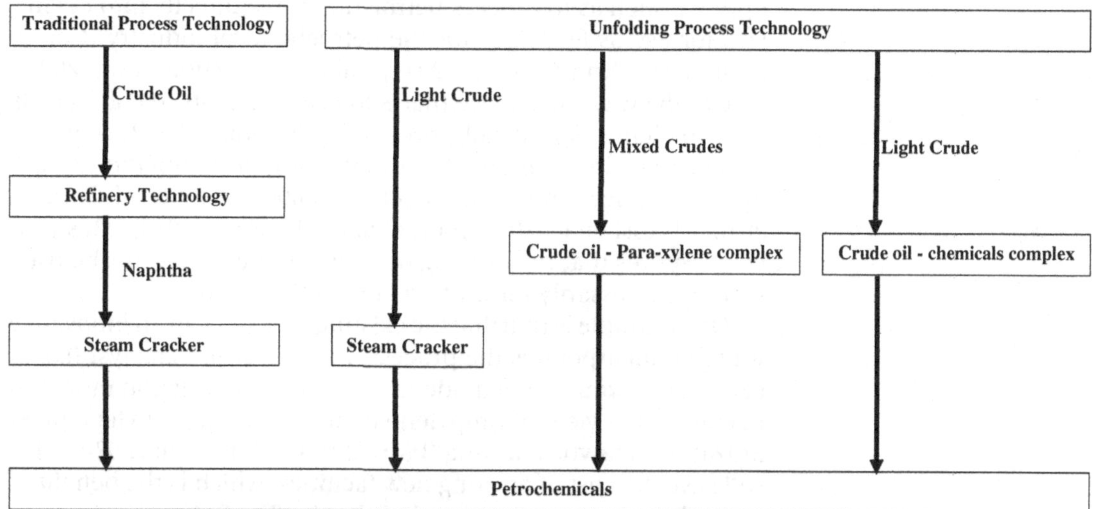

Fig. 1 Crude oil to chemicals integrated production route. Adapted from Deloitte. (2019). The future of petrochemicals: Growth surrounded by uncertainty [Online]. Deloitte Development LLC. Available from: https://www2.deloitte.com/us/en/pages/energy-and-resources/articles/base-chemicals-transform-petrochemicals-industry.html.

advantages include the real-time monitoring of chemical assets while predictive maintenance accuracy can be enhanced with cost-effective safety monitoring and improved efficiency to build confidence in the process.

3 FCC unit: Conventional FCC units with high severity to maximize propylene and ethylene

Fluid Catalytic Cracking (FCC) unit is a conversion process and workhorse of modern refineries. This unit is widely used to produce useful fuels such as gasoline, diesel, kerosine, and jet fuel for transportation and raw materials such as propylene and ethylene for petrochemical industries (Alabdullah, Rodriguez-Gomez, et al., 2021; Arbel et al., 1995; John et al., 2017b; Sertić-Bionda et al., 2009), and as sources of hydrogen, which can also be relevant to increase the petrochemical pool (Yadav et al., 2020).

The challenge of changing market because of the decreasing demand for gasoline and diesel for transportation due to the use of electric vehicles, and environmental regulations, makes operators of conventional FCC units redefine their purpose. This will lead to sluggish growth in the use of fossil fuels, after 2030 compared with the surge in the demand for chemicals and petrochemicals (Alabdullah et al., 2020; Chaudhuri & Singh, 2021). To satisfy this increasing demand, refiners are considering several options, such as to process petroleum crude directly into chemicals for use as feedstock for the petrochemical industry (Corma et al., 2017; Krumm et al., 2011; Tullo, 2019; Zhou et al., 2021). One of the ways to achieve this is to use FCC units to crack crude oil into olefins, specifically, propylene obtaining 5%–20% output at high severity operating conditions of high temperatures and long residence times (Aitani et al., 2000; Akah & Al-Ghrami, 2015; Al-Absi et al., 2020; John et al., 2018; Tullo, 2019). This process requires that the existing or conventional FCC units be used without necessarily retrofitting any section of the unit.

One example is utilizing the existing FCC units in an innovative way that incorporates the process, hardware, and catalyst (i.e., it can change from a fuel mode to a chemical production mode) to maximize the yield of propylene in an advanced propylene maximization catalytic cracking (PMCC) unit (Singh et al., 2020). This will save the cost of erecting new facilities, which is the benefit of using the current conventional FCC units. The PMCC is deigned to be a high-olefins cracker that uses the varied feedstock ranges, has high propylene selectivity catalyst with high conversion contributing about 10–12 wt% to the chemicals pool from one barrel of crude and adds 7 wt% of propylene for every unit of crude processed (Singh et al., 2020).

According to Capra (2020), propylene as a petrochemical feedstock is obtained as a by-product of thermal or catalytic operations involving coking and vis-breaking, which are mainly associated with the FCC unit processing to obtain olefins at high temperature. This is the case of high severity refining where the FCC units operate at high temperatures to favor the production of gases and the current 30% of world's propylene production (Aitani et al., 2000; Al-Absi et al., 2020; Capra, 2020; John et al., 2018; Parthasarathi & Alabduljabbar, 2014). The high-severity FCC process (HS-FCC) typically operates at temperature range of 500–800°C (Aitani et al., 2000; Akah, 2017; Akah & Al-Ghrami, 2015). Sinopec uses the combination of the HS-FCC and hydrocracker to convert vacuum gas oil (VGO) to obtain 55 wt% Olefins yield (Alabdullah et al., 2020).

Another FCC unit produces a petrochemical feedstock called Refinery-Grade Propylene (RGP). When it incorporates the use of Propane-Propylene (PP) splitters, the RGP (65–70 wt%) is converted into Chemical-Grade Propylene (CGP) (92 wt%) and Polymer-Grade Propylene (99.5 wt%) (Capra, 2020). This FCC unit retrofitting with PP splitters is profitable for the production of high-grade synthetic petrochemicals such as polypropylene (67%), cumene (6%), acrylonitrile (7%), propylene oxide (8%), and acrylic acid (4%) (Capra, 2020).

Zhou et al. (2022) present some novel FCC processes of conventional crude oil catalytic cracking process (COCC) and crude oil hierarchical catalytic cracking (CHCC) technology integrating the hierarchical gasification process to crack Daqing crude oil to produce petrochemicals. These examples show how the FCC unit is being used today to enrich and meet demand for petrochemicals.

4 Riser and regenerator mathematical model

Among the various routes to convert crude to chemicals, the use of the FCC unit is considered the most viable due to less complications such as fouling commonly experienced in the steam cracker units (Al-Khattaf & Ali, 2018). This conversion process takes place in the riser of the FCC unit, which is one of the two reactors in FCC while the other is the regenerator responsible for regenerating the coke infested catalyst due to cracking reaction. The direct crude cracking to chemicals uses the traditional or conventional FCC units. This means that the mathematical modeling of the FCC units used for crude oil conversion to chemical will still be the same as that of the conventional FCC units. The material balance, energy balance, and hydrodynamics remain the same. Also, kinetic models of the crude to chemicals conversion may be the same, although, attention should be given to the variation of feedstocks and catalyst, which can alter the kinetics of the reactions. Therefore, the area to consider is their impact on the process as Al-Khattaf and Ali (2018) stated that both catalyst and feedstock types are the key considerations to produce petrochemicals such as olefins produced from the FCC unit.

The FCC riser as one of the two reactors (Kunii & Levenspiel, 1991) is modeled as a one-dimensional plug flow system (Fernandes et al., 2007; Han & Chung, 2001a, 2001b; John et al., 2019; Nayak et al., 2005; Zhu et al., 2011), which can predict the overall performance of the riser (Theologos & Markatos, 1993). A schematic diagram of the riser unit of an FCC unit and its operating conditions as seen in the industry and literature (Han & Chung, 2001b; John et al., 2019; Kunii & Levenspiel, 1991; Shah et al., 2016) is shown (Selalame et al., 2022b) in Fig. 2.

The simulation model of the riser is made up of mass, energy, and momentum balance equations for the catalyst and gaseous phases. It also includes many other correlations and thermodynamics data. Some of the model equations along with some of their parameters used for simulating the riser are found in literature (Ali et al., 1997; Arandes et al., 2000; Fernandes et al., 2005, 2007; Han & Chung, 2001a, 2001b; Ivanchina et al., 2017; John et al., 2017a, 2017b; Landeghem et al., 1996; Rao et al., 2004; Roman et al., 2009; Selalame et al., 2022a, 2022b; Villafuerte-Macías et al., 2004; Xiong et al., 2015). The simulation is accompanied by some assumptions such as the hydrocarbon feed instantly vaporizes as it encounters the hot catalyst from the regenerator, then moves upward in thermal equilibrium with the catalyst, and there is no loss of heat from the riser (Ali et al., 1997). Selalame et al. (2022b) cautioned the overt use of such assumption since vaporization of atomized droplets in fluidized beds shows wide variations in vaporization times depending on flow conditions. It is also assumed that the cracking reactions only take place on the catalyst surface in the riser and are fast enough to justify steady-state model. The rates of dispersion and adsorption inside the catalyst particles are assumed to be negligible. Again, Selalame et al. (2022b) agreed with this only when the concentration of solids in the riser is low. These views are important and applicable to the simulation of riser units involved in the crude oil to chemicals simulation.

Riser	
Dimensions	
Height	30–40 m
Diameter	1–2 m
Operating Conditions	
Gas oil inlet T	150–300°C
Catalyst inlet T	675–750°C
Solid circulation	>250 kg/m^{2}s
Catalyst to Oil (CTO)	4–10 wt%
Dispersion steam	0–5 wt%
Pressure	150–300 kPa
Solid Residence	3–15 s
Catalyst Properties	
Average size	70 μm
Density	1200–1700 kg/m^{2}s
Typical U_{mf}	0.001 m/s
Geldart Group	A

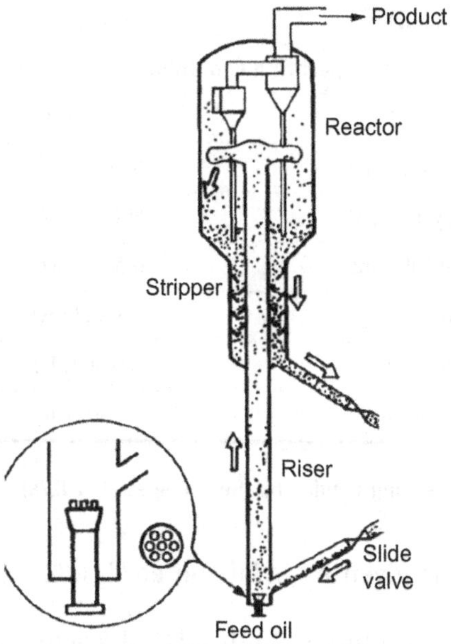

Fig. 2 Riser unit and its several dimensions and operating conditions (Selalame et al., 2022b).

Although the modeling of the FCC unit has already been thoroughly reviewed and most of the models are readily available in the literature, a comprehensive review of traditional FCC unit modeling approaches is clearly lacking in the literature, which are useful to the simulation of the riser as far as the crude oil to chemical modeling is concerned. To fill up this gap, Selalame et al. (2022b) have produced a comprehensive review of the riser modeling and analysis of the experimental data and the assumptions made in the modeling. They reviewed the current understanding of carbenium ion reaction chemistry of the riser and developed a mechanistic kinetic model of a single event kinetics methodology that utilizes an explicit description of rection pathways, contrary to the traditionally used lumped kinetics methodology. Their work is key to having an accurate modeling of the riser reactor in FCC units.

The second reactor in the FCC unit is the regenerator. Its schematic diagram and operating conditions, obtained from the

Regenerator	Dimensions	
	Height	30–40 m
	Diameter	1–2 m
	Operating Conditions	
	Gas oil inlet T	150–300°C
	Catalyst inlet T	675–750°C
	Catalyst exit T	500–500°C
	Coke entering	0.5–1.5 wt%
	Coke leaving	0.15–0.35 wt%
	Pressure	200–400 KPa
	Solid Residence	~≤ 1 h

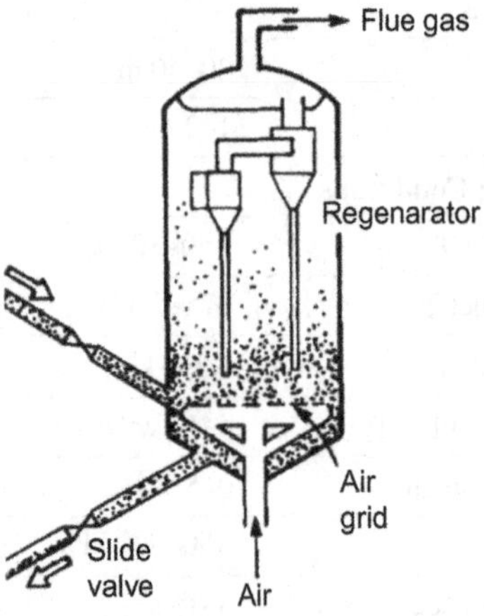

Fig. 3 Regenerator schematic diagram and its dimensions and operating conditions (Selalame et al., 2022a).

industry and literature as given by Selalame et al. (2022a), are shown in Fig. 3.

The FCC regenerator is a complex fluidized-bed reactor, which uses air for the combustions of coke deposited on the catalyst surface because of catalytic cracking of gas oil feedstock in the riser. The regenerator is regarded as a bubbling/turbulent regime fluidization bed and has very complex hydrodynamics in terms of mass, momentum, and energy transport phenomena, making it a difficult process to model (Han & Chung, 2001a; Sadeghbeigi, 2012). Due to its complex hydrodynamic nature, there are varied approaches to the modeling of the regenerator. Some of the approaches are the two-phase theory (Johnstone & Toomey, 1952), which later included the cloud and wake phases (Kunii & Levenspiel, 1969) and still in use in many modeling approaches today (Ali et al., 1997; Han & Chung, 2001b; John et al., 2017c; Kunii & Levenspiel, 1991). Selalame et al. (2022a) reviewed the modeling approaches of the regenerator and established that future models of the regenerator should make necessary adjustments to incorporate the combustion of sulfur and nitrogen-containing compounds, which are also present in coke and will improve the accuracy of the models.

The principles of the regenerator for the FCC units that process crude to chemicals will remain the same as conventional FCC regenerators. However, in some units such as the PMCC (Singh et al., 2020), it has smaller catalyst inventory and lower utility requirement, which is an advantage to the unit. A comprehensive modeling of the FCC regenerator is well documented in literature (Han & Chung, 2001a, 2001b; John, 2018; John et al., 2017c), and it is applicable to the regenerators in the crude oil to chemical units.

5 FCC catalysts and role in crude to chemical technology

One of the major challenges of producing propylene and ethylene from the FCC units and other chemical processes is that the products are mixed with several other gaseous products (Jiao et al., 2017; John et al., 2018; Knight & Mehlberg, 2011; Martin, 2022). Recently, there has been successful catalyst development that results in high selectivity to produce these useful products: propylene and ethylene. The catalyst designed for use in the PMCC unit has maximum-activity Zeolite Socony Mobil-5 (ZSM-5) zeolites with high selectivity to propylene (Martin, 2022; Singh et al., 2020), which has the potential to favor the production of propylene as a single lump, thereby overcoming the earlier challenge of having to separate propylene from gaseous product streams. The catalyst has an optimized pore structure, which allows both reactants and products to have the fastest diffusion with minimized hydrogen transfer. This helps the secondary cracking of gasoline olefins (which may not be useful in the future) to form more propylene (Singh et al., 2020).

The secondary reaction that cracks the gasoline olefins to propylene is shown in Fig. 4.

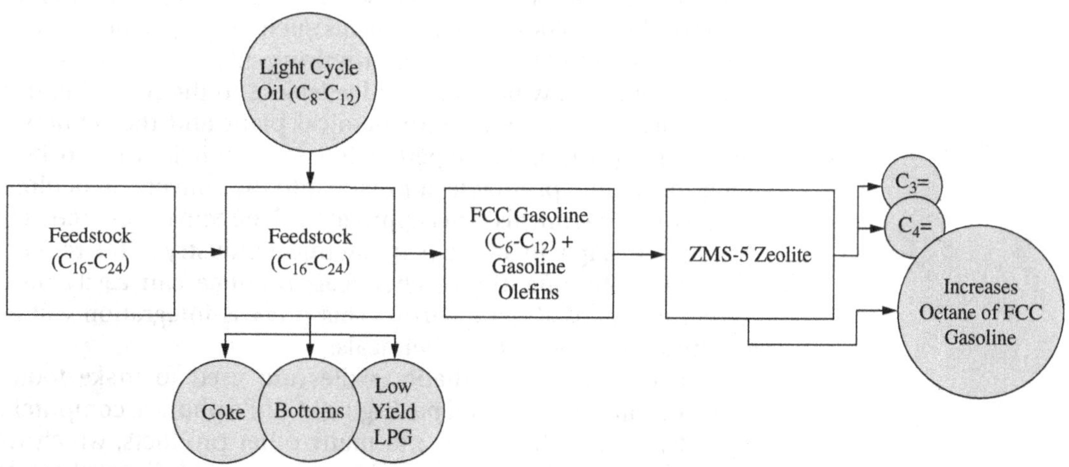

Fig. 4 High-propylene catalyst process. Adapted from Singh, R., Lai, S., Dharia, D., Hunt, D., & Cipriano, B. (2020). Conventional FCC to maximum propylene production. Hydrocarbon Processing. Houston, Texas: Catherine Watkins.

Another challenge associated with catalytic cracking is metal poisoning. Metal poisoning reduces catalyst activity and eventually reduces product conversion. The catalyst for the PMCC has advanced metal (Ni and V) trappings that are useful at high hydrogen and coke levels to protect the catalyst from being poisoned. This is good news to refiners who produce chemicals directly from crude. Singh et al. (2020) reported that the ZSM-5 zeolites yield above 19 wt%, a 3 wt% higher than the conventional catalyst as well as producing lower coke deposit on the catalyst.

6 The future of crude to chemicals

We live in a world that depends largely on petrochemical products, e.g., the cars we use for transportation are made of many petrochemical products, clothes, domestic and institutional furniture, building insulators, and road constructions all incorporate in one way or another some components of petrochemicals. Plastics as petrochemicals are considered to form a large portion of the world's materials demand. Now that the demand for fuels to power vehicles is in decline, most refineries are gradually redefining their operational objectives to produce products that satisfy customers requirement and meet both environmental and safety regulations. It is with this view that refiners use crude oil to produce feedstocks for the petrochemical industries, and this is currently dominating oil and gas discussions. Some refiners have already taken steps to build processes that convert crude directly into chemicals while others are using the existing FCC units under high-severity operating conditions to achieve the production of chemicals such as propylene and ethylene. Since the FCC unit is a promising alternative to produce these chemicals, which serve as feedstocks to the petrochemical industries, having both petrochemical plant and the refinery—FCC unit incorporated together into one unit is seen to be a long-term more profitable approach into the conversion of direct crude oil to chemicals. This approach will effectively upgrade the production capacity of the petrochemical industry to the refinery scale, and the quantity of chemicals produce can easily meet global demands (Chang, 2019). This process integration will be the future of crude oil to chemicals.

Plastics, rubber, synthetic textiles are used to make toothbrushes, carrier bags, food packaging, mobile phones, computers, carpets, clothes, furniture, and many other products, which we depend on. However, their production, use, and disposal create challenges in terms of air, water, and land pollution. These are sustainability challenges that require proactive attention. Although, the carbon in products such as plastics escapes into the atmosphere when burned or decomposed, carbon capture technologies can be used to mitigate its effect on the environment. Another approach to ensure sustainability is to build proper disposal and recycling facilities for industrial use and be made available for people to use as well. Other experts can be encouraged to come up with recycling technologies that are energy-efficient and can be integrated with both industrial and domestic use. Also, there is a need to proactively reduce the reliance on single-use plastics except for essential nonsubstitutable functions.

Other aspect of sustainable production and use of chemicals in the future will be processes involving catalytic reactions that differ from the traditional process routes, which according to (IEA, 2018) will bring about 15% of energy savings for every unit of production. Also, operators of the process plants can learn from the example of shifting from the use of coal as a feedstock to natural gas for both ammonia and methanol production in China, which consequently decreases the process emissions and energy intensity. Therefore, building sustainable processes for product development and use is another approach to the sustainability of crude to chemicals future.

References

Aitani, A., Yoshikawa, T., & Ino, T. (2000). Maximization of FCC light olefins by high severity operation and ZSM-5 addition. *Catalysis Today, 60*, 111–117.

Akah, A. (2017). Application of rare earths in fluid catalytic cracking: A review. *Journal of Rare Earths, 35*, 941–956.

Akah, A., & Al-Ghrami, M. (2015). Maximizing propylene production via FCC technology. *Applied Petrochemical Research, 5*, 377–392.

Alabdullah, M. A., Gomez, A. R., Vittenet, J., Bendjeriou-Sedjerari, A., Xu, W., Abba, I. A., & Gascon, J. (2020). A viewpoint on the refinery of the future: Catalyst and process challenges. *ACS Catalysis, 10*, 8131–8140.

Alabdullah, M., Rodriguez-Gomez, A., Shoinkhorova, T., Dikhtiarenko, A., Chowdhury, A. D., Hita, I., Kulkarni, S. R., Vittenet, J., Sarathy, S. M., Castaño, P., Bendjeriou-Sedjerari, A., Abou-Hamad, E., Zhang, W., Ali, O. S., Morales-Osorio, I., Xu, W., & Gascon, J. (2021). One-step conversion of crude oil to light olefins using a multi-zone reactor. *Nature Catalysis, 4*, 233–241.

Alabdullah, M., Shoinkhorova, T., Rodriguez-Gomez, A., Dikhtiarenko, A., Vittenet, J., Ali, O. S., Morales-Osorio, I., Xu, W., & Gascon, J. (2021). Composition-performance relationships in catalysts formulation for the direct conversion of crude oil to chemicals. *ChemCatChem, 13*, 1806–1813.

Al-Absi, A. A., Aitani, A. M., & Al-Khattaf, S. S. (2020). Thermal and catalytic cracking of whole crude oils at high severity. *Journal of Analytical and Applied Pyrolysis, 145*, 104705.

Al-Absi, A. A., & Al-Khattaf, S. S. (2018). Conversion of Arabian light crude oil to light olefins via catalytic and thermal cracking. *Energy & Fuels, 32*, 8705–8714.

Al-Ghamd, M. S. (2017). *System for conversion of crude oil to petrochemicals and fuel products integrating vacuum gas oil hydrocracking and steam cracking.* US Patent Application 16/555,794.

Ali, H., Rohani, S., & Corriou, J. P. (1997). Modelling and control of a riser type fluid catalytic cracking (FCC) unit. *Chemical Engineering Research and Design, 75*, 401–412.

Al-Khattaf, S. S., & Ali, S. A. (2018). Catalytic cracking of arab super light crude oil to light olefins: An experimental and kinetic study. *Energy & Fuels, 32*, 2234–2244.

Amghizar, I., Vandewalle, L. A., Van Geem, K. M., & Marin, G. B. (2017). New trends in olefin production. *Engineering, 3*, 171–178.

Arandes, J. M., Azkoiti, M. J., Bilbao, J., & de Lasa, H. I. (2000). Modelling FCC units under steady and unsteady state conditions. *The Canadian Journal of Chemical Engineering, 78*, 111–123.

Arbel, A., Huang, Z., Rinard, I. H., & Shinnar, R. (1995). Dynamic and control of fluidized catalytic crackers. 1. Modeling of the current generation of FCC's. *Industrial and Engineering Chemistry Research, 34*, 1228–1243.

Bourane, A., Al-Ghrami, M. S., Abba, I. A., Xu, W., & Shaik, K. M. (2016). *Process for high severity catalytic cracking of crude oil.* 14/050,708.

Bryden, K., Federspiel, M., Jr., & E. T. & Schiller, R. (2014). Processing tight oils in FCC: Issues, opportunities and flexible catalytic solutions. *American Fuel and Petrochemical Manufacturers, AFPM - AFPM Annual Meeting, 2014*(114), 127–174.

Capra, M. (2020). Unconventional improvement of propylene recovery yield at the PP splitter. In *Hydrocarbon processing*. Houston, Texas: Catherine Watkins.

Chang, R.J. 2019. Technology innovations—Crude oil to chemicals [online]. IHS Markit S&P Global. Available: https://ihsmarkit.com/research-analysis/technology-innovations-crude-oil-to-chemicals.html [Accessed 01 August 2022].

Chaudhuri, S., & Singh, R. B. (2021). Fuels and chemicals: Finding the right refinery configuration for a less predictable world. In *Hydrocarbon Processing.* Houston, Texas: Catherine Watkins.

Corma, A., Corresa, E., Mathieu, Y., Sauvanaud, L., Al-Bogami, S., Al-Ghrami, M. S., & Bourane, A. (2017). Crude oil to chemicals: light olefins from crude oil. *Catalysis Science & Technology, 7*, 12–46.

Corma, A., Sauvanaud, L., Mathieu, Y., Al-Bogami, S., Bourane, A., & Al-Ghrami, M. (2018). Direct crude oil cracking for producing chemicals: Thermal cracking modeling. *Fuel, 211*, 726–736.

Craig, J. (2020). History of oil: Regions and uses of petroleum in the classical and medieval periods. In R. Sorkhabi (Ed.), *Encyclopedia of petroleum geoscience.* Cham: Springer International Publishing.

Deloitte. 2019. The future of petrochemicals: Growth surrounded by uncertainty [Online]. Deloitte Development LLC. Available: https://www2.deloitte.com/us/en/pages/energy-and-resources/articles/base-chemicals-transform-petrochemicals-industry.html [Accessed].

Ding, L., Sitepu, H., Al-Bogami, S. A., Yami, D., Tamimi, M., Shaik, K., & Sayed, E. (2021). Effect of Zeolite-Y modification on crude-oil direct hydrocracking. *ACS Omega, 6*, 28654–28662.

Echemi 2021. Sinopec breaks through crude oil direct chemical technology [Online]. Available: https://www.echemi.com/cms/209401.html [Accessed 23 June 2022].

Fernandes, J. L., Verstraete, J. J., Pinheiro, C. I. C., Oliveira, N. M. C., & Ramôa Ribeiro, F. (2007). Dynamic modelling of an industrial R2R FCC unit. *Chemical Engineering Science, 62,* 1184–1198.

Fernandes, J., Verstraete, J.J., Pinheiro, C.C., Oliveira, N. & Ribeiro, F.R. 2005. Mechanistic dynamic modelling of an industrial FCC unit. In: Puigjaner, L.A.E., A. (Ed.) European symposium on computer aided process engineering, vol. 15. Elsevier Science B. V.

Figueirôa, S. F., Good, G. A., & Peyerl, D. (2019). *History, exploration & exploitation of oil and gas.* Springer International Publishing.

Han, I. S., & Chung, C. B. (2001a). Dynamic modeling and simulation of a fluidized catalytic cracking process. Part I: Process modeling. *Chemical Engineering Science, 56,* 1951–1971.

Han, I. S., & Chung, C. B. (2001b). Dynamic modeling and simulation of a fluidized catalytic cracking process. Part II: Property estimation and simulation. *Chemical Engineering Science, 56,* 1973–1990.

IEA. (2018). *The future of petrochemicals towards more sustainable plastics and fertilisers.* International Energy Agency.

Ivanchina, E., Ivashkina, E., & Nazarova, G. (2017). Mathematical modelling of catalytic cracking riser reactor. *Chemical Engineering Journal, 329,* 262–274.

Jenkins, S. (2014). Propylene production via propane dehydrogenation. *Chemical Engineering, 121,* 27–28.

Jiao, L., Xiong, X., Fang, X., Zang, J., Yu, H., & Liu, D. (2017). Six-lump kinetic study of propylene synthesis from methanol over HZSM-5 catalyst. *Journal of Chemical Engineering of Japan, 50,* 358–366.

John, Y. M. (2018). *Kinetic modelling simulation and optimal operation of fluid catalytic cracking of crude oil: Hydrodynamic investigation of riser gas phase compressibility factor, kinetic parameter estimation strategy and optimal yields of propylene, diesel and gasoline in fluid catalytic cracking unit.* University of Bradford.

John, Y. M., Mustafa, M. A., Patel, R., & Mujtaba, I. M. (2019). Parameter estimation of a six-lump kinetic model of an industrial fluid catalytic cracking unit. *Fuel, 235,* 1436–1454.

John, Y. M., Patel, R., & Mujtaba, I. M. (2017a). Maximization of gasoline in an industrial fluidized catalytic cracking unit. *Energy & Fuels, 31,* 5645–5661.

John, Y. M., Patel, R., & Mujtaba, I. M. (2017b). Modelling and simulation of an industrial riser in fluid catalytic cracking process. *Computers & Chemical Engineering, 106,* 730–743.

John, Y. M., Patel, R., & Mujtaba, I. M. (2017c). Optimization of fluidized catalytic cracking unit regenerator to minimize CO_2 emissions. *Chemical Engineering Transactions: The Italian Association of Chemical Engineering,* 57.

John, Y. M., Patel, R., & Mujtaba, I. M. (2018). Maximization of propylene in an industrial FCC unit. *Applied Petrochemical Research, 8,* 79–95.

Johnstone, H., & Toomey, R. (1952). Gas fluidization of solid particles. *Chemical Engineering Progress, 48,* 220.

Keusenkothen, P. F., Mccoy, J. N., Graham, J. E., & Reimann, C. D. (2009). *Process for cracking synthetic crude oil-containing feedstock.* United States patent application 11/698,514.

Knight, J., & Mehlberg, R. (2011). Maximize propylene from your FCC unit. *Hydrocarbon Process, 90,* 91–95.

Krumm, R. L., Deo, M., & Petrick, M. (2011). Direct thermal and catalytic treatment of paraffinic crude oils and heavy fractions. *Energy & Fuels, 26,* 2663–2671.

Kumar, L., Asthana, S., Laxman Newalkar, B., & Kishore Pant, K. (2022). Selective toluene methylation to p-xylene: Current status & future perspective. *Catalysis Reviews*, 1–43.

Kunii, D., & Levenspiel, O. (1969). *Fluidization engineering*. New York: Wiley.

Kunii, D., & Levenspiel, O. (1991). *Fluidization engineering*. London: Butterworth Heinemann Seris in Chemical Engineering.

Landeghem, F. V., Nevicato, D., Pitault, I., Forissier, M., Turlier, P., Derouin, C., & Bernard, J. R. (1996). Fluid catalytic cracking: modelling of an industrial riser. *Applied Catalysis A: General, 138*, 381–405.

Leprince, P. (2001). *Petroleum refining, Volume 3—Conversion processes*. Editions Technip.

Liadze, I., Macchiarelli, C., Mortimer-Lee, P., & Juanino, P. S. (2022). The economic costs of the Russia-Ukraine conflict. *NIESR Policy Paper, 32*.

Longmuir, M. (2000). Yenangyaung and its Twinza: The Burmese indigenous "Earth-Oil" industry re-examined. *Journal of Burma Studies, 5*, 17–48.

Martin, G. R. (2022). Maximising the potential of an FCCU. In *Hydrocarbon Engineering*. United Kingdom: Palladian Publications.

Mbah, R. E., & Wasum, D. (2022). Russian-Ukraine 2022 war: A review of the economic impact of Russian-Ukraine crisis on the USA, UK, Canada, and Europe. *Advances in Social Sciences Research Journal, 9*, 144–153.

Murty, K. G. (2020). History of crude oil refining. In K. G. Murty (Ed.), *Models for optimum decision making: Crude oil production and refining*. Cham: Springer International Publishing.

Nayak, S. V., Joshi, S. L., & Ranade, V. V. (2005). Modeling of vaporization and cracking of liquid oil injected in a gas–solid riser. *Chemical Engineering Science, 60*, 6049–6066.

Neely, C. J. (2022). The Russian invasion, oil and gasoline prices, and recession. In *Economic synopses*.

Parthasarathi, R. S., & Alabduljabbar, S. S. (2014). HS-FCC High-severity fluidized catalytic cracking: a newcomer to the FCC family. *Applied Petrochemical Research, 4*, 441–444.

Rao, R. M., Rengaswamy, R., Suresh, A. K., & Balaraman, K. S. (2004). Industrial experience with object-oriented modelling—FCC case study. *Chemical Engineering Research and Design, 82*, 527–552.

Rodríguez-Vallejo, D. F., Guillén-Gosálbez, G., & Chachuat, B. (2020). What is the true cost of producing propylene from methanol? The role of externalities. *ACS Sustainable Chemistry & Engineering, 8*, 3072–3081.

Roman, R., Nagy, Z. K., Cristea, M. V., & Agachi, S. P. (2009). Dynamic modelling and nonlinear model predictive control of a fluid catalytic cracking unit. *Computers & Chemical Engineering, 33*, 605–617.

Sadeghbeigi, R. (2012). *Fluid catalytic cracking handbook. An expert guide to the practical operation, design, and optimization of FCC units*. Houston, Texas: Gulf Publishing.

Selalame, T. W., Patel, R., Mujtaba, I. M., & John, Y. M. (2022a). A review of modelling of the FCC unitnmdash; Part II: The Regenerator. *Energies, 15*, 388.

Selalame, T. W., Patel, R., Mujtaba, I. M., & John, Y. M. (2022b). A review of modelling of the FCC unit. Part I: The riser. *Energies, 15*, 308.

Sertić-Bionda, K., Gomzi, Z., & Mužic, M. (2009). Modeling of gas oil catalytic cracking in a fixed bed reactor using a five-lump kinetic model. *Chemical Engineering Communications, 197*, 275–288.

Shah, M. T., Utikar, R. P., Pareek, V. K., Evans, G. M., & Joshi, J. B. (2016). Computational fluid dynamic modelling of FCC riser: A review. *Chemical Engineering Research and Design, 111*, 403–448.

Shell Global. n.d. Crude oil to chemicals [Online]. Shell Global. Available: https://www.shell.com/business-customers/catalysts-technologies/licensed-technologies/refinery-technology/crude-oil-to-chemicals.html# [Accessed 23 June 2022].

Singh, R., Lai, S., Dharia, D., Hunt, D., & Cipriano, B. (2020). Conventional FCC to maximum propylene production. In *Hydrocarbon Processing*. Houston, Texas: Catherine Watkins.

Speight, J. G. (2020a). Conventional crude oil. In *Refinery of the Future* (2nd ed.). Elsevier.

Speight, J. G. (2020b). *Refinery of the future* (2nd ed.). Elsevier.

Theologos, K. N., & Markatos, N. C. (1993). Advanced modeling of fluid catalytic cracking riser-type reactors. *AIChE Journal, 39*, 1007–1017.

Tullo, A. H. (2019). Why the future of oil is in chemicals, not fuels. In *Chemical and Engineering News*. America: American Chemical Society.

Usman, A., Siddiqui, M. A. B., Hussain, A., Aitani, A., & Al-Khattaf, S. (2017). Catalytic cracking of crude oil to light olefins and naphtha: Experimental and kinetic modeling. *Chemical Engineering Research and Design, 120*, 121–137.

Villafuerte-Macías, E. F., Aguilar, R., & Maya-Yescas, R. (2004). Towards modelling production of clean fuels: Sour gas formation in catalytic cracking. *Journal of Chemical Technology & Biotechnology, 79*, 1113–1118.

Xiong, K., Lu, C., Wang, Z., & Gao, X. (2015). Quantitative correlations of cracking performance with physiochemical properties of FCC catalysts by a novel lump kinetic modelling method. *Fuel, 161*, 113–119.

Yadav, V. G., Yadav, G. D., & Patankar, S. C. (2020). The production of fuels and chemicals in the new world: critical analysis of the choice between crude oil and biomass vis-a-vis sustainability and the environment. *Clean Technologies and Environmental Policy, 22*, 1757–1774.

Zhang, D., Zong, P., Li, J., Wang, C., Qiao, Y., & Tian, Y. (2022). Fundamental studies and pilot verification of an olefins/aromatics-rich chemical production from crude oil dehydrogenation catalytic pyrolysis process. *Fuel, 310*, 122435.

Zhou, X., Li, S., Wang, Y., Zhang, J., Zhang, Z., Wu, C., Chen, X., Feng, X., Liu, Y., Zhao, H., Yan, H., & Yang, C. (2022). Crude oil hierarchical catalytic cracking for maximizing chemicals production: Pilot-scale test, process optimization strategy, techno-economic-society-environment assessment. *Energy Conversion and Management, 253*, 115149.

Zhou, X., Sun, Z., Yan, H., Feng, X., Zhao, H., Liu, Y., Chen, X., & Yang, C. (2021). Produce petrochemicals directly from crude oil catalytic cracking, a techno-economic analysis and life cycle society-environment assessment. *Journal of Cleaner Production, 308*, 127283.

Zhu, C., Jun, Y., Patel, R., Wang, D. W., & Ho, T. C. (2011). Interactions of flow and reaction in fluid catalytic cracking risers. *AICHE Journal, 57*, 3122–3131.

PART IV

Machine learning techniques for modeling process systems

9

Hybrid model for a diesel cloud point soft-sensor

E. Turco Neto[a], Syed Ahmad Imtiaz[a], S. Ahmed[a], and R. Bhushan Gopaluni[b]

[a]*Center for Risk, Integrity and Safety Engineering (C-RISE), Department of Process Engineering, Memorial University of Newfoundland, St. John's, NL, Canada.*
[b]*Department of Chemical and Biological Engineering, University of British Columbia, Vancouver, BC, Canada*

1 Introduction

Models are at the center of process systems engineering (PSE) in various applications, including, design, control, optimization, monitoring, and prediction of properties (i.e., soft-sensor). In the early stage of PSE, mostly mechanistic models based on conservation laws (mass, energy, and momentum) complemented with various constitutive laws including equilibrium relations, reaction kinetics, and various other property relations were used to build the model. Mechanistic models are time-consuming to build and often require major assumptions as many mechanisms are poorly understood and difficult to model. Also, the execution time of the models could be an issue for on-line applications as mechanistic models involve solving ordinary differential equations (ODEs) and in some cases partial differential equations (PDEs) which take time to converge. This may limit the utility of the models for the desired application.

As process systems became more equipped with instruments and data became available, data-driven models also known as black-box models gained popularity. These models are quick to build, require minimal process knowledge, and often easy to implement. Several data-driven models received popularity in PSE, for example, various time series models, including finite impulse model (FIR) for the model predictive controller (MPC), autoregressive exogenous input (ARX), Box-Jenkins (BJ) models; partial least squares (PLS) and support vector regression (SVR) as soft sensors for predicting quality variables; principal

Modeling of Chemical Process Systems. https://doi.org/10.1016/B978-0-12-823869-1.00010-7

component analysis (PCA), independent component analysis (ICA), and canonical variate analysis (CVA) for process fault detection and diagnosis. However, these models lack the rigor of a mechanistic model. A black-box model is only valid for the range of data for which the model has been trained. The parameters of a black-box model do not have any physical significance, making the model less interpretable. As such, the reliability of these models is often questioned, and the models are avoided for system-critical applications.

There has been a significant effort to include the available system knowledge to the data-driven model and make them more interpretable. This has led to the emergence of the hybrid models, also sometimes referred to as gray-box models. The main idea for the hybrid model is to incorporate all system knowledge available in different forms to make the model more transparent as well as to improve the overall predictive ability of the model. The early work on hybrid models started in the 1990s, and since then researchers have worked in different directions to make the data-based model more transparent. A detailed review of hybrid models is beyond the scope of this book. We recommend interested readers to the following review articles Klatt and Marquardt (2009), Von Stosch et al. (2014), and Sansana et al. (2021).

1.1 Terminology related to hybrid model

There is a fair amount of ambiguity in the terminology around the hybrid model. From a systems theoretic point of view, a system is characterized as a hybrid system when it exhibits discrete event behavior along with continuously varying system behavior (Cameron & Hangos, 2001). However, from a modeling perspective, this is rarely used as a criterion to describe the hybrid model. In Von Stosch et al. (2014) and Sansana et al. (2021), researchers have tried to bring some clarity to the terminology. In the early days, models were mostly mechanistic, and empirical relations were used mainly to complement mechanistic knowledge. In a mechanistic model, in addition to conservation equations, many constitutive equations are used. These constitutive equations have a varying degree of mechanistic understanding. For example, reaction kinetics and equation of states have some physical understanding with parameters calibrated using data, while property relations for specific heat are purely empirical. Therefore, in the true sense, most mechanistic models are not entirely mechanistic. As data-driven modeling approaches gained popularity, empirical components started playing a dominant role in the

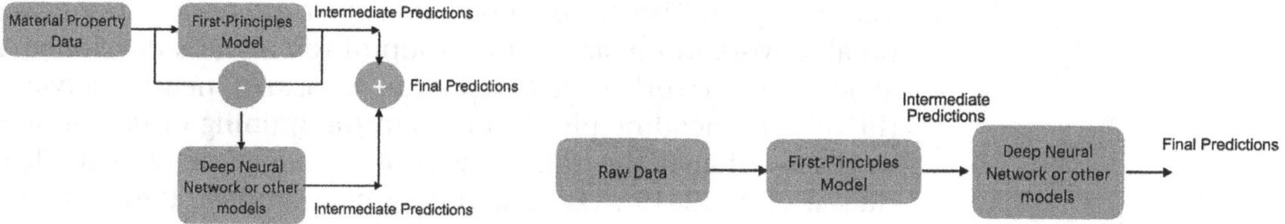

Fig. 1 A schematic showing the parallel (left) and series (right) approaches to hybrid modeling.

model. At the same time, efforts were made to bring more transparency to the model in terms of causality, interpretability, and consistency in prediction. These sets of models have been widely referred to as hybrid models.

The other trend has been to combine black box and white box submodels in series, parallel in different order to develop the system model, as shown in Fig. 1. In Von Stosch et al. (2014), researchers have further distinguished a class of models as a "semiparametric hybrid" model where part of the model will be parametric and part will be nonparametric with different levels of transparency.

In many process applications, including optimization and real-time predictions often, mechanistic models are replaced by data-driven surrogate models to generate the outputs at a higher frequency. Data are drawn from the rigorous mechanistic model to train the nonparametric model. The combination of the mechanistic and surrogate models is also referred to as a hybrid or gray-box models.

With the re-emergence of neural networks and deep neural networks, there is significant interest in the process industries to apply the models for various process systems engineering applications. However, unlike various web applications (e.g., marketing and social media), the risk associated with process industries is very high. In addition to an accurate prediction, it is important to have correct interpretation for the models. Also, these interpretations need to be robust under noise or disturbances (Ghorbani et al., 2019). One approach is to do a post hoc interpretability analysis that seeks to explain the sensitivity of prediction on different input features without diving deep into the black box. The other approach is to train the models so that the models trained neural network obeys the laws of physics. Researchers have worked in this direction since the 1990s primarily by adding the governing equations as constraints during the neural network training. A comprehensive review of physics-guided neural networks can be found in Willard et al. (2020).

Raissi et al. (2019) have gained the most attention among the notable work in the area. This group of researchers developed a general framework called a physics-informed neural network (PINN) for encoding physical laws in the training of neural networks to enhance the transparency of the neural network models (Raissi et al., 2019). Using a variety of examples, they demonstrated that the proposed data-driven algorithms can solve general nonlinear partial differential equations and construct surrogate models that obey the physical constraints. The idea here is to construct a neural network so that it is forced to follow constraints that naturally arise due to known scientific principles. For instance, you may require a model to satisfy certain mass balance or energy balance principles. These principles are added as constraints during the training of the neural network.

In some contexts, hybrid models are also referred to as "Digital Twins." Digital twins are nothing but complex models that mimic the behavior of a given system or process. The approaches described above could potentially be used to build hybrid digital twins. In addition, Fasel et al. (2022) developed a new approach to discover underlying first-principles models using data. This approach is called "Sparse identification of nonlinear systems." The algorithm uses a predetermined bank of models (both first-principles and empirical models) and a sparse optimization algorithm to identify the right combination of models that describe a given system. This algorithm is often used in combination with deep neural networks to build hybrid models (Cozad et al., 2015; Goodfellow et al., 2016; Psichogios & Ungar, 1992). One advantage of these models is that they will be partially interpretable (meaning the various parameters in the model will have physical interpretation), unlike models obtained using purely empirical approaches.

There are two ways in which first-principles models are embedded in an empirical model. The prediction from a first-principles model is augmented using an empirical model in a parallel or a series fashion. Figure below shows these approaches.

In addition to the above trends, researchers have used many ad hoc methods to incorporate system knowledge in the data-driven model with an ultimate goal to make the model more robust, causal, and interpretable, with better predictive ability over a wider range. In the following section, we will describe the development of a hybrid model for the process system. A hybrid soft-sensor was developed to predict the cloud point of dewaxed diesel fuels produced from catalytic dewaxing reactors in real-time. The soft-sensor has a mechanistic model as well as a surrogate data-based model that runs in parallel with the mechanistic model for online predictions. The mechanistic model has three modules: true boiling point (TBP) module, the reactor module, and the thermodynamic model. While the latter two modules are mechanistic, the TBP module is a more data-centric model.

2 Case study: A hybrid model for diesel cloud point prediction

Cloud point is defined as the highest temperature at which the first wax crystal forms in the fuel (Rakoczy & Morse, 2013). Cloud point is widely used in petroleum refineries to characterize diesel fuel's cold flow properties. The measurement is commonly done through visual techniques defined by the standard ASTM D2500. The visual measurement depends on the operator's skill and experience and has a high degree of uncertainty associated with the measurement. Also, the frequency of these lab measurements in a refinery is between 2 and 4 times a day, making it unsuitable for real-time control. As an alternative, a reactor model to accurately predict the cloud point values in real-time, waiving the dependence on visual measurements, is beneficial. The solid-liquid equilibrium (SLE) theory can be used to estimate the cloud point of hydrocarbon mixtures with reasonable accuracy. Therefore, integrating a mechanistic reactor model into the solid-liquid equilibrium provides a simple way to develop a soft-sensor to predict online the product cloud point from the knowledge of the reactor operating conditions and feedstock distillation data during the catalytic dewaxing operation. A surrogate model is developed in parallel to facilitate online prediction of the cloud point.

2.1 Cloud point soft-sensor—Mechanistic model

The schematic representation for the proposed cloud-point soft-sensor is shown in Fig. 2.

Each module in the sensor is described as follows:

- **TBP module**: Receives the diesel feedstock distillation data and converts into detailed mole composition distribution in the MTHS matrix form using the concepts described in Section 2.1.1;
- **Reactor module**: Reads the information regarding the process operating conditions and predicted inlet composition distribution. Then, these values are set as boundary conditions for the hydro-dewaxing model described in Sections 2.1.2 and 2.1.3 to simulate the operation and predict the outlet composition of the diesel product;

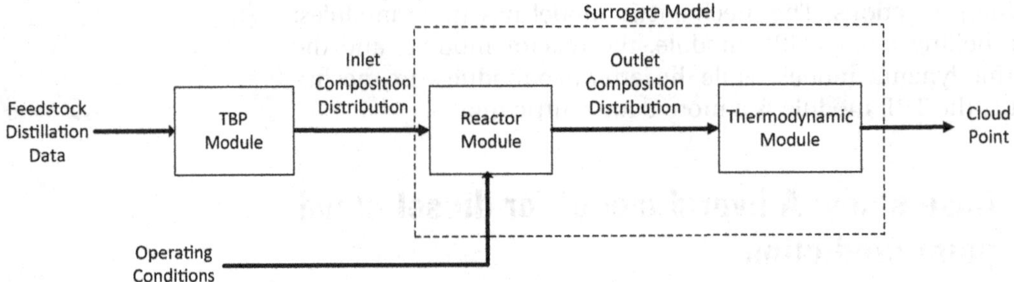

Fig. 2 Schematic representation for the proposed cloud point soft-sensor framework.

- **Thermodynamic module**: Accepts the simulated outlet composition distribution and uses this information as input for the solid-liquid equilibrium flash algorithm discussed in Section 2.1.4 to estimate the cloud point of the dewaxed diesel product.

The reactor and thermodynamic modules are easily coupled to compute the cloud point of the dewaxed diesel for a given set of operating conditions and compositions. However, the simulations can be computationally expensive due to the strong nonlinear behavior of the differential equations. Their complex functional form would also bring difficulties when implementing the sensor for online prediction purposes. To overcome these issues, a simpler surrogate model was running in parallel to the reactor and thermodynamic modules for reproducing the point prediction when the sensor is intended for on-line applications.

2.1.1 TBP module

In this work, the boundary conditions necessary to carry out the reactor simulations require detailed knowledge of the hydrocarbon mole distribution in the diesel feedstock. Even though advanced analytical techniques such as chromatography have considerably improved over the years, petroleum fractions are still being characterized using the ASTM-D86 distillation data and bulk properties such as specific gravity, cloud point, and viscosity. Hence, converting such compact and generalized information into detailed mole distribution is crucial to the soft-sensor development as the diesel properties vary from unit to unit, as the streams can be blended before being sent to the catalytic dewaxing unit. However, these streams are usually constituted by thousands of hydrocarbons differing in type, sizing, and degree of branching, and thereby, such interconversion is not straightforward.

This task can be accurately performed using a characterization technique initially proposed by Peng (1999). The approach represents the compositions of a refining stream in matrix form denominated molecular type and homologous series (MTHS). Its columns represent the type of molecules that can be possibly found in a diesel stream, such as paraffins, naphthenes, and aromatics, while the rows represent their molecular size based on the carbon number (Ahmad et al., 2011). For each molecular type and carbon number, several model compounds have been assigned based on synthetic and real diesel molecular information provided by Froment et al. (1994) and Ahmad et al. (2011). The entries in the matrix represent their overall mole fractions and are the unknown variables to be determined.

To illustrate the application of this approach to build the TBP module, the bulk information for an industrial diesel feedstock provided by a local refinery, shown in Table 1, is used.

It was observed from the industrial data that this feedstock was a blend of several heavy and light diesel streams produced from different units, such as fluid catalytic cracking (FCC) and hydrocracking. Therefore, model compounds with carbon number

Table 1 Distillation curve and bulk properties for an industrial diesel feedstock.

ASTM-D86 distillation curve (vol % distilled) (°F)	
Initial boiling point (IBP)	307.2
10%	412.3
30%	458.4
50%	493
70%	527.4
90%	564.1
Final boiling point (FBP)	617.2
Bulk physical properties	
Specific gravity	0.8435
Cloud point (°C)	−23.78
Sulfur content (ppm)	4817.9
Nitrogen content (ppm)	56.32
Paraffinic-naphthenic-aromatic (PNA) analysis	
Paraffinic content (mol%)	42.57
Naphthenic content (mol%)	30.25
Aromatic content (mol%)	27.18

ranging from C_8 to C_{24} were considered to describe this industrial feed. The average molecular weight and PNA composition values were estimated for this feedstock using correlations taken from American Petroleum Institute (1997). To obtain the entries of the MTHS matrix, an optimization procedure was used to minimize the objective function given by Eq. (1) .

$$F=\left(\frac{SG^{cal}-SG^{ind}}{SG^{ind}}\right)^2+\left(\frac{M^{cal}-M^{ind}}{M^{ind}}\right)^2+\left(\frac{x_S^{cal}-x_S^{ind}}{x_S^{ind}}\right)^2+\left(\frac{x_{Nap}^{cal}-x_{Nap}^{ind}}{x_{Nap}^{ind}}\right)^2$$
$$+\left(\frac{x_{aro}^{cal}-x_{aro}^{ind}}{x_{arom}^{ind}}\right)^2+\left(\frac{x_{par}^{cal}-x_{par}^{ind}}{x_{par}^{ind}}\right)^2+\sum_{i=1}^{7}\left(\frac{V_{TBP,i}^{cal}-V_{TBP,i}^{ind}}{V_{TBP,i}^{ind}}\right)^2 \tag{1}$$

This equation represents the deviation between the industrial values for the properties in Table 2 and those calculated using the optimal mole fraction distribution obtained from the minimization problem. The specific gravity can be evaluated by Eq. (2) (Ahmad et al., 2011).

$$\frac{1}{SG^{cal}}=\sum_{i,j}\frac{w_{i,j}}{SG_{i,j}} \tag{2}$$

The density at 60°F for each model compound was calculated using the Peng-Robinson equation of state, and their individual critical properties were obtained using Joback's group contribution method (Poling et al., 2001). The average molecular weight of the simulated diesel feed was obtained through Eq. (3) (Ahmad et al., 2011).

$$M^{cal}=\sum_{i,j}x_{i,j}M_{i,j} \tag{3}$$

Table 2 Validation results obtained from the MTHS algorithm.

Property	Industrial value	Calculated value
MeABP (°C)	249.74	256.46
Cloud point (°C)	−23.78	−23.42
Sulfur content (ppm)	4817.9	4807.96
Paraffinic content (mol%)	42.57	42.62
Naphthenic content (mol%)	30.25	30.21
Aromatic content (mol%)	27.18	27.17

Finally, the total mole fraction of paraffins, aromatics, naphthenes, and sulfur compounds can be easily computed using Eq. (4) (Wu, 2010). It is important to mention that benzothiophene and 4,6-dimethyl dibenzothiophene were assigned as model compounds to represent the sulfur content as they are extensively considered to formulate mechanistic kinetic models developed to simulate hydrotreating reactors.

$$x_j^{cal} = \sum_i x_{i,j} \tag{4}$$

The last term in Eq. (1) represents the deviations related to industrial and calculated TBP curves. First, the ASTM D-86 distillation data from Table 1 are converted into a TBP distillation curve using interconversion relationships found in the American Petroleum Institute (1997). Then, the cumulative percent volume distilled is calculated by sorting the normal boiling point temperature in ascending order with respect to the volume fraction of each model compound and applying the cumulative summation (Wu, 2010).

To solve this optimization problem, two linear constraints must be imposed to obtain a realistic mole composition distribution. The first one is based on the fact that the sum of the mole fractions in the MTHS matrix must be equal to 1 as mathematically shown by Eq. (5).

$$\sum_{i,j} x_{i,j} = 1 \tag{5}$$

The last constraint is related to the total nitrogen content in the diesel feed. As only quinoline has been assigned as a model compound to represent the nitrogen content, its composition in the matrix must be equal to the total nitrogen content provided in Table 1.

The optimization problem was solved using the interior-point algorithm available in the MATLAB® optimization toolbox, and the results are summarized in Table 2. The calculated cloud point value was obtained by applying the thermodynamic module described in Section 2.1.3. The obtained MTHS matrix is shown in Appendix.

It can be observed that the composition distribution along with selected model compounds was satisfactory to model the diesel feed with good accuracy. Fig. 3 shows a comparison between calculated and industrial TBP curves, confirming the accuracy of the methodology applied.

2.1.2 Hydro-dewaxing (HDW) reactor model

The catalytic dewaxing reactor model is formulated based on the single-event theory proposed by Froment and co-workers (Baltanas et al., 1989; Kumar, 2004; Kumar & Froment, 2007;

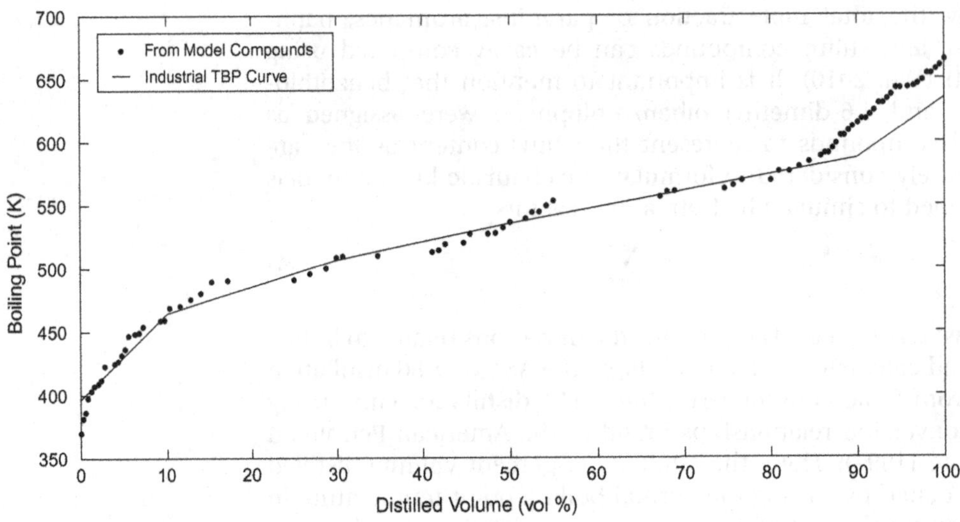

Fig. 3 Comparison between industrial and calculated TBP curves.

Svoboda et al., 1995). In this approach, the rate constant is expressed as shown by Eq. (6).

$$k = \left(\frac{\sigma_{gl}^{r}}{\sigma_{gl}^{\#}}\right)\left(\frac{k_B T}{h}\right)\exp\left(\frac{\Delta\widehat{S}^{0\#}}{R}\right)\exp\left(-\frac{\Delta H^{0\#}}{RT}\right) \quad (6)$$

The first term in this equation represents the ratio between the global symmetry numbers for reactant and activated complex, which is a denominated number of single events (n_e) and accounts for the effects related to the changes in molecular structure due to chemical reaction. The other terms are related to changes in intrinsic enthalpy and entropy and can be lumped into the so-called single event rate constant ($\tilde{k}$), which is independent of the feedstock type. Hence, Eq. (6) can be rewritten as given by Eq. (7) (Kumar, 2004).

$$k = n_e \tilde{k} \quad (7)$$

The shape-selective cracking and isomerization are the main reactions taking place within the catalytic dewaxing unit, which follow the mechanisms governed by the carbenium ion chemistry. Those reactions are usually carried out on bifunctional shape-selective catalysts such as Pt/ZSM-5, which contain metal sites responsible for the formation of intermediate olefins due to dehydrogenation reactions and acid sites where the cracking and isomerization steps take place. The common elementary steps promoted by metal and acid sites on Pt/ZSM-5 are shown in Table 3.

Table 3 Common elementary steps promoted by metal and acid sites on Pt/ZSM-5.

Dehydrogenation/hydrogenation	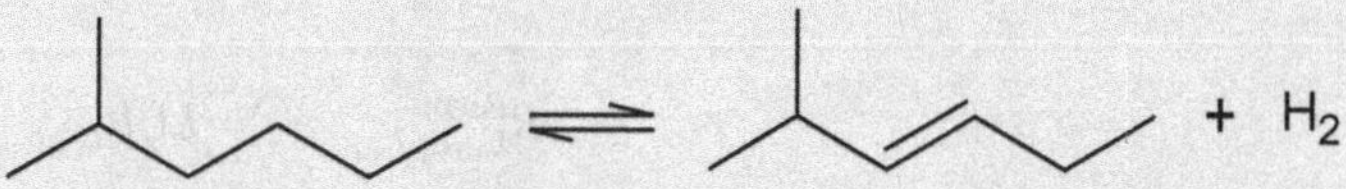 ⇌ + H_2
Protonation/deprotonation	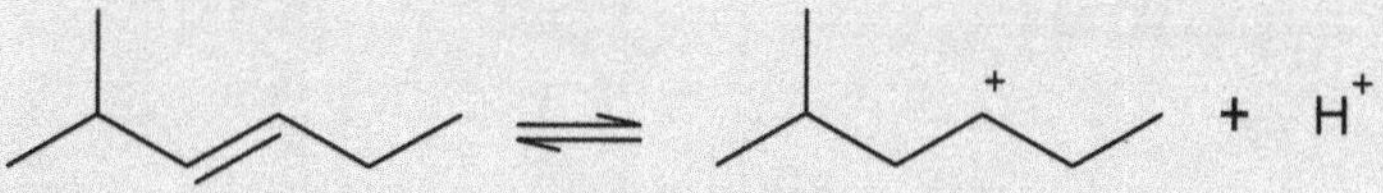 ⇌ + H^+
Hydride shift	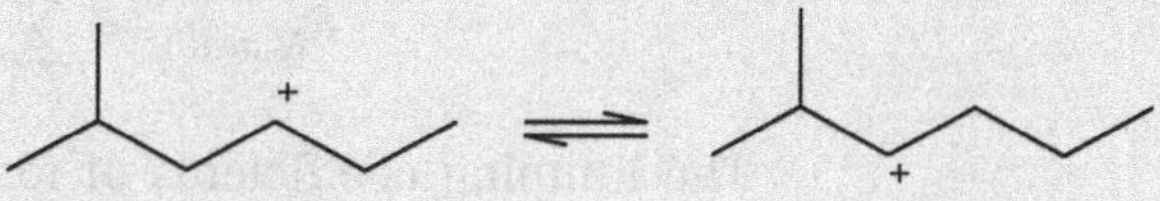 ⇌
β-scission	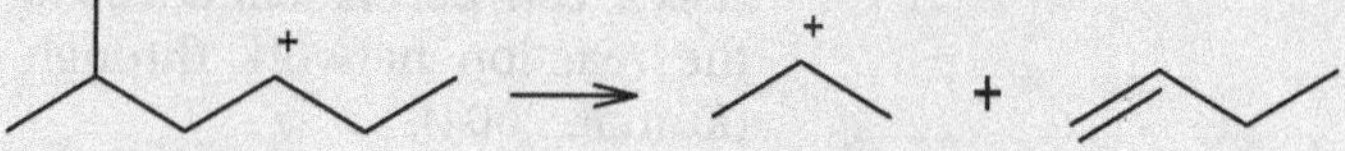 → +
PCP isomerization	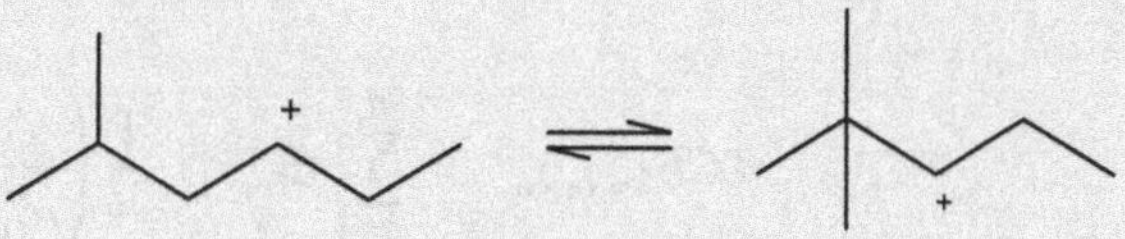 ⇌

By assuming ideal hydrocracking conditions, the reaction network proposed by Baltanas and Froment (1985) was used to obtain all the individual elementary steps assuming that the reactions taking place at the metal sites are fast enough to be considered in pseudo-equilibrium. The shape-selectivity effects were incorporated into the reaction network by eliminating reactions involving reactants or products whose molecular structures are sterically hindered in the pore-mouths and micropores of the catalyst. The methyl-shift reactions were also suppressed as shifting the methyl branches can induce the ion to move outside the pore-mouth, which is not consistent with the reaction mechanism (Laxmi Narasimhan et al., 2003). Finally, the posteriori-lumping technique is used to lump the individual compounds identified in the reaction network into common groups based on the carbon number and degree of branching, whose internal composition distribution is governed by the thermodynamic equilibrium (Vynckier & Froment, 1991). The rate expressions considered in this work to compute the net rate of formation for the components reacting in the HDW bed are given by Eqs. (8) through (10) (Kumar, 2004).

$$R^{Cons}_{L_m,l_{m,w}} = LCC_{L_m,l_{m,w}} \frac{\tilde{k}^{*}_{l}(m,w) K_{L,L_m} C^{L}_{L_m}}{C^{L}_{H_2}\left[1+\sum_{i} K_{L,L_m} C^{L}_{L_m}\right]} \tag{8}$$

$$R^{Form}_{L_m,L_k,l_{m,w}} = \sum_{k} LCF_{L_m,L_k,l_{m,w}} \frac{\tilde{k}^{*}_{l}(m,w) K_{L,L_k} C^{L}_{L_k}}{C^{L}_{H_2}\left[1+\sum_{i} K_{L,L_k} C^{L}_{L_k}\right]} \tag{9}$$

$$R_{L_m,net^{Form}} = \sum_{l=1}^{n_l}\left(R^{Form}_{L_m} - R^{Cons}_{L_m}\right) \tag{10}$$

The lumping coefficients of formation (LCF) and consumption (LCC) can be calculated using the information obtained from the reaction network through Eqs. (11) and (12), respectively (Kumar, 2004).

$$LCF_{L_m,L_k,l_{m,w}} = \sum_{q=1}^{q_T} n_{e,q} y_{i,L_k} \frac{1}{n}\left(\frac{\sigma^{gl}_{Pi}}{\sigma^{gl}_{O_{ij}}\sigma^{gl}_{H_2}}\right) \sum_{j=1}^{n} \tilde{K}^{(P_i \leftrightarrow O_{ij})}_{DH} \tilde{K}^{(O_{ij} \leftrightarrow O_r)}_{isom} \tag{11}$$

$$LCC_{Lm,l_{m,w}} = \sum_{q=1}^{q_T} n_{e,q} y_{i,Lm} \frac{1}{n}\left(\frac{\sigma^{gl}_{Pi}}{\sigma^{gl}_{O_{ij}}\sigma^{gl}_{H_2}}\right) \sum_{j=1}^{n} \tilde{K}^{(P_i \leftrightarrow O_{ij})}_{DH} \tilde{K}^{(O_{ij} \leftrightarrow O_r)}_{isom} \tag{12}$$

The single-event kinetic parameters for catalytic dewaxing on Pt/ZSM-5 were obtained in our previous work, described in Chapter 3 of this book, and are used to propose the hydro-dewaxing reactor model in this section.

It is assumed that both gas and liquid phases flow down along the hydro-dewaxing catalytic bed in plug flow pattern and that the catalyst particles are completely surrounded by liquid. The reactions involved in the catalytic dewaxing mechanism are usually endothermic; however, the overall conversion in this process usually ranges from low to moderate values to avoid yield losses. Therefore, it is assumed that the hydro-dewaxing bed is nearly isothermal and the continuity equations required to obtain the composition profiles for a given lump in both gas and liquid phases are expressed by Eqs. (13) and (14), respectively (Kumar & Froment, 2007).

$$\frac{1}{A}\frac{dF^{G}_{i}}{dz} = -k_{o,i} a_v\left(\frac{C^{G}_{i}}{K^{C,VLE}_{i}} - C^{L}_{i}\right) \qquad \text{at } z=0, F^{G}_{i} = F^{G}_{i,0} \tag{13}$$

$$\frac{1}{A}\frac{dF_i^L}{dz} = k_{o,i}a_v\left(\frac{C_i^G}{K_i^{C,VLE}} - C_i^L\right) + R_{i,net}^{form} \quad \text{at } z = 0, F_i^L = F_{i,0}^L \quad (14)$$

By considering the two-film theory to model gas-liquid interphase mass transfer, the overall mass transfer coefficient can be calculated based on the individual mass transfer film coefficients for both liquid and gas sides using Eq. (15) (Kumar, 2004).

$$\frac{1}{k_{o,i}a_v} = \frac{1}{K_i^{C,VLE}k_G a_v} + \frac{1}{k_L a_v} \quad (15)$$

These coefficients were estimated using the correlations proposed by Reiss (1967) and Sato et al. (1972), respectively. The equilibrium coefficient in concentration basis expressed by Eq. (6) can be written as a function of the true vapor-liquid equilibrium constant, which is calculated using the Peng-Robinson equation of state.

$$K_i^{C,VLE} = K_i^{VLE}\frac{\rho_{\text{molar}}^G}{\rho_{\text{molar}}^L} \quad (16)$$

This set of differential equations were solved using the Backward Differentiation Formula (BDF) available in MATLAB® to obtain the composition profiles along the HDW bed, and the model was checked in our previous work (Turco Neto, 2016).

2.1.3 Hydrodesulfurization reactor model

In this work, the commercial CoMo/Al_2O_3 was selected as the catalyst to be used for hydrodesulfurization (HDS) in order to eliminate the sulfur compounds that can poison the dewaxing catalyst while producing ultra-low sulfur diesel (ULSD). This catalyst contains two types of active sites (Van Parijs & Froment, 1986):

- The σ-sites where hydrogenolysis reactions take place, which are responsible for extracting the sulfur atom from the reactant molecular structure;
- The τ-sites where hydrogenation reactions take place, which are responsible for the saturation of the aromatic rings found in the molecular structure of sulfur compounds;

The model compounds used to represent the overall sulfur content in the diesel feedstock were dibenzothiophene and 4,6-dimethyldibenzothiophene. The reaction mechanism governing the conversion of dibenzothiophene to cyclohexylbenzene and H_2S was proposed by Vanrysselberghe and Froment (1996) as shown in Fig. 4.

Likewise, the mechanism for the hydrodesulfurization of 4,6-dimethyldibenzothiophene, which is converted into bicyclohexyl,

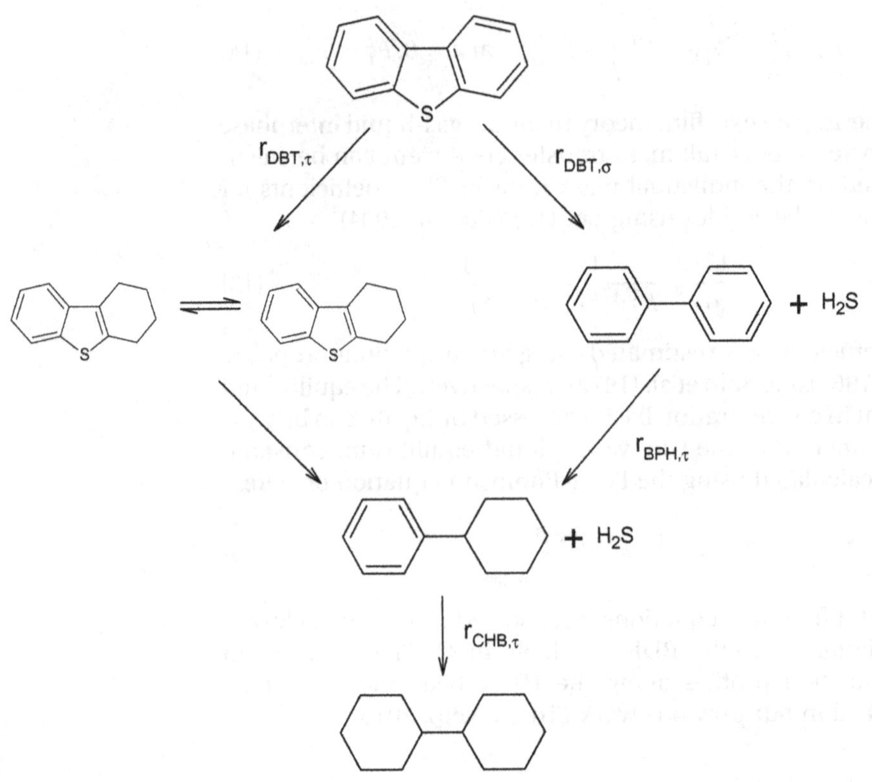

Fig. 4 Mechanism for hydrodesulfurization of dibenzothiophene (Vanrysselberghe & Froment, 1996).

3,3′-dimethylbicyclophenyl, and H_2S, was proposed by Vanrysselberghe et al. (1998) and is illustrated by Fig. 5.

According to Vanrysselberghe et al. (1998), the technical limitations in producing ULSD can be explained by the slow conversion of 4,6-dimethyldibenzothiophene, as the methyl groups attached to its aromatic rings block the sulfur atom from being adsorbed at the active sites of the catalyst. These reactions were used as the basis to obtain the values of the net rate of consumption for each component involved in the HDS reaction mechanism. The rate equations and kinetic parameters are going to be discussed in the next section.

Kinetic parameters and rate equations

The Hougen-Watson rate equations and kinetic parameters related to the previously discussed hydrodesulfurization mechanisms were taken from several sources available in the literature. The rate equations as well as the rate and adsorption equilibrium constants are shown in Tables 4 through 7.

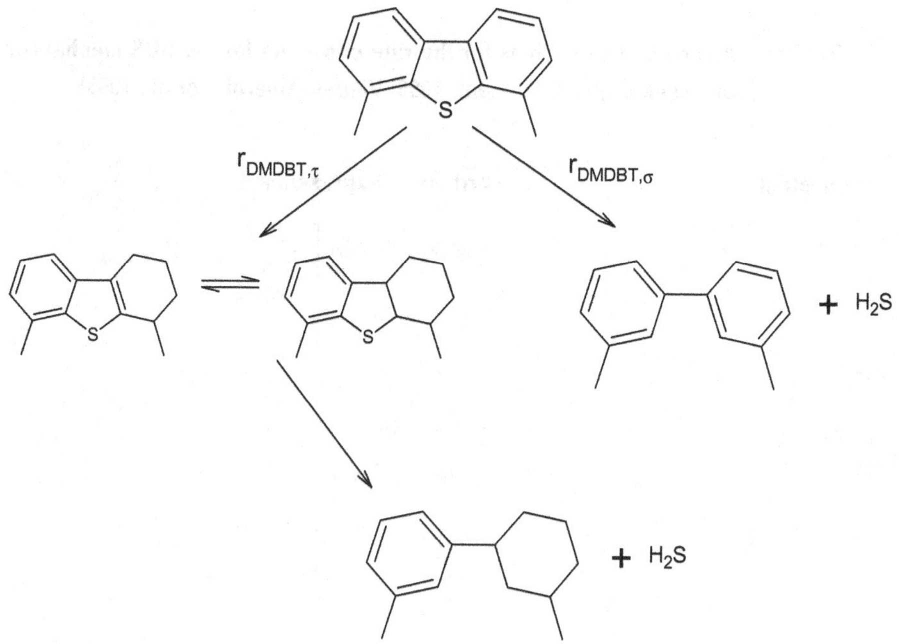

Fig. 5 Mechanism for hydrodesulfurization of 4,6-dimethyldibenzothiophene (Vanrysselberghe et al., 1998).

Table 4 Hougen-Watson rate equations for the main reactions taking place in the HDS process (Vanrysselberghe & Froment, 1996; Vanrysselberghe et al., 1998).

Reaction type	Hougen-Watson rate equation
Hydrogenolysis of dibenzothiophene	$r_{DBT,\sigma} = \frac{k_{DBT,\sigma} K_{DBT,\sigma} K_{H,\sigma} C_{DBT} C_{H_2}}{DEN_\sigma}$ (17)
Hydrogenation of dibenzothiophene	$r_{DBT,\tau} = \frac{k_{DBT,\tau} K_{DBT,\tau} K_{H,\tau} C_{DBT} C_{H_2}}{DEN_\tau}$ (18)
Hydrogenation of biphenyl	$r_{BPH,\tau} = \frac{k_{BPH,\tau} K_{BPH,\tau} K_{H,\tau} C_{BPH} C_{H_2}}{DEN_\tau}$ (19)
Hydrogenation of cyclohexylbenzene	$r_{CHB,\tau} = \frac{k_{CHB,\tau} K_{CHB,\tau} K_{H,\tau} C_{CHB} C_{H_2}}{DEN_\tau}$ (20)
Hydrogenolysis of 4,6-dimethyl-dibenzothiophene	$r_{DMBT,\sigma} = \frac{k_{DMBT,\sigma} K_{DMBT,\sigma} K_{H,\sigma} C_{DMBT} C_{H_2}}{DEN_\sigma}$ (21)
Hydrogenation of 4,6-dimethyl-dibenzothiophene	$r_{DMBT,\tau} = \frac{k_{DMBT,\tau} K_{DMBT,\tau} K_{H,\tau} C_{DMBT} C_{H_2}}{DEN_\tau}$ (22)

Table 5 Arrhenius expressions for the rate constants for the HDS mechanisms (Vanrysselberghe & Froment, 1996; Vanrysselberghe et al., 1998).

Rate constant	Arrhenius expression	
$k_{DBT,\sigma}$	$2.44336 \times 10^{10} \exp\left[-\frac{122,770}{RT}\right]$	(23)
$k_{DBT,\tau}$	$2.86757 \times 10^{16} \exp\left[-\frac{186,190}{RT}\right]$	(24)
$k_{BPH,\tau}$	$3.4112 \times 10^{23} \exp\left[-\frac{255,714}{RT}\right]$	(25)
$k_{CHB,\tau}K_{CHB,\tau}$	3.38631×10^{-1}	
$k_{DMDBT,\sigma}$	$6.4456 \times 10^{7} \exp\left[-\frac{106,223}{RT}\right]$	(26)
$k_{DMDBT,\tau}$	$3.68208 \times 10^{27} \exp\left[-\frac{299,042}{RT}\right]$	(27)

Table 6 Van't Hoff expressions for the adsorption equilibrium constants for the HDS mechanisms (Vanrysselberghe & Froment, 1996; Vanrysselberghe et al., 1998).

Adsorption equilibrium constant	Van't Hoff expression	
$K_{H,\sigma}$	$3.36312 \times 10^{-11} \exp\left[\frac{113,232}{RT}\right]$	(28)
$K_{H,\tau}$	$1.40255 \times 10^{-15} \exp\left[\frac{142,693}{RT}\right]$	(29)
$K_{H_2S,\sigma}$	$1.47118 \times 10^{-8} \exp\left[\frac{105,670}{RT}\right]$	(30)
$K_{DBT,\sigma}$	7.5686×10^{1}	
$K_{DBT,\tau}$	$2.50395 \times 10^{-7} \exp\left[\frac{76,840}{RT}\right]$	(31)
$K_{BPH,\tau}$	$4.96685 \times 10^{-4} \exp\left[\frac{37,899}{RT}\right]$	(32)
$K_{DMBT,\sigma}$	1.80397×10^{1}	
$K_{DMBT,\tau}$	$1.58733 \times 10^{-8} \exp\left[\frac{90,485}{RT}\right]$	(33)

Table 7 Stoichiometric table for hydrodesulfurization reactions.

	$r_{DBT,\sigma}$	$r_{DBT,\tau}$	$r_{BPH,\tau}$	$r_{CHB,\tau}$	$r_{DMBT,\sigma}$	$r_{DMBT,\tau}$
H_2	1	1	0	0	1	1
H_2S	−2	−5	−3	−3	−2	−5
DBT	−1	−1	0	0	0	0
BPH	1	0	−1	0	0	0
CHB	0	1	1	−1	0	0
DMDBT	0	0	0	0	−1	−1
DMBPH	0	0	0	0	1	0
MCHT	0	0	0	0	0	1

The denominator in the rate equations can be expressed by Eqs. (34) and (35) (Vanrysselberghe & Froment, 1996; Vanrysselberghe et al., 1998)

$$DEN_\sigma = \left(\begin{array}{l} 1 + K_{DBT,\sigma}C_{DBT} + K_{BPH,\sigma}C_{BPH} + \sqrt{K_{H,\sigma}C_{H_2}} + K_{H_2S,\sigma}C_{H_2S} + \\ K_{DMDBT,\sigma}C_{DMDBT} \end{array} \right)^3 \tag{34}$$

$$DEN_\tau = \left(1 + K_{DBT,\tau}C_{DBT} + K_{BPH,\tau}C_{BPH} + \sqrt{K_{H,\tau}C_{H_2}} + K_{DMDBT,\tau}C_{DMDBT}\right)^3 \tag{35}$$

Hence, by analyzing the reaction mechanisms and rate expressions, the stoichiometric table required to calculate the net rate of consumption for the several compounds involved in the hydrodesulfurization process can be built as presented in Table 7.

The entries in Table 7 represent the stoichiometric coefficients for the different components reacting through the pathway indicated by reaction rate. Negative and positive numbers are related to reactants and products, respectively, while 0 is used to indicate that a compound is not reacting in that pathway.

Model equations

The continuity and energy equations required to describe the conversion process in a HDS catalytic bed is similar to those presented to model the catalytic dewaxing one. To simplify the problem, the diffusion limitations in the catalyst pellet as well as the resistance to mass transport from the bulk of liquid phase to the surface of the pellet are not considered. Hence, the continuity equation for a given compound in liquid phase accounting for interphase mass transfer and reaction kinetics is expressed by Eq. (36) (Froment et al., 1994).

$$\frac{1}{A}\frac{dF_i^L}{dz} = k_{o,i}a_v\left(\frac{C_i^G}{K_i^{C,VLE}} - C_i^L\right) - R_{i,HDS}^{net}(1-\varepsilon_b)$$
$$\text{at } z=0, F_i^L = F_{i,0}^L \tag{36}$$

Also, the continuity equation for the same component in gas phase is identical to the one presented by Eq. (13). The net rate of consumption expressed by Eq. (37) can be calculated using stoichiometry table and the Hougen-Watson rate expressions (Boesen, 2010).

$$R_{i,HDS}^{net} = -\rho_P \sum_{j=1}^{N_R} S[i,j]r_j \tag{37}$$

The energy equations used to obtain the temperature distributions for gas and liquid phases are expressed by Eqs. (38) and (39), respectively, which account for the interphase convective heat transport and the heat effects due to the exothermic reactions (Froment et al., 1994). Also, it is assumed that the temperature at the gas-liquid interface is equal to the one at liquid phase as the heat generated in the catalyst due to the HDS reactions is quickly transferred to this phase.

$$u_G\rho_G C_{pG}\frac{dT_G}{dz} = h_G a_v(T_L - T_G) + \sum_{i=1}^{N} N_i a_v C_{piG}(T_G - T_L)$$
$$\text{at } z=0, T_G = T_{G,0} \tag{38}$$

$$u_L\rho_L C_{pL}\frac{dT_L}{dz} = \rho_B \sum_{j=1}^{N_R} r_j(-\Delta H_{Rj}) + \sum_{i=1}^{N} N_i a_v \Delta H_{vi}$$
$$\text{at } z=0, T_L = T_{L,0} \tag{39}$$

The gas phase heat capacity for each model compound was estimated applying the correlations given by Poling et al. (2001), which were then used along with the Peng-Robinson equation of state to calculate the heat capacity for the gas mixture. The interphase heat transfer coefficient was evaluated using the Chilton-Colburn analogy (Froment et al., 2010). Finally, the heat of vaporization and formation for each component were obtained using the Joback's group contribution method (Poling et al., 2001). The composition and temperature profiles along the HDS bed were obtained by solving a set of differential equations represented by Eqs. (36) though (39), which were implemented and solved in MATLAB® using the backwards differentiation formula (BDF) approach.

2.1.4 Solid-liquid equilibrium thermodynamic model

When the cloud point of a diesel mixture is reached, the first solidified paraffinic crystals are in thermodynamic equilibrium with the liquid phase. Therefore, the phase equilibrium criterion given by Eq. (40) must be satisfied.

$$\hat{f}_i^L = \hat{f}_i^S \tag{40}$$

The fugacity of a given component can be written in terms of its activity coefficient and fugacity at pure state. Therefore, Eq. (41) can be rearranged into Eq. (43) (Ghanaei et al., 2007).

$$x_i^L \gamma_i^L f_{i,\text{pure}}^L = x_i^S \gamma_i^S f_{i,\text{pure}}^S \tag{41}$$

The solid-liquid equilibrium constant can be calculated by rewriting Eq. (41) into Eq. (42), which determines how a particular compound is distributed between the phases (Ghanaei et al., 2007).

$$K_i^{SL} = \frac{x_i^S}{x_i^L} = \frac{\gamma_i^L f_{i,\text{pure}}^L}{\gamma_i^S f_{i,\text{pure}}^S} \tag{42}$$

As confirmed by Coutinho et al. (2000), isoparaffinic, aromatic, and naphthenic compounds act as solvents and have almost a negligible impact on the cloud point temperature. Therefore, their composition can be lumped and represented by a single isoparaffinic pseudo-component selected to match the average molecular weight of the diesel fuel. Also, the solid phase is assumed to be only consisted of normal paraffins, and the pseudo-component is avoided in this phase by setting its fugacity at pure state in Eq. (42) to a large value as suggested by Pedersen (1995).

The fugacity ratio of each paraffinic component can be expressed as function of the fusion properties using an expression proposed by Coutinho et al. (2000) and then modified by Ghanaei et al. (2007) represented by Eq. (43).

$$\ln \frac{f_{i,\text{pure}}^S}{f_{i,\text{pure}}^L} = \frac{\Delta H_i^f}{RT_i^f}\left(1 - \frac{T_i^f}{T}\right) + \frac{1}{RT}\int_T^{T_i^f} \Delta C_{p,i}^{LS} dT - \frac{1}{R}\int_T^{T_i^f} \Delta C_{p,i}^{LS}\frac{dT}{T} + \frac{1}{RT}\int_P^{P_i^f} \Delta V_i dP \tag{43}$$

This equation represents the changes in enthalpy and entropy during the solidification process of a pure compound. The fusion enthalpy and temperature as well as the phase change heat capacity were estimated using the Eqs. (44) through (46) proposed by Won (1989) and Broadhurst (1962).

$$\Delta H_i^f = 0.1426 M_i T_i^f \quad (44)$$

$$T_i^f = \begin{cases} 0.040C_n^3 - 2.2133C_n^2 + 46.197C_n - 45.777 & C_n < C_{16} \\ 0.0028C_n^3 - 0.3185C_n^2 + 13.559C_n + 143.15 & C_n \geq C_{16} \end{cases} \quad (45)$$

$$\Delta C_{P,i}^{LS} = 0.3033 M_i T_i^f - 4.635 \times 10^{-4} M_i T_i^f \quad (46)$$

The last term in Eq. (43) is Poynting correction and represents the effect of pressure in the thermodynamic model (Ghanaei et al., 2007). In this work, this term has been neglected as the cloud point measurements are usually made at atmospheric pressure.

As the diesel product leaving the catalytic dewaxing reactor consists of several hydrocarbons differing in terms of size and degree of branching, nonideal behavior is expected for both solid and liquid phases. Therefore, the activity coefficients for the components in liquid phase can be evaluated using Eq. (47).

$$\ln \gamma_i^L = \ln \gamma_i^{res} + \ln \gamma_i^{fv} \quad (47)$$

The first term in Eq. (47) was evaluated using the UNIFAC model, which accounts for energetic interactions among the molecules. The second one was calculated using Flory-free volume equation represented by Eq. (48) and describes the effect of molecular size on the activity coefficient (Coutinho et al., 2000).

$$\ln \gamma_i^{fv} = \ln \frac{\varphi_i}{x_i^L} + 1 - \frac{\varphi_i}{x_i^L} \quad (48)$$

where

$$\varphi_i = \frac{x_i^L\left(V_i^{1/3} - V_{Wi}^{1/3}\right)}{\sum_j x_j^L\left(V_j^{1/3} - V_{Wj}^{1/3}\right)} \quad (49)$$

The parameter V_{Wi} is the van der Walls volume for a pure compound and can be determined by the group contribution method proposed by Bondi (1964). The activity coefficients for the compounds in solid phase are estimated using the predictive Wilson model given by Eq. (50).

$$\ln \gamma_i^S = 1 - \ln\left(\sum_i x_j^S \Lambda_{ij}\right) - \sum_k \frac{x_k^S \Lambda_{kj}}{\sum_j x_j^S \Lambda_{kj}} \quad (50)$$

where

$$\Lambda_{ij} = \exp\left(-\frac{\lambda_{ij} - \lambda_{ii}}{RT}\right) \quad (51)$$

Eq. (51) represents the characteristic energy parameter that is a function of the interaction energy for a pair of similar or different molecules, which can be evaluated by Eqs. (52) and (53), respectively (Ghanaei et al., 2007).

$$\lambda_{ii} = -\frac{2}{6}(\Delta H_{\text{sub},i} - RT) \tag{52}$$

$$\lambda_{ij} = \lambda_{ji} = \lambda_{jj} \tag{53}$$

The index "*j*" in Eq. (52) represents the smallest normal alkane in the pair being compared. The sublimation enthalpy can be estimated using Eq. (54) (Ghanaei et al., 2007).

$$\Delta H_{\text{sub},i} = \Delta H_{\text{vap},i} + \Delta H_i^f + \Delta H_{tr,i} \tag{54}$$

The heat of vaporization in Eq. (56) was evaluated using the correlation proposed by Morgan and Kobayashi (1994) given by Eqs. (55) through (61).

$$x = 1 - \frac{T}{T_C} \tag{55}$$

$$\begin{aligned}\Delta H_v^0 = {} & 5.2804x^{0.3333} + 12.865x^{0.8333} + 1.171x^{1.2083} - 13.116x \\ & + 0.4858x^2 - 1.088x^3\end{aligned} \tag{56}$$

$$\begin{aligned}\Delta H_v^1 = {} & 0.80022x^{0.3333} + 273.23x^{0.8333} + 465.08x^{1.2083} - 638.5x \\ & -145.12x^2 + 74.049x^3\end{aligned} \tag{57}$$

$$\begin{aligned}\Delta H_v^2 = {} & 7.2543x^{0.3333} - 346.45x^{0.8333} - 610.48x^{1.2083} + 839.89x \\ & + 160.05x^2 - 50.71x^3\end{aligned} \tag{58}$$

$$\frac{\Delta H_{\text{vap}}}{RT_C} = \Delta H_v^0 + \omega \Delta H_v^1 + \omega^2 \Delta H_v^2 \tag{59}$$

Finally, the transition enthalpy in Eq. (54) can be calculated using the correlation formulated by Broadhurst (1962) represented by Eq. (60).

$$\Delta H_{tr} = \begin{cases} 0.0009C_n^3 + 0.0011C_n^2 + 0.1668C_n + 3.693 & C_n < C_{19} \\ -0.0032C_n^3 + 0.2353C_n^2 - 3.912C_n + 25.261 & C_n \geq C_{19} \end{cases} \tag{60}$$

The solid-liquid flash algorithm proposed by Ghanaei et al. (2007) represented by the block diagram in Fig. 6 was implemented in MATLAB® to perform the phase equilibrium calculations required to calculate the cloud point temperature.

The two convergence criteria used in this algorithm are expressed by Eqs. (61) and (62).

$$\text{crit}_1 = \left| \sum_i x_i^S - 1 \right| \tag{61}$$

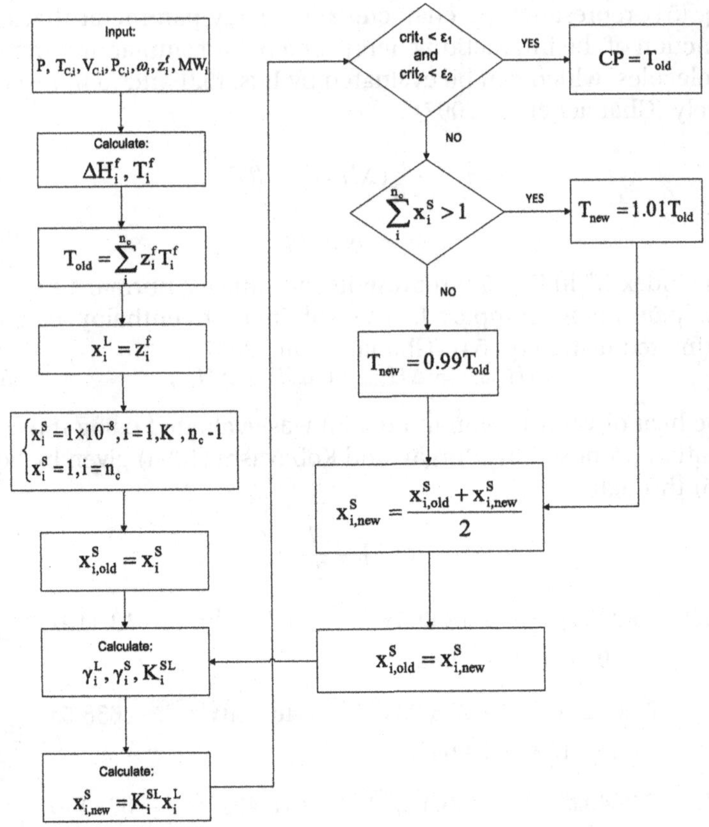

Fig. 6 Solid-liquid equilibrium calculation algorithm (Ghanaei et al., 2007).

Table 8 Validation of the solid-liquid equilibrium algorithm for cloud point estimation.

Diesel feedstock	Experimental Cloud point (Coutinho et al., 2000)	Calculated Cloud point
Nondesulfurized	4.3°C	4.48°C
Desulfurized	5.5°C	4.86°C

$$\text{crit}_2 = \max\left|\frac{x_{i,\text{new}}^S - x_{i,\text{old}}^S}{x_{i,\text{old}}^S}\right| \quad (62)$$

To check if the algorithm was correctly implemented, the experimental values of cloud point for two diesel feedstocks presented by Coutinho et al. (2000) were compared to the calculated ones shown in Table 8.

Besides the measurement errors related to the experimental data, the deviations between experimental and calculated cloud point values can also be explained based on the accuracy of the correlations used to estimate the fusion properties of the pure compounds.

2.2 Surrogate model

The mechanistic model described in the previous section can be used to easily design, simulate, and analyze the reactor operation based on changes in different process variables. However, the high complexity of the differential equations and thermodynamic calculations can make its implementation difficult to perform online cloud point estimation. Also, the reactor simulations can be time-consuming as the reactor model is highly nonlinear. To overcome these drawbacks, a surrogate model with simpler functional form and capable of reproducing the results with good accuracy is developed in the following sections.

There are several approaches to developing a surrogate model. In the one-shot process, all the experimental design points are used at once to build the surrogate model. The main drawback of this methodology lies in the fact that no information is available about the cloud point response before the reactor and thermodynamic simulations are carried out. Thereby, the ideal size for the design is yet to be discovered. Therefore, the design could either be oversampled, and the advantages of using the surrogate model would be lost as solving the differential equations can be expensive or undersampled, resulting in an inaccurate surrogate model as the design would not properly cover the design space (van der Herten et al., 2015).

In the sequential design approach, additional points are iteratively added to an initial design of experiments based on the information obtained from the surrogate model built at the previous iteration (Crombecq et al., 2011). In the context of this work, these extra samples must be added at regions where the cloud point response is nonlinear, as those are usually harder to be described. The iterations are carried out until a convergence criterion based on the accuracy of the proposed surrogate model is reached. This technique does not require prior knowledge about the size of the design; however, it faces a trade-off problem related to the exploration and exploitation features (van der Herten et al., 2015).

The exploration feature induces the sequential design strategy to cover the model domain as uniformly as possible, resulting in a global surrogate model capable of identifying all important regions containing nonlinearities or local optima in the surface response. However, this methodology would fail to accurately describing those locations as it would not place enough points at those (van der Herten et al., 2015). On the other hand, the exploitation capability would focus only on the nonlinear regions already identified at previous iterations and miss other ones not explored, resulting in a surrogate model that is only good for local predictions (Crombecq et al., 2011).

Therefore, building an online soft sensor capable of providing global predictions and accurately describing highly nonlinear regions in the cloud point response is one of the main goals of this work. In this context, the fuzzy sequential design strategy proposed by van der Herten et al. (2015) was found to be suitable for achieving this objective.

2.2.1 Design of computer experiments

The first step in applying the sequential design strategy is to create an initial design of experiments considering the changes in the most important process variables affecting the product cloud point response. This procedure provides initial knowledge about its functional form using only a few solutions obtained by using the original sensor framework. The LHSV, reactor inlet temperature, and pressure are selected as the main process variables in this work as they strongly affect the conversion profiles, especially in the HDW beds responsible for reducing the diesel product's cloud point.

Also, the effect of the feed paraffinic composition distribution must be incorporated into the online soft-sensor design due to its impact on both solid-liquid equilibrium behavior and HDW conversion. However, it is not possible to consider the composition of all 24 normal paraffinic compounds due to dimensionality constraints. To overcome this issue, only the mole fractions for the heavy normal paraffins ranging from C_{20} to C_{24} are chosen for the analysis as they have a major impact on the cloud point response.

As an application example, the ranges for the process parameters chosen to build the surrogate model to replace the reactor and thermodynamic modules in the current sensor framework are shown in Table 9.

The initial experiment design chosen to initialize the sequential design algorithm was a "max-min" Latin Hypercube containing 35 sample points. Once cloud point responses were obtained for all design points, the hyperparameters of the Gaussian process surrogate model were estimated by maximizing the likelihood function using the MATLAB® Statistical and Machine Learning Toolbox. At each iteration, the root relative square error (RRSE)

Table 9 Process variables and ranges adopted to develop the diesel cloud point predictor.

Process variable	Range
Mole fraction of n-C_{20}	$7.22\times10^{-5}\le x_{n-C_{20}}^{inlet}\le1.8\times10^{-3}$
Mole fraction of n-C_{21}	$6.9\times10^{-5}\le x_{n-C_{21}}^{inlet}\le1.8\times10^{-3}$
Mole fraction of n-C_{22}	$7.6\times10^{-5}\le x_{n-C_{22}}^{inlet}\le1.9\times10^{-3}$
Mole fraction of n-C_{23}	$3.9\times10^{-5}\le x_{n-C_{23}}^{inlet}\le3.2\times10^{-3}$
Mole fraction of n-C_{24}	$1.3\times10^{-4}\le x_{n-C_{24}}^{inlet}\le3.2\times10^{-3}$
Inlet pressure (bar)	$45\le P_{reactor}\le70$
Reactor inlet temperature (K)	$593.15\le T_{in}\le633.15$
LHSV (h^{-1})	$2\le LHSV\le5$

given by Eq. (63) was calculated by comparing the predicted cloud point values evaluated from the surrogate model with those obtained from the original sensor integrating the mechanistic models using a validation data set containing 30 data points different than those adopted for training purposes.

$$RRSE = \sqrt{\frac{\sum_i (y_i - \hat{y}_i)^2}{\sum_i (y_i - \bar{y})^2}} \tag{63}$$

If the RRSE is higher than the threshold value set as 0.07, the obtained surrogate model is used by the FLOLA-Voronoi algorithm to obtain information about the neighborhood surrounding the cloud point response evaluated at each design point, and the degree of nonlinearity is measured at those locations. At this stage, a new point is augmented to the current design considering the exploration and exploitation trade-offs. Finally, the cloud point for this new sample is evaluated using the integrated reactor and thermodynamic models, and the hyperparameters are re-estimated. The iterations are carried out until the convergence criterion is reached. The block diagram representing this iterative procedure is shown in Fig. 7.

2.2.2 *Nonlinearity tracking and exploration capability*

As building a surrogate model to reproduce the original sensor response involves a high dimensional design space, it would be difficult to show if highly nonlinear or unexplored regions are actually

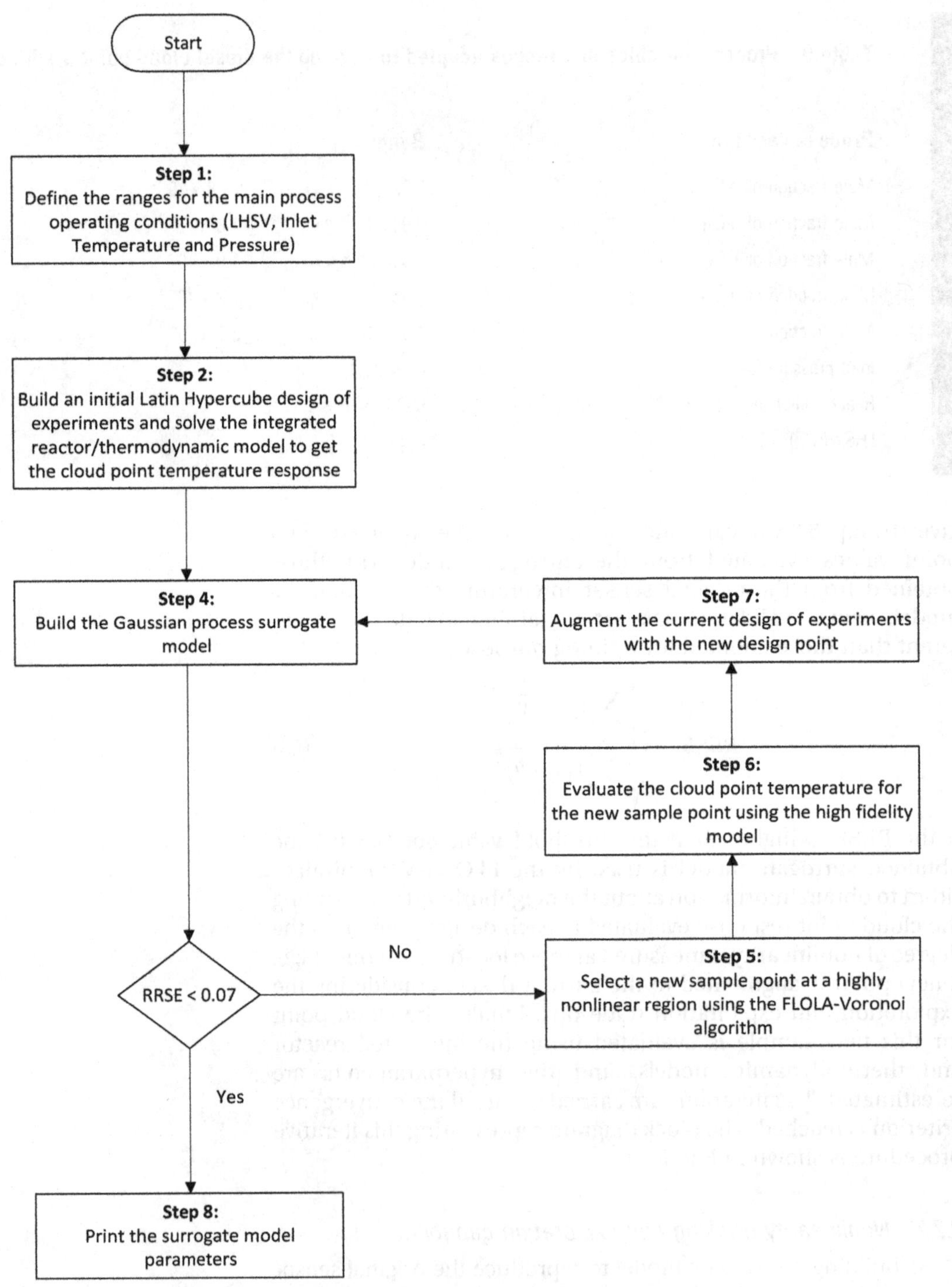

Fig. 7 Overall algorithm for the cloud point surrogate model development.

being tracked by the overall algorithm presented in Fig. 7. To check if the sequential design strategy adopted in this work was correctly implemented, the algorithm was tested using the benchmark function "Peaks," whose functional form is expressed by Eq. (64).

$$y = 3(1-x_1)^2 \exp\left[-x_1^2 - \left(x_2^2+1\right)^2\right] \\ -10\left(\frac{x_1}{5} - x_1^2 - x_2^2\right)\exp\left(-x_1^2 - x_2^2\right) \\ -\frac{1}{3}\exp\left[\left(x_1^2+1\right)^2 - x_2^2\right] \quad x_1, x_2 \in [-5, 5] \tag{64}$$

The procedure started with a Latin hypercube design with 15 points augmented with a 2^2 factorial design (van der Herten et al., 2015). The convergence threshold was set to 0.05, as the low dimensionality of the problem allows a tighter convergence criterion. The comparison between the actual function and surrogate model responses is presented in Fig. 8.

It can be observed that an almost perfect match between actual and predicted responses was obtained by using the FLOLA-Voronoi algorithm. To check how the additional points were placed in the design space, the initial and final experiment designs were plotted and compared, as shown in Fig. 9.

It can be observed that the density of points is higher at the center of the design space where the nonlinear peaks are located, and they are spread in such a way that all of those regions are explored. Also, some additional points are placed outside the nonlinear regions because the Voronoi cells tend to be large and close to the boundaries and the exploration feature of the algorithm tries to reduce their sizes as much as it can. Therefore, it can be concluded that the chosen fuzzy parameters and sequential

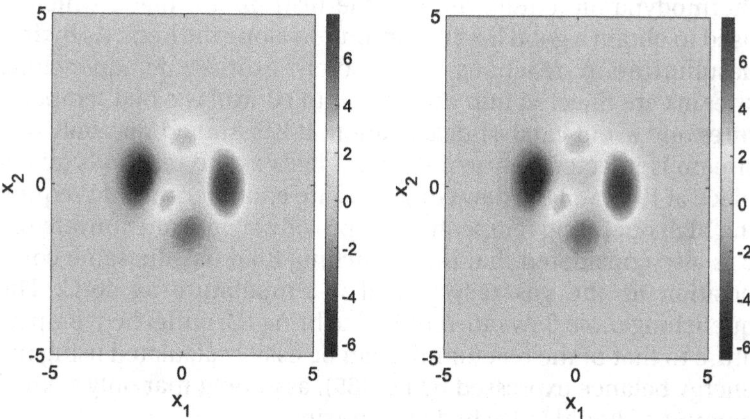

Fig. 8 Contour plots for the actual "Peaks" function (left) and surrogate model (right).

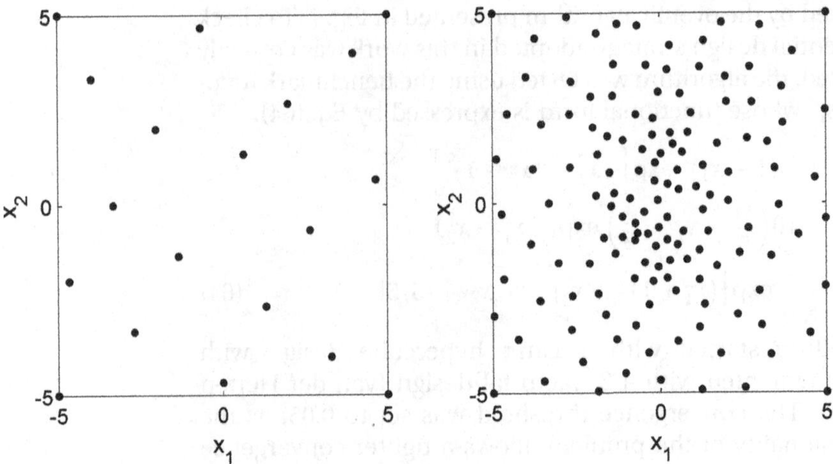

Fig. 9 Comparison between initial (left) and final (right) experiment designs.

design algorithm satisfy the conditions necessary to track nonlinearity while exploring the whole design space, and it is suitable to be applied in this work.

3 Results

3.1 Industrial case study

The proposed cloud point soft-sensor is developed for the industrial reaction unit shown in Fig. 10, which consists of two multibed reactors with two HDS and one HDW section each.

In a typical operation, the gas recycle and diesel streams are first mixed, and it is assumed that the resulting mixture reaches thermodynamic equilibrium at the distribution plate, which is used to obtain a good liquid distribution along the beds. As hydrodesulfurization reactions are severely exothermic, quenching streams are injected into the system to control the bed temperatures and avoid catalyst deactivation. It is assumed that only one stream is injected between the HDS beds as the reactions taking place at the catalytic dewaxing ones are endothermic and require the high operating temperature to provide significant conversion. It is also considered that the quenching fluid has the same composition as the gas recycle and a temperature as 40°C. The quenching mole flow rate required to bring the outlet bed temperature to that of the reactor inlet can be easily calculated using the energy balance expressed by Eq. (65), assuming that only a small amount of liquid is flashed after mixing.

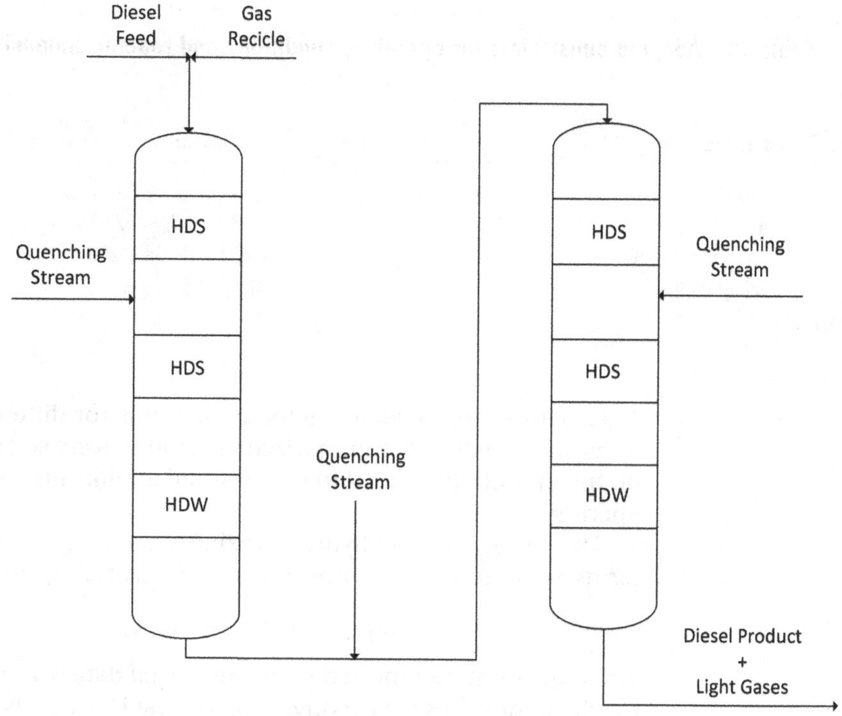

Fig. 10 Schematic representation of the industrial reactor unit under study.

$$F_{\text{quench}}^{G} = \frac{\left[F_{\text{total}}^{G} C_{pG}\left(T_{G}^{\text{out}} - 40\right) + F_{\text{total}}^{L} C_{pL}\left(T_{L}^{\text{out}} - 40\right)\right]}{C_{pG}\left(T_{\text{in}} - 40\right)} \tag{65}$$

The values for the reactor operating conditions and bed lengths were not provided by the local refinery where a similar reactor is applied; however, they can be estimated from target values of product sulfur concentration and cloud point. Hence, it can be done by minimizing the objective function represented by Eq. (66).

$$F = \left(\frac{C_{S}^{\text{ind}} - C_{S}^{\text{cal}}}{C_{S}^{\text{ind}}}\right)^{2} + \left(\frac{CP^{\text{ind}} - CP^{\text{cal}}}{CP^{\text{ind}}}\right)^{2} \tag{66}$$

As this equation can be highly nonlinear, bound constraints must be applied to the decision variables chosen as inlet pressure, temperature, H_2/Oil ratio, and bed lengths to obtain a realistic solution. These constraints were chosen based on common industrial values reported by Satterfield (1975). Also, the industrial value of liquid hourly space velocity (LHSV) was found to be 3.25.

Table 10 Adopted constraints for operating conditions and reactor dimensions.

Process variable	Constraint
Inlet pressure (bar)	$40 \leq P_{reactor} \leq 60$
Inlet temperature (K)	$603.15 \leq T_{in} \leq 623.15$
Hydrogen-to-oil ratio (scf/bbl)	$1000 \leq H_2/Oil \leq 2500$
Catalytic bed length (m)	$3 \leq L_i \leq 6$
Reactor diameter (m)	$1 \leq D \leq 3$

After performing several reactor simulations for different conditions, the constraints summarized in Table 10 were selected based on the limitations related to vapor-liquid equilibrium and reactor operation.

The performance of hydrodewaxing reactors is usually given in terms of cloud point improvement, represented by Eq. (67).

$$CP_{imp} = |CP_{feed} - CP_{prod}| \tag{67}$$

The improvement reported in the industrial data was around −3° C, which indicates small conversions in the HDW beds. For a better application example, the optimization procedure applied in this section was performed using the same industrial distillation data and sulfur content reported for the diesel feedstock described in Section 2.1.1 while assuming the feed and product cloud points as −20°C and −35°C, respectively. The gas recycle composition was taken directly from the industrial data, which is not reported in this work due to proprietary reasons. The calculated values of outlet sulfur concentration and cloud point for the dewaxed diesel product were obtained by applying the developed soft-sensor containing both reactor and thermodynamic modules described in Section 2.1. The constrained interior-point algorithm available in the MATLAB® Optimization Toolbox was used to solve the minimization problem. The estimated reactor dimensions and operating conditions are shown in Table 11.

By simulating the reactor operation using the proposed sensor and the previous dimensions and operating conditions, the calculated outlet sulfur concentration and cloud point temperature were checked with the target values as shown in Table 12.

The calculated sulfur content is very close to the target one while the cloud point is deviated by approximately 2.5°C. It can be explained based on conflicting objectives between the hydrodesulfurization and hydrodewaxing processes in terms of

Table 11 Reactor dimensions and operating conditions for the industrial reactor unit.

Process parameter	Value
1st HDS bed length (m)	5.04
2nd HDS bed length (m)	5
3rd HDS bed length (m)	5
4th HDS bed length (m)	5.03
1st HDW bed length (m)	4.84
2nd HDW bed length (m)	4.90
Reactor diameter (m)	2.06
Reactor pressure (bar)	54
Reactor inlet temperature (K)	612.84
Hydrogen-to-oil ratio (scf/bbl)	1320

Table 12 Comparison between target and calculated values for the output variables.

Outlet variable	Target value	Calculated value
Sulfur content (ppm)	4.8	4.81
Cloud point (°C)	−35	−32.47

pressure. The rate equations for HDS and HDW are directly and inversely proportional to the hydrogen solubility in liquid phase, respectively. Hence, decreasing the hydrogen concentration in that phase by lowering the pressure within the reaction system would decrease the sulfur conversion but significantly increase the cloud point improvement.

The temperature profile obtained by simulating the industrial unit using the estimated dimensions and operating conditions is shown in Fig. 11.

It was observed from the simulation that the temperature difference between gas and liquid phases were very small, meaning that the resistance for interphase heat transfer was not significant, and for this reason, only the temperature profile for the liquid phase is shown in Fig. 11. The regions of temperature raise correspond to the HDS beds, while the flat ones are related to the isothermal HDW sections. Also, the quenching locations can be identified by the regions of sudden temperature drop. The overall

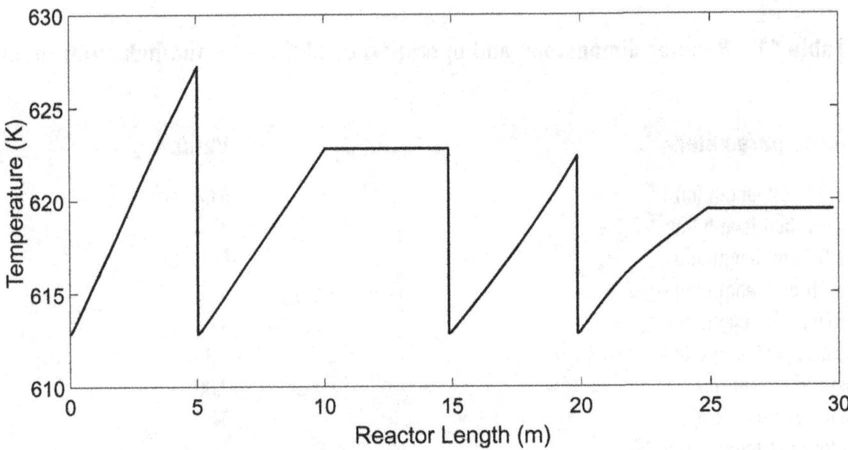

Fig. 11 Simulated temperature profile across the industrial unit.

profile satisfies the common technical constraints imposed in industry once the temperature raise in each HDS section is below the maximum allowed value of 30°C to preserve the catalyst activity. Finally, the temperature profile for the last HDS bed presents more nonlinearity than the other ones as the amount of sulfur compounds is very low in that section, and the heat effects due to evaporation become comparable to those related to the exothermic reactions.

To assess the performance of the reactor in producing ULSD, the simulated sulfur concentration profile as well as the evolution of the total mole flow rate for the individual sulfur compounds along the industrial unit are shown in Figs. 12 and 13, respectively.

Fig. 12 shows that the operating conditions and reactor dimensions found in this section led to an industrial reaction unit that satisfactorily reaches the target sulfur concentration to produce diesel with ultra low sulfur content while reducing the cloud point to desired values. The first reactor reduces the sulfur concentration from 4807.96 to 2000 ppm and the second one from 2000 to 4.81 ppm. By analyzing Figs. 12 and 13, it can be observed that hydrodesulfurization of dibenzothiophene controls the processes up to the 3rd HDS bed. At the last HDS section, that compound reaches full conversion, and the slow conversion of 4,6-dimethyldibenzothiophene becomes the main limiting mechanism, which explains the technical limitations commonly found in refineries when attempting to further reduce the sulfur content below 8 ppm identified as ULSD region.

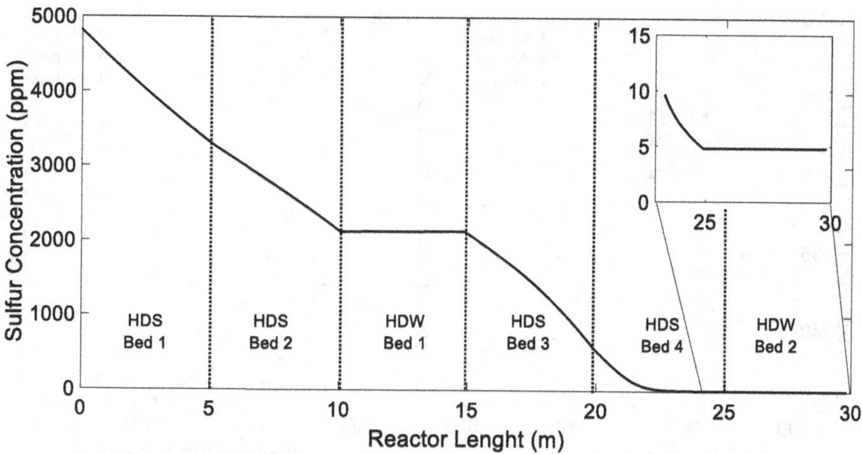

Fig. 12 Simulated sulfur concentration profile along the industrial reactor.

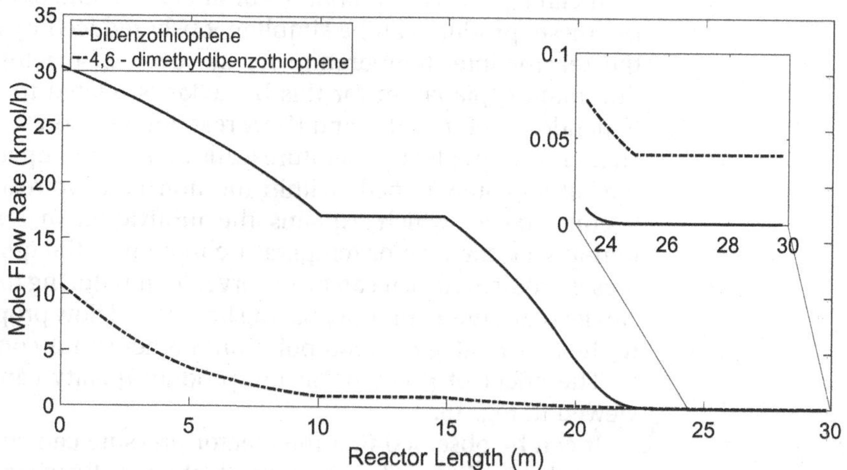

Fig. 13 Mole flow rate evolution for the sulfur compounds along the industrial reactor.

To further illustrate the application of the proposed soft-sensor, a sensitivity analysis study involving the changes in LHSV, reactor inlet temperature, and pressure was performed as the hydrodewaxing conversion are mostly affected by those variables based on the results obtained in our previous work. Fig. 14 shows the effect of temperature on the product cloud point and the sulfur content for three different reactor pressures keeping the LHSV at its nominal value of $3.25\,h^{-1}$.

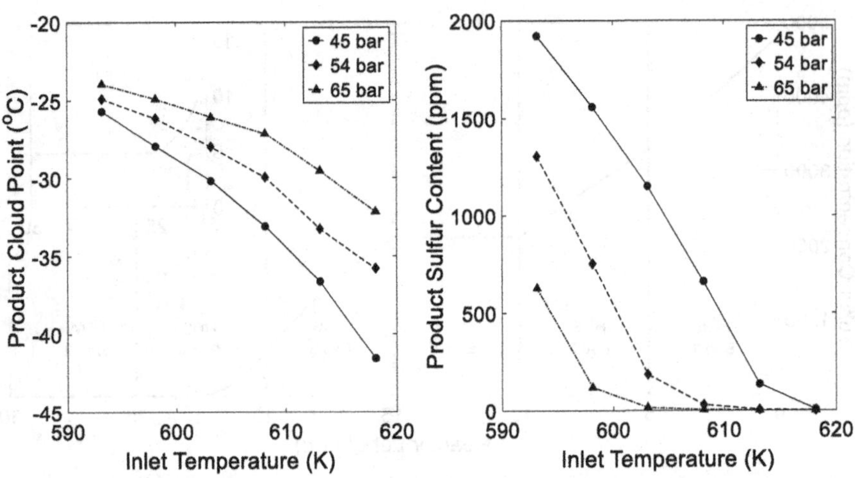

Fig. 14 Effect of inlet temperature on the product cloud point and sulfur content at different pressures for LHSV=3.25 h^{-1}.

It can be observed that both sulfur content and cloud point of the diesel product can be simultaneously reduced by increasing the reactor inlet temperature for all levels of reactor pressure. The main explanation for this behavior is related to the strong dependence of the HDS and HDW reaction rates on temperature. Higher reactor inlet temperature leads to higher temperature profiles at all catalytic beds within the unit and the reaction rates become faster, which explains the motivation in refineries to manipulate the reactor temperature to control the quality of the diesel product. Also, it can be observed that reducing the pressure can lead to a diesel product having better cold flow properties due to the lower value of cloud point but higher sulfur content.

The effect of pressure on the product quality can be better viewed in Fig. 15.

It can be observed that the reactor pressure can considerably affect the quality of the diesel product specially when operating at high inlet temperatures, and the conflicting objectives between the HDW and HDS mechanisms seem to be related to this variable. If the reactor inlet temperature is kept at 613 K and pressure is reduced, the unit produces diesel having better cold flow properties but richer in sulfur compounds. As previously discussed, decreasing the pressure reduces the solubility of hydrogen in liquid phase, and consequently, the reaction rates in the HDW beds become faster, while the conversion of sulfur compounds slows down at the HDS ones. Therefore, the reactor pressure must be carefully selected to produce diesel fuels that meet the desired specifications related to sulfur concentration and cloud point.

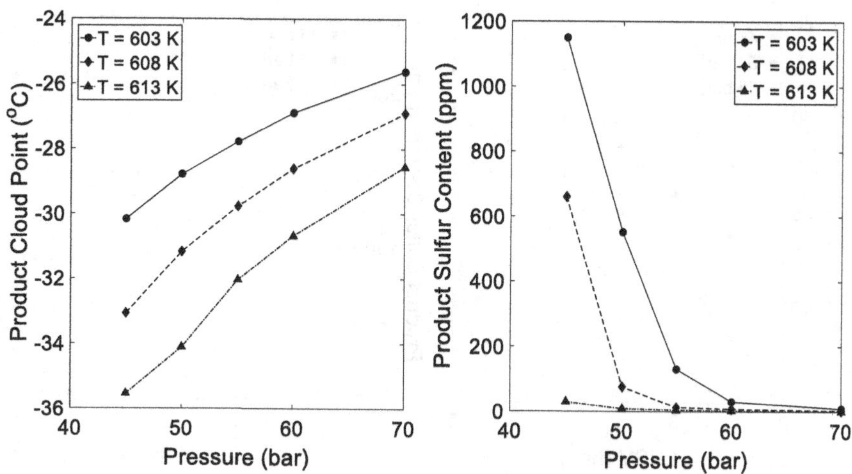

Fig. 15 Effect of inlet pressure on the product cloud point and sulfur content at different temperatures for LHSV = 3.25 h^{-1}.

The effect of LHSV on the product quality was also studied based on changes in temperature or pressure as shown in Figs. 16 and 17, respectively.

It can be noted in both figures that decreasing the LHSV indeed reduces the values of both product cloud point and sulfur concentration for all different inlet temperature and pressure conditions as the sulfur and heavy normal paraffinic compounds spend more

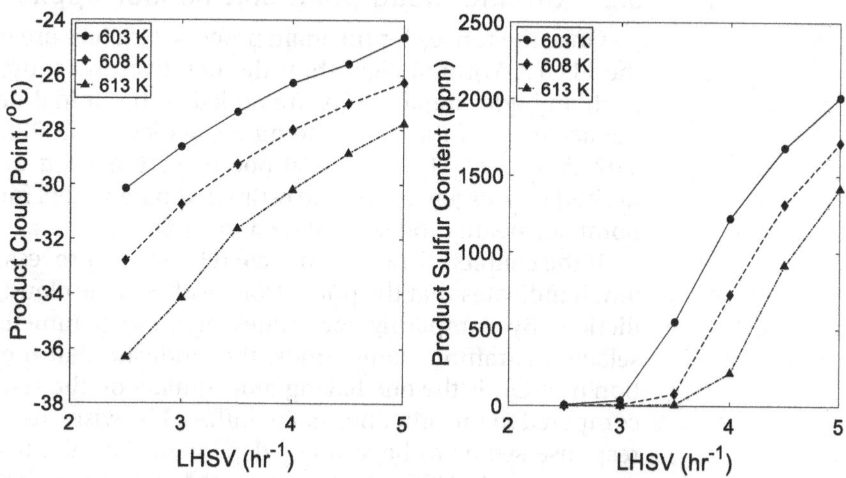

Fig. 16 Effect of LHSV on the product cloud point and sulfur content at different temperature conditions at $P_{reactor}$ = 54 bar.

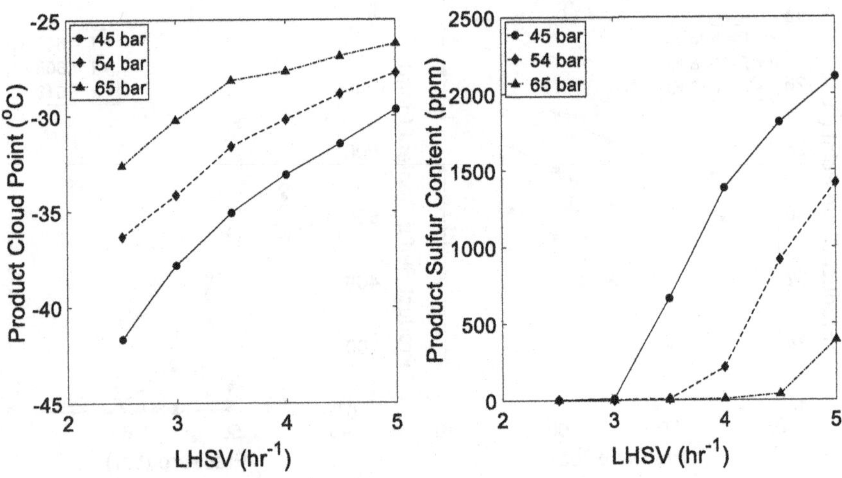

Fig. 17 Effect of LHSV on the product cloud point and sulfur content at different pressure conditions at $T_{in} = 613$K.

time reacting within the catalytic beds due to the higher residence time. The interaction between LHSV and reactor pressure must also be carefully analyzed. It can be observed from Fig. 17 that the effect of LHSV on the reactor performance is strongly affected by the trade-off between the HDS and HDW objectives when the reactor pressure is changed, and such interaction mostly affects the sulfur concentration of the diesel product.

3.2 On-line cloud point soft-sensor application

Once the ranges for the main process variables are included in the FLOLA-Voronoi algorithm, the iterative procedure is initialized, and additional points are added to the initial design until the accuracy of the surrogate model achieves the RRSE value of 0.07. A total of 117 additional points were needed to reach the desired convergence, and the estimated parameters for the cloud point surrogate model are shown in Table 13.

If the characteristic length scale related to a process variable is low, it indicates that the prior strongly affects the cloud point prediction. By comparing the values of these parameters for the selected paraffinic compounds, they indicate that the composition of n-C_{24} is the one having most impact on the sensor output compared to the other normal paraffins. Likewise, the cloud point response seems to be strongly dependent on inlet temperature, pressure, and LHSV, which is expected as these variables have a major impact on the reactor conversion.

Also, the composition of n-C_{20} is shown to be slightly more important than that for n-C_{21}. It can be related to the limitations

Table 13 Estimated hyperparameters of the cloud point surrogate model.

Parameter	Value
$\widehat{\mu}$	−39.5969
$\widehat{\sigma}$	0.5111
σ_f	17.6824
$\theta_{n-C_{20}}$	47.8151
$\theta_{n-C_{21}}$	69.1604
$\theta_{n-C_{22}}$	26.0629
$\theta_{n-C_{23}}$	26.0071
$\theta_{n-C_{24}}$	4.4899
θ_P	5.5723
θ_T	2.6925
θ_{LHSV}	1.8166

of the predictive Wilson model used to describe the solid phase behavior in the thermodynamic module. As reported by Coutinho and Ruffier-Meray (1997), heavy paraffinic compounds can form multiple solid phases due to differences in the solidification process, and this phenomenon cannot be well-described by that thermodynamic model. However, it can be observed that the difference between those parameters is not very large, which also can be linked to their experimental observation that the formation of multiple solid phases does not affect the cloud point of the mixture much (Coutinho & Ruffier-Meray, 1997).

To check if using the surrogate model would affect the accuracy of the sensor, its response was compared with that evaluated using the original sensor framework integrating the reactor and thermodynamic modules for both training and validation data sets, as shown in Figs. 18 and 19.

It can be observed in Figs. 18 and 19 that the surrogate model can accurately reproduce the response obtained from the original sensor for both validation and training data sets. To get a quantitative view of the accuracy of the proposed surrogate model, Fig. 20 shows the residuals between their responses for both training and validation data sets.

The maximum residual observed is approximately 1.5°C when adopting an RRSE of 0.07 as convergence criteria for the FLOLA-Voronoi algorithm. Therefore, replacing the reactor and thermodynamic modules with the surrogate model does not affect the accuracy of the sensor and results in a sensor framework that has a more straightforward formulation and can be easily

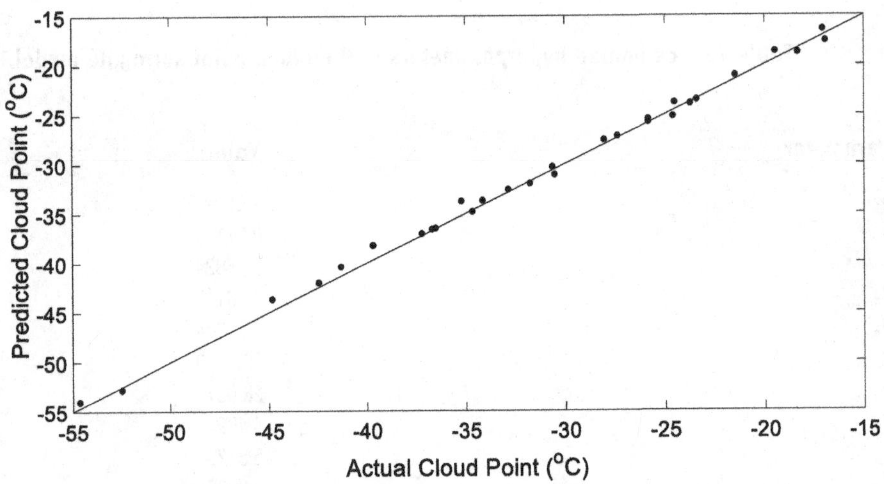

Fig. 18 Parity plot comparing the cloud point response for the product diesel obtained from the surrogate model (predicted) and detailed model (actual) for the validation data set.

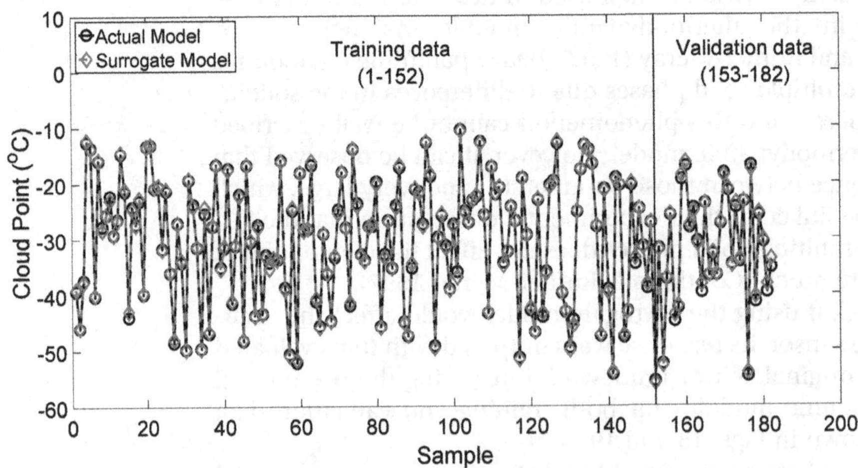

Fig. 19 Comparison between the surrogate model and original soft-sensor responses for both training and validation data sets.

implemented when intended to be used for online cloud point predictions.

To check if nonlinear regions affect the accuracy of the surrogate model, the RRSE values for both training and validation data sets were also calculated for a metamodel built using the "max-min" Latin hypercube having the same size as the final design obtained from the sequential approach. This comparison is shown in Fig. 21.

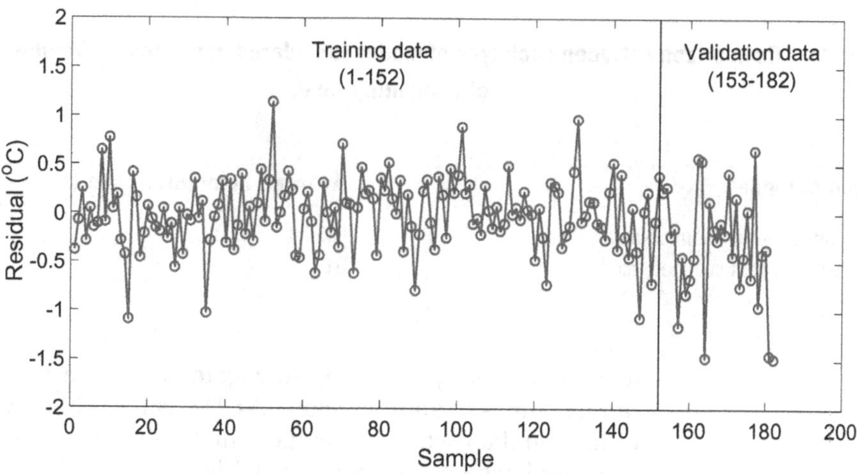

Fig. 20 Residuals between the responses obtained from surrogate and reactor model-based soft-sensors.

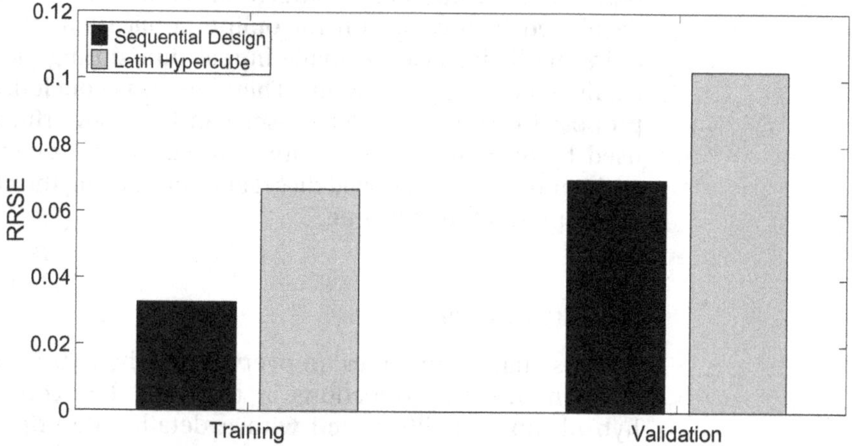

Fig. 21 Comparison between FLOLA-Voronoi and Latin hypercube designs in terms of surrogate model accuracy.

The RRSE values obtained applying the FLOLA-Voronoi algorithm are approximately 50% and 30% lower than those observed using the "max-min" Latin hypercube for both training and validation data sets. As the Latin hypercube is a pure explorative design, it attempts to explore the design space as uniformly as possible. A surrogate model built using this design may not correctly describe highly nonlinear regions. Therefore, the accuracy of the soft sensor is affected by the nonlinear behavior of the cloud point response and must be handled appropriately.

Finally, to assess the performance of the cloud point soft-sensor in terms of computing time, the sensors integrating

Table 14 Comparison between each type of model considered in the sensor framework in terms of computing time.

Type of model	Average computing time (s)
Integrated reactor/thermodynamic model	635.79
Gaussian process surrogate model	0.03

reactor/thermodynamic and surrogate models, respectively, were run five times in an Intel Xeon 3.5 GHz with 32 GB RAM using the validation data set. The average time to obtain the solutions was measured, which is reported in Table 14.

Table 14 shows a drastic reduction in computing time when replacing the reactor and thermodynamic modules with the surrogate model. Another advantage of using the surrogate model developed in this section for online applications is that cloud point predictions can be made in real-time during the industrial catalytic dewaxing operation. Therefore, it is concluded that the proposed cloud point soft-sensor can be a powerful tool to be used by oil refineries attempting to gain better quality control of the produced winterized diesel and monitoring the cloud point of the product in real-time.

4 Summary

This chapter provides an overview of the hybrid model and different research directions in the area. The concept of the hybrid model is illustrated with a detailed description of the development of a soft sensor that was proposed to predict the cloud point of dewaxed diesel fuels produced from catalytic dewaxing units. Its framework integrates three different modules, one data-based and two mechanistic, to carry out the predictions of the cloud point of diesel. In parallel, a surrogate model was developed to predict the cloud point in real-time for advanced control applications. The results presented in this work confirm that the proposed hybrid model can effectively replace the visual procedures commonly adopted in petroleum refineries to measure the cloud point of diesel fuels processed in catalytic dewaxing units. Also, it can be easily implemented to control, design, and optimize the hydro-dewaxing units without depending on costly and time-consuming experiments carried out in pilot reactors.

Appendix

	NP	IP	MA	DA	TA	A_A	A_AA	A_N	AN	MN	DN	TN	N_N	SI	SII	NI
C8	0.0321	0.0067	0.0054	*	*	*	*	*	*	*	*	*	*	*	*	*
C9	0.0042	0.0073	*	*	*	*	*	*	0.00525	0.00903	*	*	*	*	*	0.07725
C10	0.0019	0.0119	*	0.1099	*	*	*	*	*	*	0.0267	*	*	*	*	*
C11	0.0034	0.0131	*	*	*	*	*	*	*	*	*	*	*	*	*	*
C12	1.8E-5	0.0371	*	*	0.00246	0.0380	*	0.0400	*	*	*	*	0.0765	0.0193	*	*
C13	0.0059	0.0790	*	*	*	*	*	*	0.0225	*	*	*	*	*	*	*
C14	0.0010	0.0052	*	*	*	*	*	*	*	*	*	0.1218	*	*	0.00691	*
C15	0.0003	0.0068	0.028	*	*	*	*	*	*	0.0449	*	*	*	*	*	*
C16	0.0004	0.0159	*	*	*	*	0.0275	*	*	*	*	*	*	*	*	*
C17	0.0016	0.0240	*	*	*	*	*	*	*	*	*	*	*	*	*	*
C18	0.0004	0.0367	*	*	*	*	*	*	*	*	*	*	*	*	*	*
C19	0.0004	0.0056	*	*	*	*	*	*	*	*	*	*	*	*	*	*
C20	0.0004	0.0054	*	*	*	*	*	*	*	*	*	*	*	*	*	*
C21	0.0003	0.0052	*	*	*	*	*	*	*	*	*	*	*	*	*	*
C22	0.0004	0.0050	*	*	*	*	*	*	*	*	*	*	*	*	*	*
C23	0.0002	0.0047	*	*	*	*	*	*	*	*	*	*	*	*	*	*
C24	0.0006	0.0044	*	*	*	*	*	*	*	*	*	*	*	*	*	*

A_A, A_AA, biphenyl type compounds; A_N, AN, naphthene-aromatic type compounds; DA, Di-aromatics; DN, Di-naphthenes; IP, isoparaffins; MA, mono-aromatics; MN, mono-naphthenes; N_N, bicyclohexyl compounds; NI, nitrogen compounds; NP, normal paraffins; SI, nonsubstituted dibenzothiophene; SII, substituted dibenzothiophene; TA, tri-aromatics; TN, tri-naphthenes.

References

Ahmad, M. I., Zhang, N., & Jobson, M. (2011). Molecular components-based representation of petroleum fractions. *Chemical Engineering Research and Design, 89*(4), 410–420.

American Petroleum Institute. (1997). *Technical data book, petroleum refining.* Washington: API.

Baltanas, M. A., & Froment, G. F. (1985). Computer generation of reaction networks and calculation of product distributions in the hydroisomerization and hydrocracking of paraffins on Pt-containing bifunctional catalysts. *Computers & Chemical Engineering, 9*(1), 71–81.

Baltanas, M. A., Van Raemdonck, K. K., Froment, G. F., & Mohedas, S. R. (1989). Fundamental kinetic modeling of hydroisomerization and hydrocracking on noble metal-loaded faujasites. 1. Rate parameters for hydroisomerization. *Industrial & Engineering Chemistry Research, 28*(7), 899–910.

Boesen, R. R. (2010). *Investigation and modelling of diesel hydrotreating reactions.* Lyngby: Technical University of Denmark.

Bondi, A. (1964). van der Walls volumes and radii. *The Journal of Physical Chemistry, 68*(3), 441–451.

Broadhurst, M. G. (1962). An analysis of the solid phase behavior of the normal paraffins. *Journal of Research of the National Bureau of Standards, 66A*(3), 241–249.

Cameron, I. T., & Hangos, K. (2001). *Process modelling and model analysis.* Elsevier.

Coutinho, J. A. P., Dauphin, C., & Daridon, J. L. (2000). Measurements and modelling of wax formation in diesel fuels. *Fuel, 79*(6), 607–616.

Coutinho, J. A. P., & Ruffier-Meray, V. (1997). Experimental measurements and thermodynamic modeling of paraffinic wax formation in undercooled solutions. *Industrial & Engineering Chemistry Research, 36*(11), 4977–4983.

Cozad, A., Sahinidis, N. V., & Miller, D. C. (2015). A combined first-principles and data-driven approach to model building. *Computers & Chemical Engineering, 73*, 116–127.

Crombecq, K., Gorissen, D., Deschrijver, D., & Dhaene, T. (2011). A novel hybrid sequential design stretegy for global surrogate modeling of computer experiments. *SIAM Journal on Scientific Computing, 33*(4), 1948–1974.

Fasel, U., Kutz, J. N., Brunton, B. W., & Brunton, S. L. (2022). Ensemble-SINDy: Robust sparse model discovery in the low-data, high-noise limit, with active learning and control. *Proceedings of the Royal Society A, 478*(2260), 20210904.

Froment, G. F., De Wilde, J., & Bischoff, K. B. (2010). *Chemical reactor analysis and design.* Hoboken, NJ: Willey.

Froment, G. F., Depauw, G. A., & Vanrysselberghe, V. (1994). Kinetic modeling and reactor simulation in hydrodesulfurization of fractions. *Industrial & Engineering Chemistry Research, 33*(12), 2975–2988.

Ghanaei, E., Esmaeilzadeh, F., & Fathi Kaljahi, J. (2007). A new predictive thermodynamic model in the wax formation phenomena at high pressure condition. *Fluid Phase Equilibria, 254*(1-2), 126–137.

Ghorbani, A., Abid, A., & Zou, J. (2019, July). Interpretation of neural networks is fragile. In *Proceedings of the AAAI conference on artificial intelligence (Vol. 33, No. 01, pp. 3681–3688).*

Goodfellow, I., Bengio, Y., & Courville, A. (2016). *Deep learning.* MIT Press.

Klatt, K. U., & Marquardt, W. (2009). Perspectives for process systems engineering—Personal views from academia and industry. *Computers & Chemical Engineering, 33*(3), 536–550.

Kumar, H. (2004). *Single event kinetic modeling of the hydrocracking of paraffins.* M. S. thesis Texas A&M University, College Station.

Kumar, H., & Froment, G. F. (2007). A generalized mechanistic kinetic model for the hydroisomerization and hydrocracking of long-chain paraffins. *Industrial & Engineering Chemistry Research, 46*(12), 4075–4090.

Laxmi Narasimhan, C. S., Thybaut, J. W., Marin, G. B., Jacobs, P. A., Martens, J. A., Denayer, J. F., & Baron, G. V. (2003). Kinetic modeling of pore mouth catalysis in the hydroconversion of n-octane on Pt-H-ZSM-22. *Journal of Catalysis, 220*(2), 399–413.

Morgan, D. L., & Kobayashi, R. (1994). Extension of pitzer CSP models for vapor pressures and heats of vaporization to long-chain hydrocarbons. *Fluid Phase Equilibria, 94*(15), 51–87.

Pedersen, K. S. (1995). Prediction of cloud point temperatures and amound of wax formation. *SPE Production & Facilities, 10*(1), 46–49.

Peng, B. (1999). *Molecular modeling of refining processes.* Ph.D. dissertation Manchester: The University of Manchester.

Poling, B. E., Prausnitz, J. M., & O'Connell, J. P. (2001). *The properties of gases and liquids.* New York: McGraw-Hill.

Psichogios, D. C., & Ungar, L. H. (1992). A hybrid neural network—First principles approach to process modeling. *AIChE Journal, 38*(10), 1499–1511.

Raissi, M., Perdikaris, P., & Karniadakis, G. E. (2019). Physics-informed neural networks: A deep learning framework for solving forward and inverse problems involving nonlinear partial differential equations. *Journal of Computational Physics, 378,* 686–707.

Rakoczy, R. A., & Morse, P. M. (2013). Consider catalytic dewaxing as a tool to improve diesel cold flow properties. *Hydrocarbon Processing,* 67–69.

Reiss, L. P. (1967). Cocurrent gas-liquid contacting in packed columns. *Industrial & Engineering Chemistry Process Design and Development,* 6(4), 486–499.

Sansana, J., Joswiak, M. N., Castillo, I., Wang, Z., Rendall, R., Chiang, L. H., & Reis, M. S. (2021). Recent trends on hybrid modelling for Industry 4.0. *Computers & Chemical Engineering, 151,* 107365.

Sato, Y., Hirose, H., Takahashi, F., & Toda, M. (1972). *Proceedings of the first pacific chemical engineering congress, October 10–14.* Kyoto, Japan.

Satterfield, C. N. (1975). Trickle-bed reactors. *AIChE Journal, 21*(2), 209–228.

Svoboda, G. D., Vynckier, E., Debrabandere, B., & Froment, G. F. (1995). Single-event rate parameters for paraffin hydrocracking on a Pt/US-Y zeolite. *Industrial & Engineering Chemistry Research, 34*(11), 3793–3800.

Turco Neto, E. (2016). *Single event kinetic modelling of catalytic dewaxing on commercial Pt/ZSM-5.* MEng thesis Memorial University of Newfoundland and Labrador.

van der Herten, J., Couckuyt, I., Deschrijver, D., & Dhaene, T. (2015). A fuzzy hybrid sequential design strategy for global surrogate modeling of high-dimensional computer experiments. *SIAM Journal on Scientific Computing, 37*(2), A1020–A1039.

Van Parijs, I. A., & Froment, G. F. (1986). Kinetics of hydrodesulfurization on a $CoMo/Al_2O_3$ catalyst. 1. Kinetics of the hydrogenolysis of benzothiophene. *Industrial and Engineering Chemistry Product Research and Development, 25*(3), 437–443.

Vanrysselberghe, V., & Froment, G. F. (1996). Hydrodesulfurization of dibenzothiophene on a $CoMo/Al_2O_3$ catalyst: Reaction network and kinetics. *Industrial & Engineering Chemistry Research, 35*(10), 3311–3318.

Vanrysselberghe, V., Gall, R. L., & Froment, G. F. (1998). Hydrodesulfurization of 4-methyldibenzothiophene and 4,6-dimethyldibenzothiophene on a $CoMo/Al_2O_3$ catalyst: Reaction network and kinetics. *Industrial & Engineering Chemistry Research, 37*(4), 1235–1242.

Von Stosch, M., Oliveira, R., Peres, J., & de Azevedo, S. F. (2014). Hybrid semi-parametric modelling in process systems engineering: Past, present and future. *Computers & Chemical Engineering, 60*, 86–101.

Vynckier, E., & Froment, G. F. (1991). Modeling of the kinetics of complex processes based upon elementary steps. In G. Astarita, & S. I. Sandler (Eds.), *Kinetic and thermodynamic lumping of multicomponent mixtures* (pp. 131–161). Amsterdam: Elsevier Publishers B.V.

Willard, J., Jia, X., Xu, S., Steinbach, M., & Kumar, V. (2020). Integrating physics-based modelling with machine learning: A survey. *arXiv preprint arXiv:2003.04919, 1*(1), 1–34.

Won, K. W. (1989). Thermodynamic calculation of cloud point temperatures and wax phase composition of refined hydrocarbon mixtures. *Fluid Phase Equilibria, 53*, 377–396.

Wu, Y. (2010). *Molecular management for refining operations.* Ph.D. Dissertation Manchester: The University of Manchester.

10

Large-scale process models using deep learning

R. Bhushan Gopaluni, Liang Cao, and Yankai Cao
Department of Chemical and Biological Engineering, University of British Columbia, Vancouver, BC, Canada

1 Large-scale system modeling challenges

With the stringent requirements for product quality and cost, the complexity and automation degree of the process are continuously growing (Ding, 2014). As the scale of plants grows, it is important to enhance the process's safety, reliability, and robustness. Over the last decades, we have seen tremendous improvement in industrial data, computational power, and major theoretical advances in machine learning (Bishop, 2006). How to successfully adapt the data-driven approach to the industrial process has been the focus of the industry and academia (Ge et al., 2017; Gopaluni et al., 2020). Data collected from large-scale processes often retain unique characteristics, which pose great challenges for process modeling (Cao et al., 2020; Fan et al., 2014; Gopaluni, 2008; Qin, 2014; Sun et al., 2019; Wang et al., 2020; Wasikowski & Chen, 2010; Zhou et al., 2014; Zhu et al., 2021).

The first challenge is that modern processes are characterized by many variables such as temperature, pressure, flow rate, etc. (Qin, 2014). The complex structural mechanisms and unknown nonlinear relationships between the variables, making it difficult to build accurate and general mathematical models (Fan et al., 2014; Gopaluni, 2008).

The second challenge is that modern processes have irregular sampling and multiple timescales (Cao et al., 2020; Zhou et al., 2014). Process variables are generally sampled on the order of minutes or even seconds, while key variables are sampled even more slowly, primarily measured hourly or daily. Fig. 1 shows an example of this challenge. The classical approach is to first

Modeling of Chemical Process Systems. https://doi.org/10.1016/B978-0-12-823869-1.00009-0

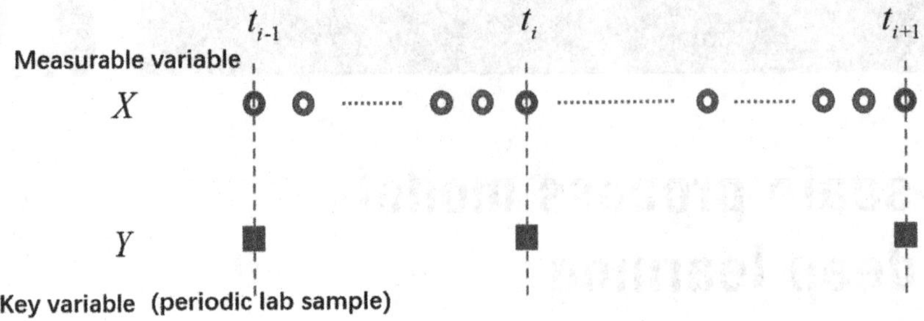

Fig. 1 Irregular sampling and multiple timescale.

downsample the process variables, align them with the key variables, and then transform the modeling problem into a general regression modeling problem, which can be solved using sophisticated techniques such as multivariate statistical analysis and machine learning. However, only a tiny fraction of the process variable data is used to build regression models, resulting in a large amount of process variable data that are rich in information being directly discarded and not fully utilized.

The third challenge is that most modern processes are implemented through closed-loop control, producing relatively steady-state data with little volatility, and few representative samples (Sun et al., 2019; Wang et al., 2020; Wasikowski & Chen, 2010; Zhu et al., 2021). Similar data contain relatively sparse information, so it is difficult to find valuable information. Additionally, many of these data sets have missing values and are noisy, which increases the difficulty of mining data information.

2 Motivation for deep learning algorithms

Before the advent of deep learning (also known as deep structured learning) (Arel et al., 2010; LeCun et al., 2015; Schmidhuber, 2015), most machine learning algorithms relied on shallow-structured architectures, such as Gaussian mixture models (Permuter et al., 2006), extreme learning machines (Huang et al., 2006), and support vector machines (Cortes & Vapnik, 1995). These architectures typically contain at most one or two layers of nonlinear feature transformations. Shallow architectures solve many simple or well-defined problems, but their limited modeling and representational capabilities often perform poorly when dealing with more complex situations. Therefore, deep architectures with multiple layers of nonlinear transformations are required to extract informative latent representations from a large number of inputs.

Deep learning is a subfield of machine learning based on artificial neural networks inspired by studying brain structure and function (Rosenblatt, 1958). Although artificial neural networks were invented in the 1950s, the limitation of computational complexity and the insufficiency of some theories or algorithms have prevented them from reaching their potential in academia and industry. In recent years, advances in computing technology, new methods of unsupervised pretraining, and more efficient training and optimization methods, the power of deep learning has been unleashed and played a vital role in the current era of artificial intelligence.

High-dimensional, nonlinear, and dynamic characteristics exist widely in the process industry. Deep learning has robust nonlinear fitting and dynamic modeling capabilities and can effectively analyze high-dimensional data. It has been widely studied in process modeling and process control. Generally speaking, the learning ability of a model is related to its complexity. If the complexity of a learning model can be enhanced, its expression ability can often be improved. Deep learning enhances the model's capacity for expression by deepening the neural network model. Compared with traditional machine learning, deep learning attempts to automatically extract good latent features or multiple levels of representation through the training of deep neural networks, eliminating the need for manual feature engineering. This can save significant time and effort in the modeling process, especially when dealing with high-dimensional and complex data.

3 Deep learning methods

Deep learning methods can be broadly classified into two main categories according to the model architecture and usage intentions: supervised learning (labeled data available) and unsupervised learning/generative model (no available information about labeled data). In the following sections, we will introduce these two methods and highlight some widely used techniques in large-scale process modeling.

3.1 Supervised deep learning

Supervised learning is commonly used for tasks such as classification, regression, and object detection, where the target labels are well-defined. For supervised deep learning methods, generally speaking, there are three basic models, namely multilayer

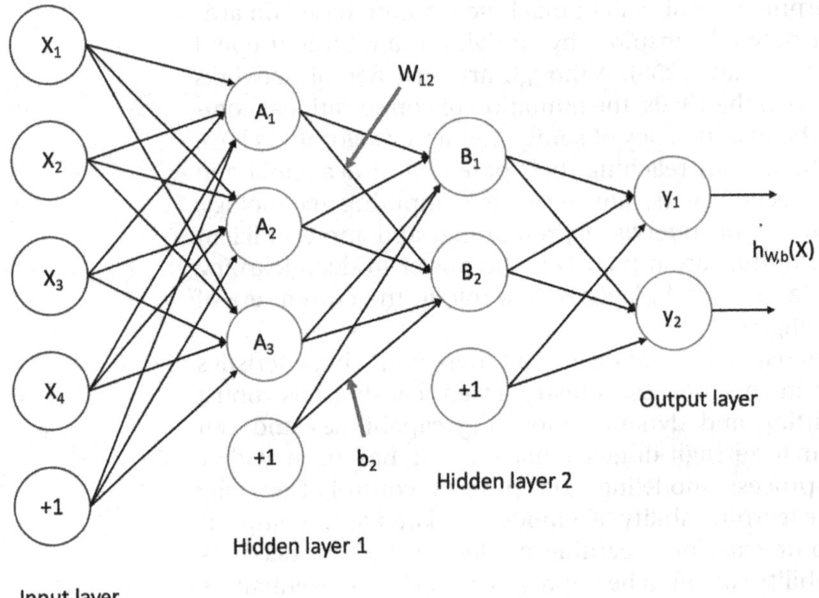

Fig. 2 A schematic diagram of MLP.

perceptron (MLP) (Imseng et al., 2013; Mohamed et al., 2012; Pinto et al., 2011), convolutional neural network (CNN) (Kavukcuoglu et al., 2010; Krizhevsky et al., 2012; Zeiler et al., 2011), and recurrent neural network (RNN) (Bengio et al., 2013; Graves et al., 2013; Hermans & Schrauwen, 2013).

MLP is a traditional neural network architecture that including an input layer, one or more hidden layers, and an output layer. Backpropagation (Rumelhart et al., 1986) is a widely used algorithm for training MLP. Fig. 2 gives a schematic example of MLP. For a single neuron in the neural network, it can be computed as follows:

$$h_{w,b} = f\left(w^T x + b\right),$$
$$f(z) = \frac{1}{1 + e^{-z}} \tag{1}$$

w, b are the parameters of this neuron and $f(\bullet)$ is the activation function. Each neuron in the network receives input from neurons in the previous layer, applies an activation function to the weighted sum of inputs, and passes the output to neurons in the next layer.

CNN is another powerful supervised deep learning model, which is very effective in computer vision and image recognition.

Unlike MLP, which is a fully connected network, CNN is designed to capture spatial and temporal relationships in data, making it particularly suitable for processing structured data. CNN generally consists of multiple modules stacked to form a deep model; each module consists of a convolutional layer and a pooling layer. The weight sharing in the convolutional layer, together with pooling schemes to reduce the dimensions of data, enables CNN to extract local repetitive features and reduce computation cost, which are the cores of CNN. Fig. 3 gives a schematic example of CNN.

The third supervised deep learning model is RNN, which has made breakthroughs in processing sequential data tasks (speech recognition, voice recognition, time series prediction, natural language processing, etc.). It has become the default configuration for sequence modeling tasks with its powerful ability to capture the temporal dependencies in data. An RNN is a type of network structure used to process sequential data, which was first proposed in the 1980s(Rumelhart et al., 1986). In traditional machine learning methods, sequential modeling usually uses hidden Markov model (Fine et al., 1998) and conditional random field (Chatzis & Demiris, 2013). In recent years, RNN can handle sequential information and capture information between long-distance samples. In addition, the architecture of RNN allows it to use the hidden node state to store valuable historical information in the sequence so that the network can learn the information in the entire sequence. Fig. 4 gives a schematic example of RNN. As we can see, the same neural network gets reused many times with the current input data X_t and previously hidden state h_{t-1} to get output O_t and new hidden state h_t. Each layer of RNN has the same parameters and the loops allow information to

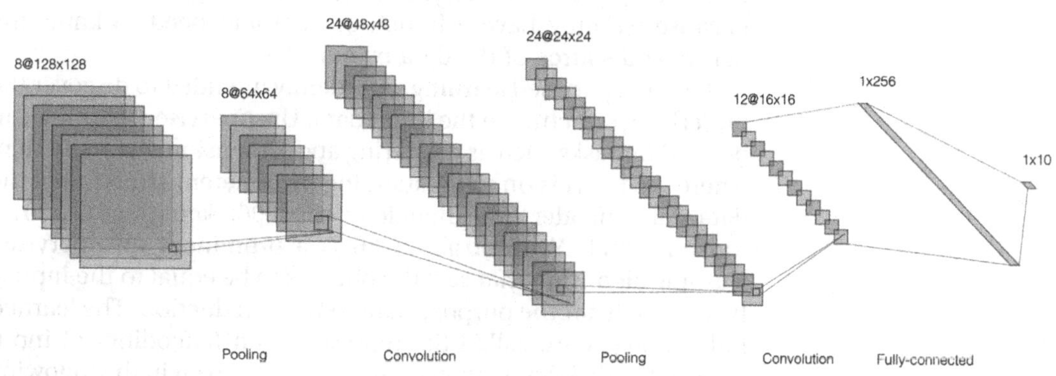

Fig. 3 A schematic diagram of CNN.

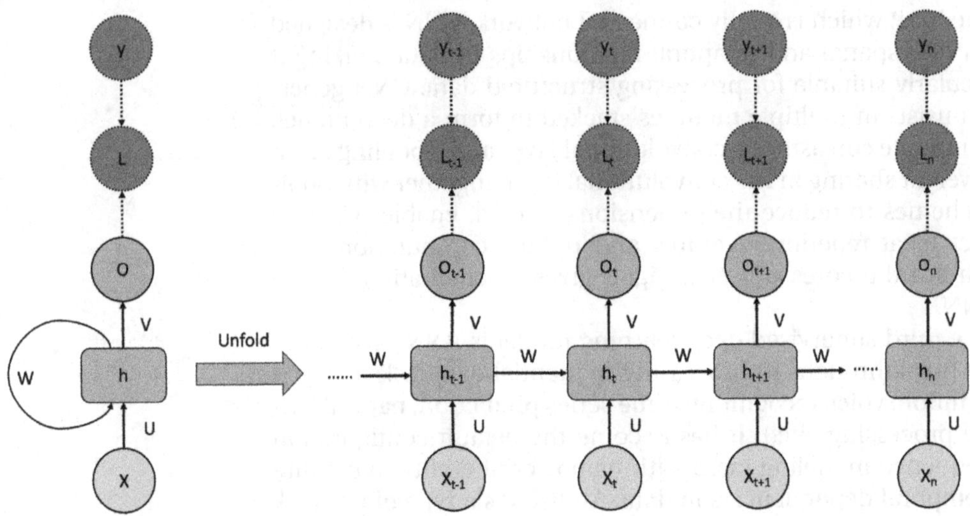

Fig. 4 A schematic diagram of RNN.

persist across time steps, making it ideal for modeling sequential data. The following section will give a detailed introduction to RNN, and we will use RNN to model large-scale chemical processes.

3.2 Unsupervised deep learning

To achieve a deeper level of understanding of the world, it is not enough for machines to simply recognize patterns in data. It is equally important for machines to understand how data are produced without any explicit labels or supervision. However, learning how data are produced is a much more complex and challenging problem than just an understanding of the data itself because we only have a bunch of data and need to know the nature and source of the data by ourselves.

For unsupervised learning, it is mainly intended to discover the underlying structure in the input data. Unsupervised learning can be used for tasks such as clustering and dimensionality reduction, where the focus is on understanding the inherent structure of the data. Autoencoder (AE) (Bengio et al., 2006; Sainath et al., 2012; Vincent, 2011; Yuan et al., 2020) is a prominent unsupervised learning algorithm that sets the outputs to be equal to the inputs. It is typically for the purpose of dimension reduction. The learned hidden layers are called the representation (encoding) of input data. A detailed description of the AE will be given in the following section.

For generative model (Ranzato et al., 2011), it is a subfield of unsupervised learning and mainly intended to characterize joint statistical distributions and generate new data that has similar characteristics to the training data. More formally, given a dataset *X*, generative models capture the joint probability *P*(*X*). With the distribution of the generative model, we can have the machine sample the distribution and create new samples that are similar to the original dataset. In recent years, generative models that use neural networks have gradually become the mainstream methods. There are two main types of generative models that are popular in academia, namely variational autoencoder (VAE) (Bengio et al., 2013; Hsu et al., 2017; Lin et al., 2020) and generative adversarial network (GAN) (Goodfellow et al., 2014). Compared with GAN's idea of adversarial training between a generator and a discriminator, VAE uses variational inference to solve the model optimization problem more elegantly. As VAEs are better at modeling continuous data and capturing the underlying structure of the data. In the context of large-scale chemical process modeling, VAEs can be a useful tool. A detailed description of the VAE is given in the following section.

4 Exploring key deep learning methods in large-scale process modeling

In the following section, we will take a closer look at RNN, LSTM, AE and VAE, which are key building blocks of modern deep learning systems. We will analyze the theoretical foundations of these techniques, highlight their practical benefits, and apply them to real-world complex modeling problems.

4.1 Recurrent neural network

The key advantage of RNN is its ability to capture long-term dependencies in sequential data, which is critical for modeling large-scale chemical processes. The following section will give a detailed introduction to RNN. The RNN designs a repetitive structure with a time-dependent relationship to process sequential data. RNN can map input sequences to output sequences with the same length (or different lengths with input sequences). In Fig. 4, one input sequence of the network is $x_t, x_{t+1}, \ldots, x_n$, the input, hidden state, output, target output, and loss at time *t* are x_t, h_t, o_t, y_t, and L_t, respectively. *U* is the weight matrix from the input to the hidden state, *W* is the connection weight matrix between different hidden states, and *V* is the weight matrix from the hidden state to the output. For the basic RNN, the calculation can be given as follows:

- The hidden state h_t is jointly determined by the input x_t at the current moment and the hidden state h_{t-1} at the previous moment.

$$h_t = \delta_1(Ux_t + Wh_{t-1} + b) \tag{2}$$

  where δ_1 is the activation function, b is the bias.
- The calculation formula for the output state o_t is

$$o_t = \delta_2(Vh_t + c) \tag{3}$$

  where δ_2 is the activation function and c is the bias.
- The training loss over the sequence can be defined as the sum of the losses at different time steps:

$$L = \sum_{i=t}^{n} L_t = \sum_{i=t}^{n} Loss(o_i, y_i) \tag{4}$$

It is worth noting that RNN performs the same operation for input, which means the recurrent structure will share parameters (such as the weight matrix U, V, W); this sharing mechanism can significantly reduce the number of parameters that the network needs to learn and enable the network to model the long-term dependencies in the sequence. RNN performs excellently in processing sequence data because it has a long-term memory function, which can compress data and obtain a long-term data representation. On the one hand, considering that the state at different time is dependent on each other, so the state information at each time needs to be stored, which results in a large memory consumption during the entire training process; on the other hand, the whole algorithm cannot be calculated in parallel, resulting in the slow implementation of the algorithm. Therefore, in the training process of RNN, only the gradient of k time steps is usually backpropagated (truncated backpropagation). However, it should be noted that truncated backpropagation may affect the generalization performance of the model because it may not capture long-term dependencies. As the number of layers increases, the gradients may "vanish" (tend to zero) or "explode" (tend to infinity). These problems make optimization difficult and ultimately make the network unable to remember sequence information entered long ago (Basodi et al., 2020; Bengio et al., 1994). To address these issues, long short-term memory network (LSTM) was proposed by Hochreiter and Schmidhuber (1997).

4.2 Long short-term memory network

Due to the vanishing gradient and exploding gradient problem of basic RNN, it often fails to achieve the expected effect in practical tasks. LSTM is one of the most famous extensions of RNN. It can not only effectively process short-term information but also memorize valuable long-term information, thereby reducing the training difficulty of RNN and improving the learning ability of the network (Yuan et al., 2020).

Compared with the traditional RNN, LSTM is still based on the input x_t and the hidden state h_{t-1} to calculate h_t, but the internal structure is more carefully designed. As shown in Fig. 5, the internal structure of LSTM includesinput gate i_t, forget gate f_t, output gate o_t, and cell state c_t. The input gate controls how much the new information is updated to the cell; the forget gate controls how much the information of the previous step is forgotten in the cell; and the output gate controls how much the information is passed. In the classical LSTM, the updating formula of the cell at the tth step is

$$f_t = \sigma(W_f x_t + U_f h_{t-1} + b_f) \tag{5}$$

$$i_t = \sigma(W_i x_t + U_i h_{t-1} + b_i) \tag{6}$$

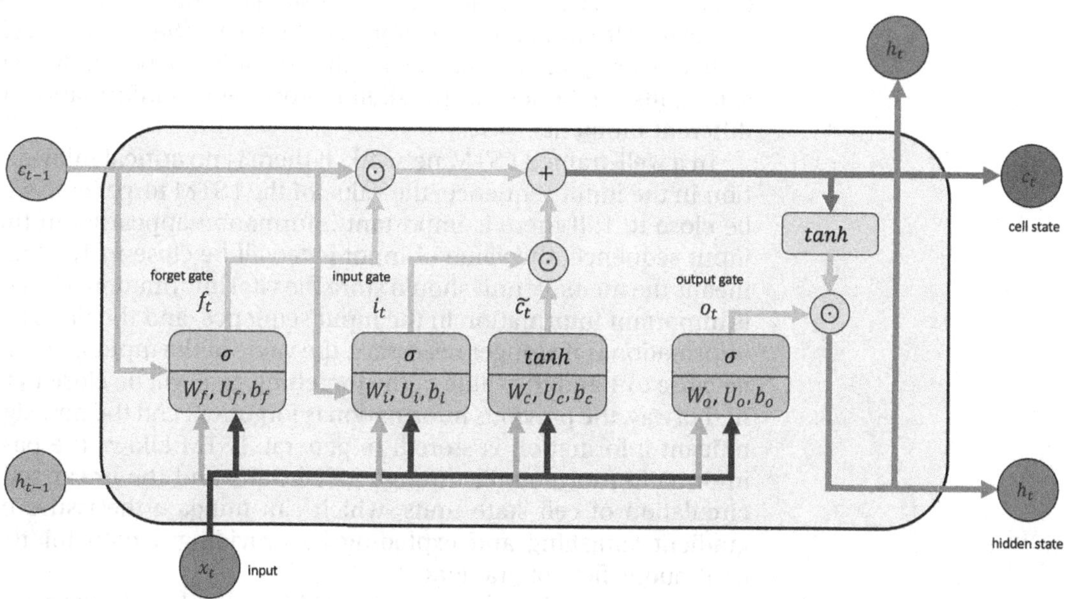

Fig. 5 A basic cell of LSTM.

$$o_t = \sigma(W_o x_t + U_o h_{t-1} + b_o) \tag{7}$$

$$\widetilde{c}_t = \tanh\left(W_c x_t + U_c h_{t-1} + b_c\right) \tag{8}$$

$$c_t = f_t \odot c_{t-1} + i_t \odot \widetilde{c}_t \tag{9}$$

$$h_t = o_t \odot \tanh(c_t) \tag{10}$$

where i_t is obtained by linear transformation of input x_t and the hidden state h_{t-1}, and then by activation function σ. The result of the input gate i_t is a vector, where each element is a real number between 0 and 1 to control the amount of information flowing through the gate in each dimension. W_i, U_i, and b_i are the parameters of the input gate, which needs to be learned in the training process. The calculation of the forgetting gate f_t and the output gate o_t is similar to the input gate i_t, and they have their own parameters W, U, and b. The transition of the cell state from the last moment c_{t-1} to the current moment c_t is jointly controlled by the input gate, candidate cell state and the forget gate. The hidden state h_t is determined by output gate o_t and cell state c_t. The input gate determines how much of the input information at the candidate cell state $\sim c_t$ is stored in the cell state. The forget gate determines how much of the cell state information at the previous cell state c_{t-1} is forgotten, and the final cell state unit is obtained by adding two gate signals. It can be seen that in the LSTM, there is a linear self-circulation relationship between the internal cell state units c_{t-1} and c_t. The linear self-circulation between the cell state units can be seen as the sliding processing of information at different moments.

In a well-trained LSTM network, if there is no critical information in the input sequence, the value of the LSTM forget gate will be close to 1; if there is important information appearing in the input sequence, the value of input gate will be close to 1, which means the memory unit should store the vital information; if there is important information in the input sequence, and the previous information is no longer necessary, the value of the input gate will be close to 1 and the value of the forgetting gate will be close to 0. In this way, the previous information is forgotten, and the new significant information is stored. In general, LSTM allows the past information to re-enter through gated units and the linear self-circulation of cell state units, which can mitigate the issues of gradient vanishing and exploding by providing a path for the continuous flow of gradient.

The successful application of LSTM in several fields has made researchers aware of the effectiveness of gate units. Recall that

LSTM is not only designed to be sensitive to short-term memory but also to capture valuable long-term memory. So, can we design an RNN with only two gate units, one for short-term memory and the other for long-term memory? This is precisely the idea of the gated recurrent unit GRU (Vincent et al., 2010). GRU has only two gates, one for controlling the amount of information to be forgotten (reset gate) and another for controlling the amount of new information to be added (update gate). Compared to LSTM, GRU has fewer parameters and is easier to compute and implement. Overall, RNN has been developed over the years, and understanding its development history and common variants allows us to choose the best network structure based on the practical research problem.

4.3 Autoencoder

In large-scale chemical process modeling, autoencoder can extract useful features and remove noise from the data, thus helping to build more accurate and robust models. AE is often used for learning a representation or encoding of the original data by setting the outputs equal to the inputs. As shown in Fig. 6, the

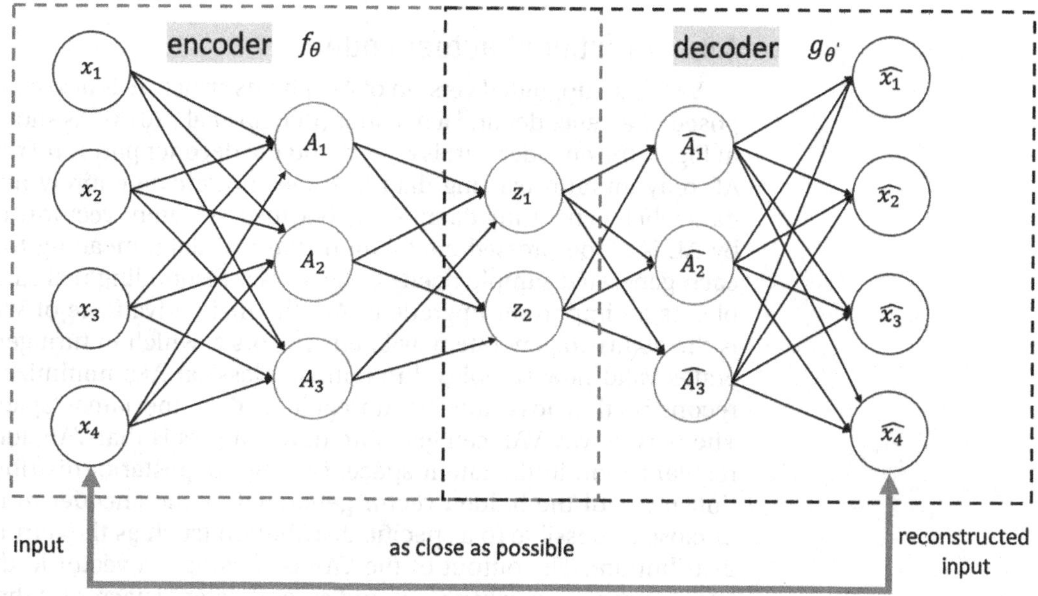

Fig. 6 An example of the AE.

standard AE consists of two parts: encoder and decoder. First, the encoder compresses the real data x into a hidden vector z in the low-dimensional hidden space ($z = f_\theta(x)$), which can be regarded as the essence of the input; and then the decoder decompresses the hidden vector z to get the generated data $\hat{x}$ ($\hat{x} = g_{\theta'}(z)$). In the training process, the generated data $\hat{x}$ are compared with the real data x, and the parameters of the encoder θ and decoder θ' are updated in the direction of reducing the difference between $\hat{x}$ and x, which can be given as follows:

$$\begin{aligned}\theta^*, \theta'^* &= \arg\min_{\theta,\theta'} \frac{1}{n}\sum_{i=1}^{n} L\left(x^i \hat{x}^i\right) \\ &= \arg\min_{\theta,\theta'} \frac{1}{n}\sum_{i=1}^{n} L(x^i g_{\theta'}(f_\theta(x^i)))\end{aligned} \tag{11}$$

which L is a loss function such as mean squared error or cross-entropy. The ultimate goal is to make the hidden vector z capture the essence of the real data x as much as possible so that the reconstructed data $\hat{x}$ is as close to the real data x as possible. AE can be applied to denoising and dimension reduction and would be an excellent tool for processing high-dimensional and noisy industrial process data.

4.4 Variational autoencoder

VAE is an upgraded version of AE, and its structure is also composed of an encoder and a decoder (Bengio et al., 2013). As shown in Fig. 7, the encoder part is $q_\phi(z|x)$ and the decoder part is $p_\theta(x|z)$. AE only imitates existing data but does not create entirely new data when generating data; this is because the latent vector used by AE is a compressed encoding of the real data, meaning that each generated sample needs to have a corresponding real sample. As an important upgrade of AE, the main advantage of VAE is the ability to generate new latent vectors z, which in turn generates valid new samples. In contrast, classical AEs minimize a reconstruction loss and are unregularized in the latent space. The reason why VAE can generate new samples is that VAE adds regularization in the latent space, forcing the posterior distribution $p(z|x)$ of the hidden vector generated by the encoder to be as close as possible to a specific distribution (such as the normal distribution). The output of the VAE is no longer a vector in the latent space but a continuous and smooth latent space, and then the latent vector z is sampled from the normal distribution of the latent space. Due to the randomness of the sampling operation,

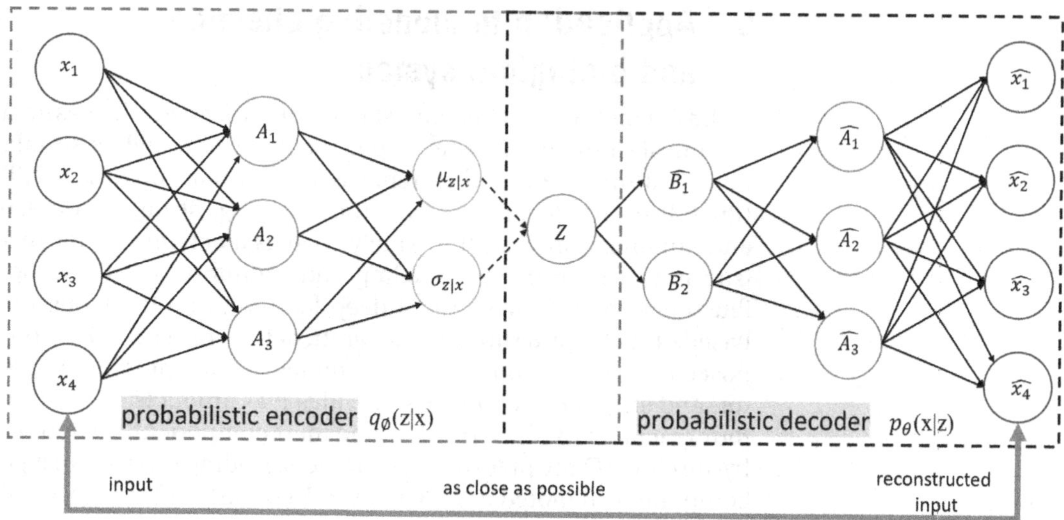

Fig. 7 An example of the VAE.

the output obtained by VAE is no longer deterministic, which creates new valid data that have similar characteristics to the original data, but with some variations.

In the encoder network, we need to make an inference (calculate the posterior $p(z|x)$). For high-dimensional latent variables, $p(x)$ is intractable, which leads to $p(z|x)$ intractable. One of the effective solutions for this intractable posterior is called deterministic approximation (also called variational inference), which $p(z|x)$ is approximated with closest distribution $q_\phi(z|x)$ from tractable probabilistic models (Bekker et al., 2015). The optimization objective of VAE mainly includes two parts, namely the reconstruction error and the regularizer on the posterior distribution $q_\phi(z|x)$. The loss function L (also called variational lower bound) could be represented as

$$L(\theta, \phi, x) = -D_{KL}[q_\phi(z|x) \parallel p_\theta(z)] + E_{q_\phi(z|x)}[\log p_\theta(x|z)] \quad (12)$$

In this loss function, the first term is the regularizer, which makes approximation inference term $q_\phi(z|x)$ close to the prior $p_\theta(z)$. The $D_{KL}[q_\phi(z|x) \| p_\theta(z)]$ means the information lost when using $p_\theta(z)$ to approximate $q_\phi(z|x)$ (Zeng et al., 2014). The second term is conceptually the reconstruction error. We aim to learn the parameters ϕ and θ via backpropagation. VAE can be applied to learn and represent complex industrial chemical process dynamics, as well as generate new process conditions or scenarios. This can provide valuable insights in modeling large-scale industrial chemical processes.

5 Application to modeling chemical and biological systems

Based on the actual chemical reaction processes, the Eastman Chemical Company has developed a chemical modeling simulation platform called the Tennessee Eastman process (TEP) (Russell et al., 2000). TEP is a time varying, strongly coupled, and nonlinear process. It is widely used as a benchmark for realistic industrial process modeling and monitoring. This section illustrates the effectiveness of deep learning as a modeling tool based on TEP. Fig. 8 shows the diagram of the TEP. The TEP is composed of five major units: reactor, condenser, compressor, separator, and stripper. The process provides two products (*G*, *H*) from four reactants (*A*, *C*, *D*, *E*). Also, one inert component (*B*) and one by-product (*F*) are present in the process, leading to a total of eight components denoted as *A*, *B*, *C*, *D*, *E*, *F*, *G*, and *H*. There are 52 different variables in this process, among which 33 can be measured in real time, while another 19 must be analyzed, respectively.

The data set used in the present work is available at https://doi.org/10.7910/DVN/6C3JR1. There are 20 faults in the data set; IDV (1) fault (A/C feed ratio changes, B composition is constant) is considered to validate the effectiveness of deep learning algorithms. In this case study, we present a hybrid anomaly detection method that utilizes an AE with LSTM encoder and decoder modules to represent time series data. Both the AE and LSTM units do not require labeled anomalies for training. The basic idea is to build an LSTM-AE model on the normal data and then use this model to calculate reconstruction error. If the reconstruction error is high, we will label it as an anomaly. The objective of the LSTM-AE model is to predict an anomaly k time steps before it occurs.

For time-series data $X = \{x_1, x_2, \ldots, x_N\}$ with N samples, and each sample contains m features, where $x_i \in R^m$ is the ith sample. At time $t(L < t < N)$, the input is S_t with L past samples $S_t = [x_{t-L+1}, \ldots, x_t]$, the output is $y_{t+k} \in \{0, 1\}$ with 1 indicating anomaly occurs at time $t + k$. In this case study, the encoder has two LSTM layers with 32 and 16 units and one dense layer with 8 units. The decoder has the same structure as the encoder. The moving window size is $L = 20$, and we will try to predict the anomaly $k = 5$ time steps in advance. We split the data into training, validation, and test set.

The following figures show how we can use a reconstruction error for anomaly detection. In Fig. 9, we calculate the reconstruction error on the entire training data with both normal and anomaly samples. The orange (light gray in print versions) and blue

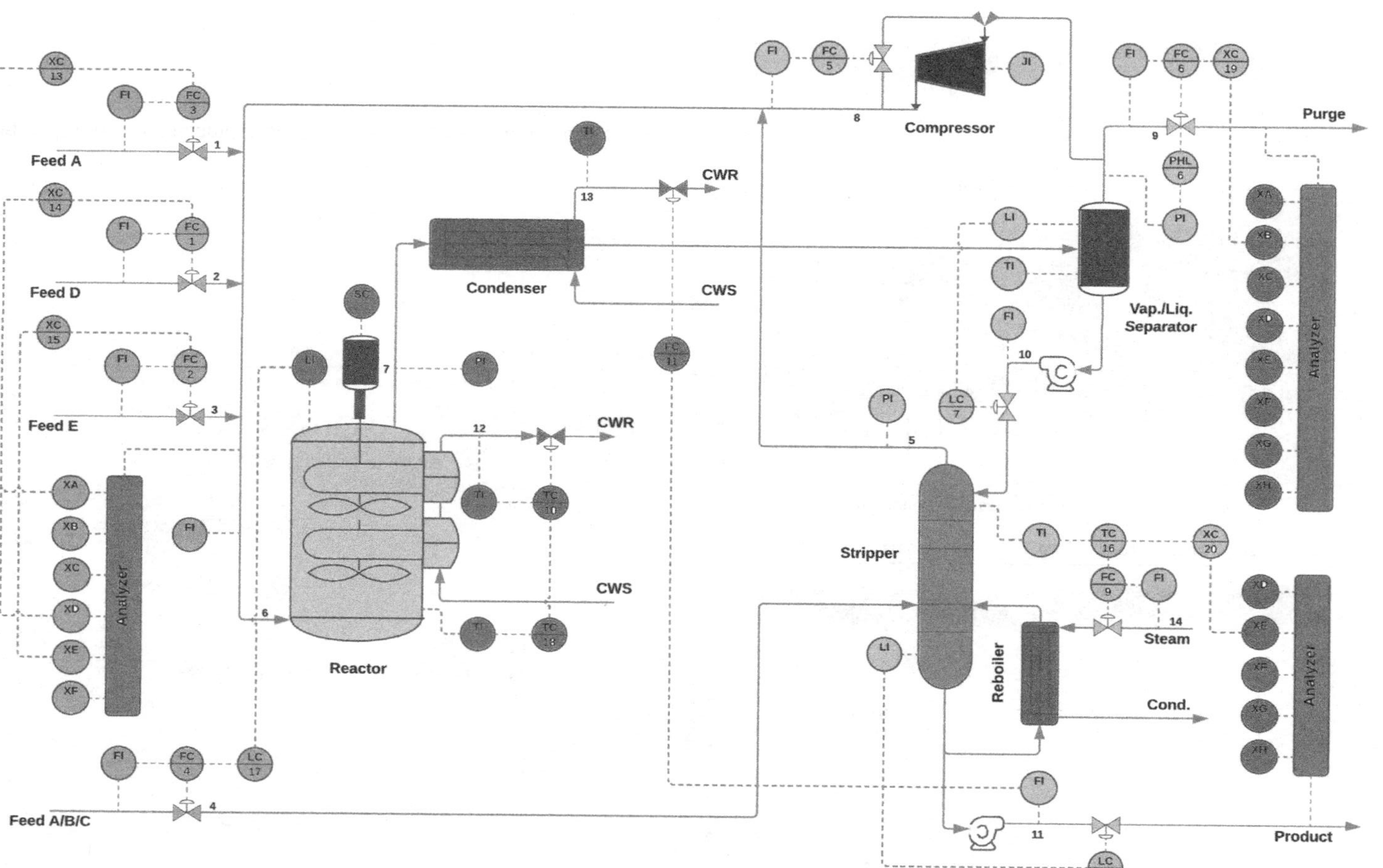

Fig. 8 TEP flowchart.

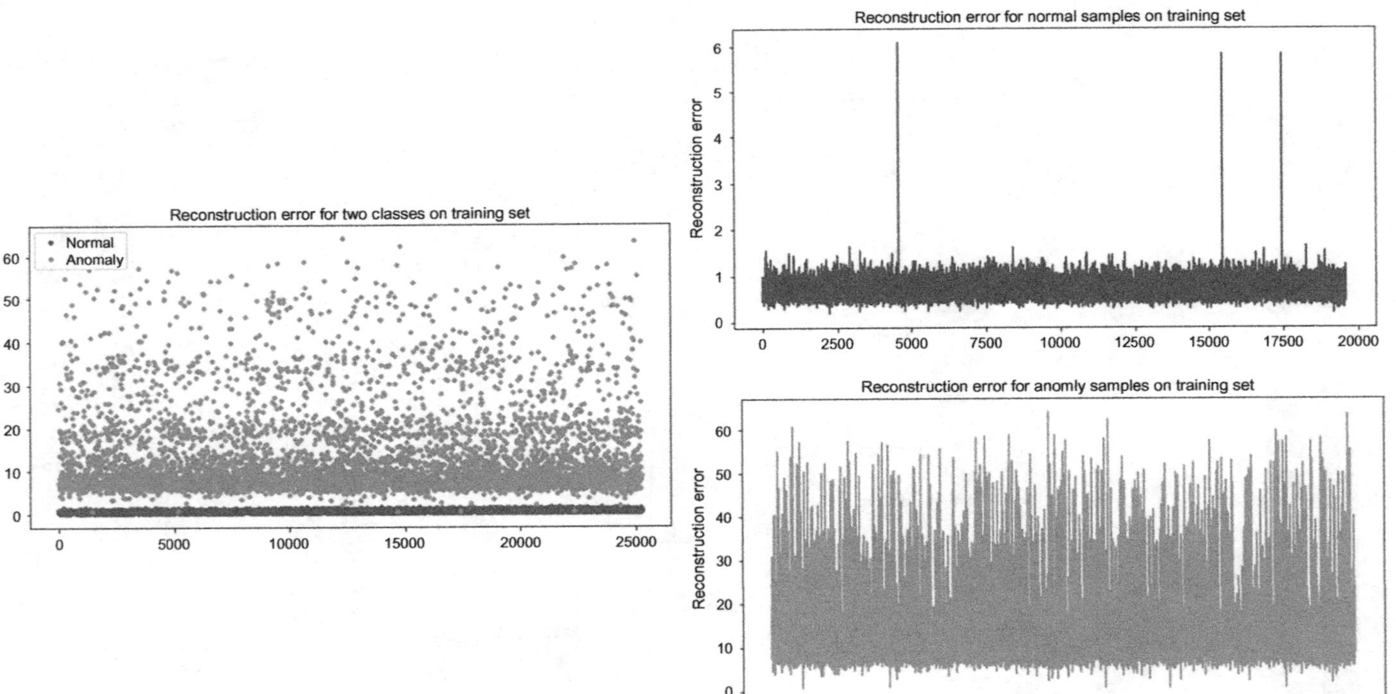

Fig. 9 The reconstruction error of LSTM-AE model on training set.

(dark gray in print versions) dot represents the anomaly and normal samples, respectively. The reconstruction error serves as the anomaly detection score for the model on the training set. The reconstruction error of normal samples should be much smaller than anomaly samples; as shown in Fig. 9, almost all anomaly samples have a more significant reconstruction error than normal samples.

As mentioned before, if the reconstruction error of one sample is high, it will be classified as an anomaly. To be more precise, we will use the validation set to determine the classification threshold. The ideal threshold should be the value of the intersection of precision and recall. From the results of Fig. 10, we can get that the optimal threshold is 2.

Fig. 11 shows the performance of the LSTM-AE model on the test set, the blue dot (dark gray in print versions) above the threshold line represents the false positive, and the orange dot (light gray in print versions) below the threshold line represents the false negative. As we can see, almost all samples are correctly classified in the test data set. Fig. 12 further shows the confusion matrix of classification. We could correctly classify 3479 anomaly samples out of a total of 3482 anomaly samples. Considering these classifications are five time steps ahead, this is a pretty good result.

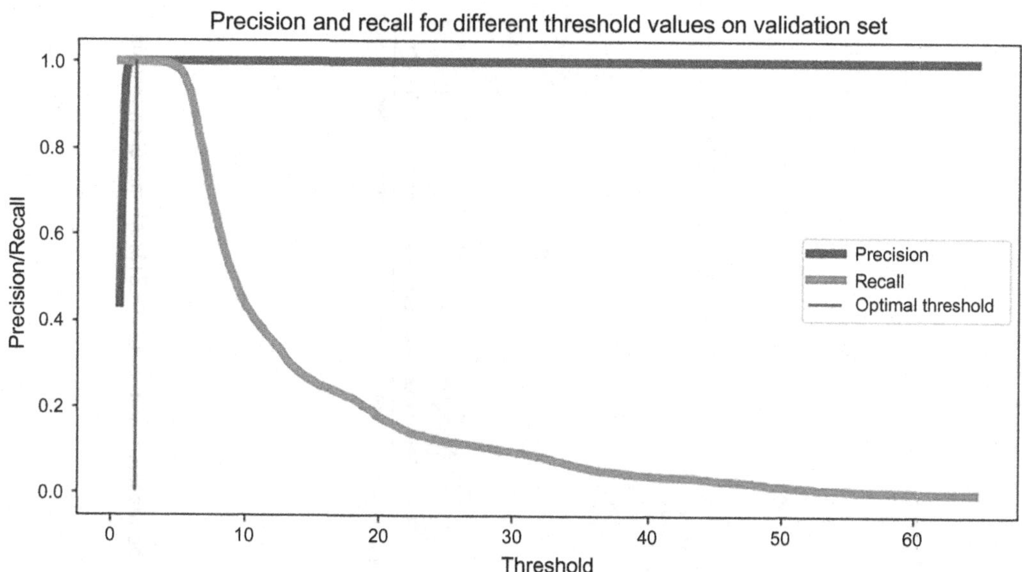

Fig. 10 The precision and recall of LSTM-AE model for different threshold values.

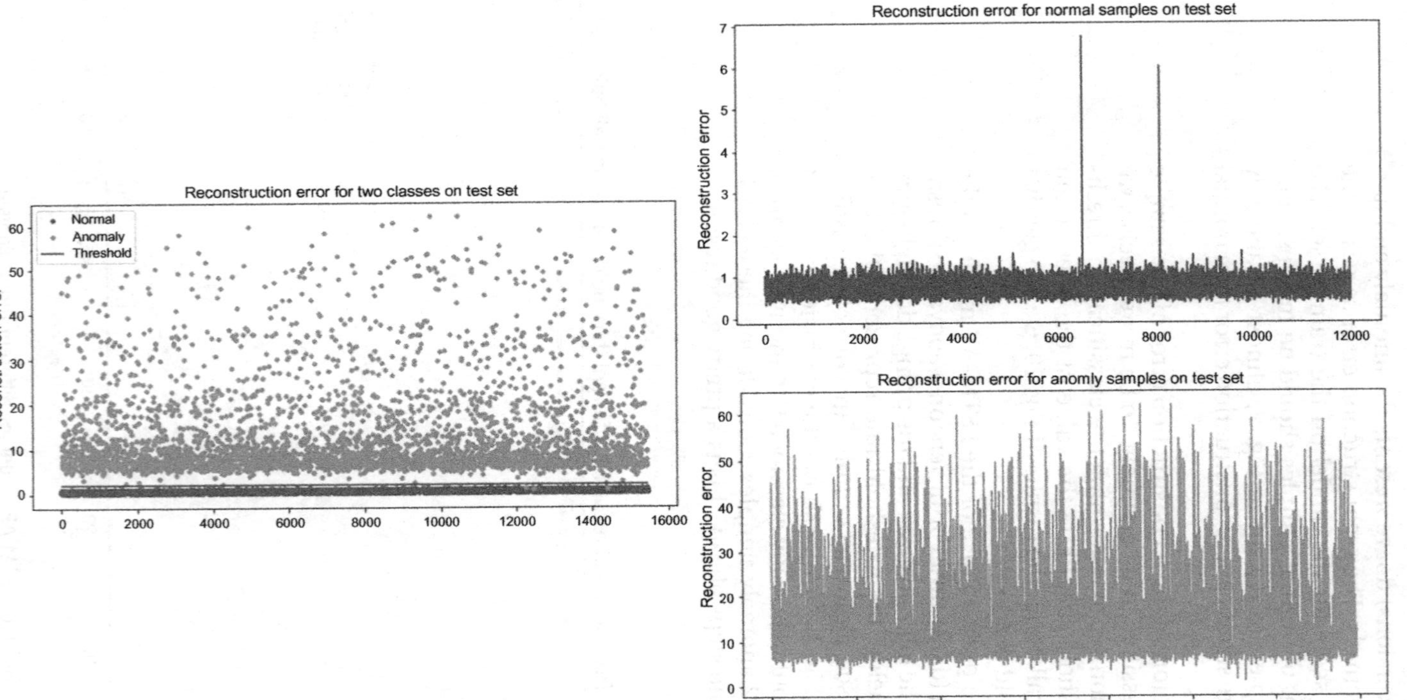

Fig. 11 The reconstruction error of LSTM-AE model on test set.

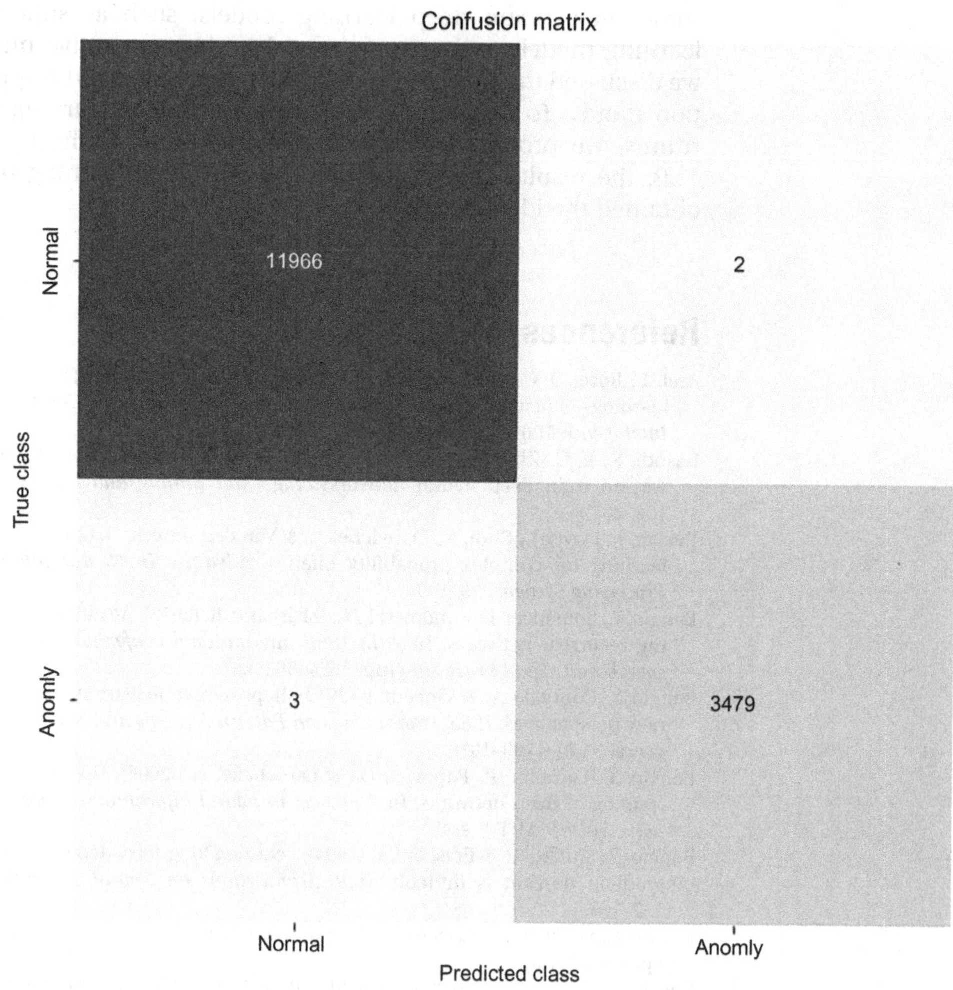

Fig. 12 Confusion matrix of LSTM-AE model on test set.

6 Summary

With the development of hardware and software, the complexity and the automation degree of industrial processes are continuously growing, which leads to poor performance of traditional modeling methods on large-scale processes. However, rapid progress in deep learning enables new opportunities for a comprehensive modeling and analysis of the large-scale processes. In this chapter, we first analyzed the necessity and significance of deep learning in large-scale process modeling. By introducing some mainstream deep learning models, such as supervised learning models and unsupervised learning/generative models, we discussed the pros and cons of deep learning and the application trends. To validate the effectiveness of deep learning algorithms, we proposed an LSTM-AE model and apply it to the TEP; the results show that the proposed deep learning model obtained the ideal performance.

References

Arel, I., Rose, D. C., & Karnowski, T. P. (2010). Research frontier: Deep machine learning—A new frontier in artificial intelligence research. *IEEE Computational Intelligence Magazine, 5*(4), 13–18.

Basodi, S., Ji, C., Zhang, H., & Pan, Y. (2020). Gradient amplification: An efficient way to train deep neural networks. *Big Data Mining and Analytics, 3*(3), 196–207.

Bekker, J., Davis, J., Choi, A., Darwiche, A., & Van den Broeck, G. (2015). Tractable learning for complex probability queries. *Advances in Neural Information Processing Systems, 28*, 1–9.

Bengio, Y., Boulanger-Lewandowski, N., & Pascanu, R. (2013). Advances in optimizing recurrent networks. In *2013 IEEE International conference on acoustics, speech and signal processing* (pp. 8624–8628).

Bengio, Y., Courville, A., & Vincent, P. (2013). Representation learning: A review and new perspectives. *IEEE Transactions on Pattern Analysis and Machine Intelligence, 35*(8), 1798–1828.

Bengio, Y., Lamblin, P., Popovici, D., & Larochelle, H. (2006). Greedy layer-wise training of deep networks. In *Advances in neural information processing systems: Vol. 19*. MIT Press.

Bengio, Y., Simard, P., & Frasconi, P. (1994). Learning long-term dependencies with gradient descent is difficult. *IEEE Transactions on Neural Networks, 5*(2), 157–166.

Bishop, C. M. (2006). *Pattern recognition and machine learning*. Berlin, Heidelberg: Springer-Verlag.

Cao, L., Yu, F., Yang, F., Cao, Y., & Gopaluni, R. B. (2020). Data-driven dynamic inferential sensors based on causality analysis. *Control Engineering Practice, 104*, 104626.

Chatzis, S. P., & Demiris, Y. (2013). The infinite-order conditional random field model for sequential data modeling. *IEEE Transactions on Pattern Analysis and Machine Intelligence, 35*(6), 1523–1534.

Cortes, C., & Vapnik, V. (1995). Support-vector networks. *Machine Learning, 20*, 273–297.

Ding, S. X. (2014). Data-driven design of fault diagnosis and fault-tolerant control systems. In *Advances in industrial control*.. London: Springer. https://books.google.ca/books?id=77y8BAAAQBAJ.

Fan, J., Qin, S. J., & Wang, Y. (2014). Online monitoring of nonlinear multivariate industrial processes using filtering KICA-PCA. *Control Engineering Practice, 22*, 205–216.

Fine, S., Singer, Y., & Tishby, N. (1998). The hierarchical hidden Markov model: Analysis and applications. *Machine Learning, 32*(1), 41–62.

Ge, Z., Song, Z., Ding, S. X., & Huang, B. (2017). Data mining and analytics in the process industry: The role of machine learning. *IEEE Access, 5*, 20590–20616. https://doi.org/10.1109/ACCESS.2017.2756872.

Goodfellow, I., Pouget-Abadie, J., Mirza, M., Xu, B., Warde-Farley, D., Ozair, S., ... Bengio, Y. (2014). Generative adversarial nets. *Advances in Neural Information Processing Systems, 27*, 1–9.

Gopaluni, R. B. (2008). Identification of nonlinear processes with known model structure under missing observations. *IFAC Proceedings Volumes, 41*(2), 6478–6483.

Gopaluni, R. B., Tulsyan, A., Chachuat, B., Huang, B., Lee, J. M., Amjad, F., ... Lawrence, N. P. (2020). Modern machine learning tools for monitoring and control of industrial processes: A survey. *IFAC-PapersOnLine, 53*(2), 218–229.

Graves, A., Mohamed, A.-R., & Hinton, G. (2013). Speech recognition with deep recurrent neural networks. In *2013 IEEE International conference on acoustics, speech and signal processing* (pp. 6645–6649).

Hermans, M., & Schrauwen, B. (2013). Training and analysing deep recurrent neural networks. In *Advances in neural information processing systems: Vol. 26.*

Hochreiter, S., & Schmidhuber, J. (1997). Long short-term memory. *Neural Computation, 9*(8), 1735–1780.

Hsu, W.-N., Zhang, Y., & Glass, J. (2017). Unsupervised domain adaptation for robust speech recognition via variational autoencoder-based data augmentation. In *2017 IEEE automatic speech recognition and understanding workshop (ASRU)* (pp. 16–23).

Huang, G.-B., Zhu, Q.-Y., & Siew, C.-K. (2006). Extreme learning machine: Theory and applications. *Neurocomputing, 70*(1), 489–501.

Imseng, D., Motlicek, P., Garner, P. N., & Bourlard, H. (2013). Impact of deep MLP architecture on different acoustic modeling techniques for under-resourced speech recognition. In *2013 IEEE workshop on automatic speech recognition and understanding* (pp. 332–337).

Kavukcuoglu, K., Sermanet, P., Boureau, Y.-L., Gregor, K., Mathieu, M., & Cun, Y. (2010). Learning convolutional feature hierarchies for visual recognition. In J. Lafferty, C. Williams, J. Shawe-Taylor, R. Zemel, & A. Culotta (Eds.), *Advances in neural information processing systems: Vol. 23.* Curran Associates, Inc. (Eds.).

Krizhevsky, A., Sutskever, I., & Hinton, G. E. (2012). ImageNet classification with deep convolutional neural networks. In F. Pereira, C. J. Burges, L. Bottou, & K. Q. Weinberger (Eds.), *Advances in neural information processing systems: Vol. 25.* Curran Associates, Inc. (Eds.).

LeCun, Y., Bengio, Y., & Hinton, G. (2015). Deep learning. *Nature, 521*, 436–444.

Lin, S., Clark, R., Birke, R., Schönborn, S., Trigoni, N., & Roberts, S. (2020). Anomaly detection for time series using VAE-LSTM hybrid model. In *2020 IEEE International conference on acoustics, speech and signal processing (ICASSP)* (pp. 4322–4326).

Mohamed, A.-R., Dahl, G. E., & Hinton, G. (2012). Acoustic modeling using deep belief networks. *IEEE Transactions on Audio, Speech, and Language Processing, 20*(1), 14–22.

Permuter, H., Francos, J., & Jermyn, I. (2006). A study of Gaussian mixture models of color and texture features for image classification and segmentation. *Pattern Recognition, 39*(4), 695–706.

Pinto, J., Garimella, S., Magimai-Doss, M., Hermansky, H., & Bourlard, H. (2011). Analysis of MLP-based hierarchical phoneme posterior probability estimator. *IEEE Transactions on Audio, Speech, and Language Processing, 19*(2), 225–241.

Qin, S. J. (2014). Process data analytics in the era of big data. *AIChE Journal, 60*(9), 3092–3100.

Ranzato, M., Susskind, J., Mnih, V., & Hinton, G. (2011). On deep generative models with applications to recognition. In *CVPR 2011* (pp. 2857–2864).

Rosenblatt, F. (1958). The perceptron: A probabilistic model for information storage and organization in the brain. *Psychological Review, 65*, 386–408.

Rumelhart, D., Hinton, G., & Williams, R. (1986). Learning representations by back-propagating errors. *Nature, 323*, 533–536.

Russell, E. L., Chiang, L. H., & Braatz, R. D. (2000). *Data-driven methods for fault detection and diagnosis in chemical processes*. London: Springer.

Sainath, T. N., Kingsbury, B., & Ramabhadran, B. (2012). Auto-encoder bottleneck features using deep belief networks. In *2012 IEEE International conference on acoustics, speech and signal processing (ICASSP)* (pp. 4153–4156).

Schmidhuber, J. (2015). Deep learning in neural networks: An overview. *Neural Networks, 61*, 85–117.

Sun, Q., Liu, Y., Chua, T.-S., & Schiele, B. (2019). Meta-transfer learning for few-shot learning. In *Proceedings of the IEEE/CVF conference on computer vision and pattern recognition (CVPR)*. June.

Vincent, P. (2011). A connection between score matching and denoising autoencoders. *Neural Computation, 23*(7), 1661–1674.

Vincent, P., Larochelle, H., Lajoie, I., Bengio, Y., & Manzagol, P. (2010). Stacked denoising autoencoders: Learning useful representations in a deep network with a local denoising criterion. *Journal of Machine Learning Research, 11*, 3371–3408.

Wang, Y., Yao, Q., Kwok, J. T., & Ni, L. M. (2020). Generalizing from a few examples: A survey on few-shot learning. *ACM Computing Surveys, 53*(3), 1–34.

Wasikowski, M., & Chen, X.-W. (2010). Combating the small sample class imbalance problem using feature selection. *IEEE Transactions on Knowledge and Data Engineering, 22*(10), 1388–1400. https://doi.org/10.1109/TKDE.2009.187.

Yuan, X., Li, L., & Wang, Y. (2020). Nonlinear dynamic soft sensor modeling with supervised long short-term memory network. *IEEE Transactions on Industrial Informatics, 16*(5), 3168–3176. https://doi.org/10.1109/TII.2019.2902129.

Yuan, X., Ou, C., Wang, Y., Yang, C., & Gui, W. (2020). A novel semi-supervised pre-training strategy for deep networks and its application for quality variable prediction in industrial processes. *Chemical Engineering Science, 217*, 115509.

Zeiler, M. D., Taylor, G. W., & Fergus, R. (2011). Adaptive deconvolutional networks for mid and high level feature learning. In *2011 International conference on computer vision* (pp. 2018–2025).

Zeng, J., Kruger, U., Geluk, J., Wang, X., & Xie, L. (2014). Detecting abnormal situations using the Kullback-Leibler divergence. *Automatica, 50*(11), 2777–2786. Nov.

Zhou, L., Song, Z., Chen, J., Ge, Z., & Li, Z. (2014). Process-quality monitoring using semi-supervised probability latent variable regression models. *IFAC Proceedings Volumes, 47*(3), 8272–8277.

Zhu, Q.-X., Hou, K.-R., Chen, Z.-S., Gao, Z.-S., Xu, Y., & He, Y.-L. (2021). Novel virtual sample generation using conditional GAN for developing soft sensor with small data. *Engineering Applications of Artificial Intelligence, 106*, 104497.

Index

Note: Page numbers followed by *f* indicate figures, *t* indicate tables, and *s* indicate schemes.

Printed in the United States
by Baker & Taylor Publisher Services